# XVIII International Symposium in Honour of Academician M. A. Usov: Problems of Geology and Subsurface Development 2014



IOP Conference Series: Earth and Environmental Science
Volume 21

Tomsk, Russia
7 - 11 April 2014

ISBN: 978-1-63439-962-3
ISSN: 1755-1307

**Printed from e-media with permission by:**

Curran Associates, Inc.
57 Morehouse Lane
Red Hook, NY 12571



Printed by Curran Associates, Inc. (2015)

For permission requests, please contact the Institute of Physics
at the address below.

Institute of Physics
Dirac House, Temple Back
Bristol BS1 6BE UK

Phone: 44 1 17 929 7481
Fax: 44 1 17 920 0979

techtracking@iop.org

**Additional copies of this publication are available from:**

Curran Associates, Inc.
57 Morehouse Lane
Red Hook, NY 12571 USA
Phone: 845-758-0400
Fax: 845-758-2634
Email: curran@proceedings.com
Web: www.proceedings.com

# Table of contents

## Volume 21

**XVIII International Scientific Symposium in Honour of Academician M. A. Usov: Problems of Geology and Subsurface Development**
**7–11 April 2014, Tomsk, Russia**

**Accepted papers received: 18 July 2014**
**Published online: 28 August 2014**

## Preface



## Papers

### Stratigraphy and Regional Geology



### Mineralogy, Geochemistry and Petrography

**Mineral Deposits. Metallogeny**

**Petroleum Geology**



**Exploration Geophysics, Mineral Exploration**



**Hydrogeology and Engineering Geology**

**High Technologies in Oil and Gas Development**



**High Technologies in Raw Hydrocarbon and Mineral Conversion**



**High Technologies in Oil and Gas Well Drilling**

**Geoecology, Environmental Protection**



**Comprehensive Utilization of Mineral Resources**



**Minerals and Hydrocarbons: Economics and Policies. Mining Laws**

IOP Conf. Series: Earth and Environmental Science **21** (2014) 011001 doi:10.1088/1755-1315/21/1/011001

# Preface

XVIII International Scientific Symposium in honor of Academician M.A. Usov "Problems of Geology and Subsurface Development" (for students and young scientists) was organized under the guidance of the Ministry of Education and Science of the Russian Federation and the Russian Foundation for Basic Research. Being one of the oldest technical higher education institutions which trains specialists who contribute to scientific research in geosciences, The Institute of Natural Resources of National Research Tomsk Polytechnic University (TPU INR) was chosen to hold the symposium. In 2014 The Institute of Natural Resources celebrated its 113th anniversary. It was founded in 1901 by V.A. Obruchev, the first geologist in Siberia, member of USSR Academy of Sciences, Hero of Socialist Labor, and the first Laureate of the Lenin Prize. He was recognized all over the world as a prominent scientist in the area of geology.

INR is the first institute of geological education and geosciences in the Asian part of Russia. Siberian Mining and Geological Schola, established by V.A. Obruchev and M.A. Usov, has been retaining its significance for discovery, exploration and development of mineral resources not only in Siberia, in the Far East and North-East of the country, but also in Central Asia.

There are a lot of outstanding scientists, engineers and manufacturers among alumni of The Institute of Natural Resources. The institute is proud of M.A. Usov, the student and first postgraduate of V.A. Obruchev, first professor and academician in Siberia, whose name is associated with the development of the mining industry in Siberia; Academician K.I. Satpaev, the founder and first president of the Academy of Sciences of Kazakhstan; Professor N.N. Urvantsev, the discoverer of the unique Norilsk ore deposits in the north of East Siberia and Professor M.K. Korovin, who considered West Siberia deposits to be prospective for oil-gas exploration. There are over 35 000 graduates of the institute and more than 450 of them became explorers of mineral deposits including one Nobel laureate, 50 laureates of the Lenin and State Prizes, more than 250 researchers with DSc and PhD, 15 academicians and corresponding members of the USSR Academy of Sciences and five Heroes of Socialist Labor.

Within the scope of the symposium there were 21 panels and workshop, being held over four days. The symposium was unique because it covered all scientific fields of geology and subsurface development: methods of prospecting and exploration for minerals and hydrocarbons, including space geological research and geoinformation systems in geology, as well as the ecological problems and integrated use of mineral resources, land management, natural resource law and economics. The students and young scientists of Russia, foreign and CIS countries participated in the symposium. The investigations presented at the symposium shed light on the latest achievements made by means of modern techniques and original methods of interpretation; the results of experimental studies and computer technologies in geology, oil and gas production and geoecology; the analysis of theoretical and experimental studies on various geological problems and environmental protection.

The reports consider the vital issues and the latest achievements of stratigraphy, paleontology,

IOP Conf. Series: Earth and Environmental Science **21** (2014) 011001 doi:10.1088/1755-1315/21/1/011001

tectonics, historical and regional geology, mineralogy, geochemistry, petrography, lithology, metallogeny, hydrogeology and engineering geology, geophysics, petroleum geology, oil and gas field development and processing of hydrocarbon and mineral resources, geoinformation systems (GIS) in geology, space geological research, oilfield equipment upgrading, modern techniques of mineral exploration, production, transportation and storage of oil and gas, drilling, mining engineering, geoecology, hydrogeoecology, environmental protection engineering, integrated use of mineral resources, land management, mining and natural resources law, and economical problems of mineral resources sectors in Russia and CIS countries. There is a special panel for those who make reports in English and German.

The scientific significance of the reports is explained by new concepts and original ideas suggested by the authors. A number of studies introduce fundamentally new discoveries. The findings of the young scientists' investigations, in both fundamental sciences and experimental studies are significant for practical application, and further investigation might lead to writing a thesis for scientific degree. The authors of some reports gained patents and licenses for their discoveries.

Within the scope of XVIII International Scientific Symposium there were 970 reports (including poster presentations) made by 1195 authors, among whom 293 and 90 came from other cities and foreign countries respectively.

90 participants from foreign countries and 293 of those who came from different cities of Russia.

There were **1195 students and young scientists** from Russia, CIS and foreign countries, who applied for the symposium with **970 reports**, including **293 applications sent from other cities. Numbers of foreign participants** equaled to **32**, made up by **representatives of 16 higher education institutions, scientific centres and industrial enterprises of 12 different cities. CIS countries** were represented by **58 participants from 30 higher education institutions, scientific centres and industrial enterprises of 20 cities. The total number of Russian participants** equaled to **206**, who represented **78 higher education institutions, scientific centres and industrial companies of 48 cities.** There were **677 reports made by students and researchers from Tomsk**, including **647** declared by those **of TPU. 73 participants** presented their reports **in English and German.**

**970 reports** made within the scope of **21 panels and the round table** embraced **827 presented by the speakers** (including **677** by those **from TPU**) and **143 poster presentations** analyzed by the experts. **Nonresidents** took an active part in the symposium having presented **150 reports**.

The reports of the symposium were made by **students (655 participants), post-graduate students (180), researchers (82), engineers (30), young teachers (18) and schoolchildren (5).**

The organizations represented by the participants were **higher education institutions (607), Russian Academy of Sciences (22), National Academy of Sciences (8), Research Institutes (18), industrial enterprises (10) and schools (5).**

The speakers who made **827 reports were students of bachelor and master degree programs (547, including 490 from TPU), post-graduate students and young researchers (130, including 80 from TPU) and five schoolchildren.** The amount of **those who attended different panels** over the **four days** of the symposium consisted of **2010 people.**

**Nine foreign countries** (except for CIS) were represented by **32 reports** made by participants from Germany, France, China, Italy, Poland, Ecuador, Iraq, Vietnam and Mongolia.

**Nine CIS countries** were represented by **58 reports** made by participants from National Academies of Sciences and Universities of Ukraine, Belarus, Kazakhstan, Latvia, Azerbaijan, Armenia, Uzbekistan, Kyrgyzstan and Tajikistan.

Russian participants came from various areas of the country: in the east from Sakhalin, Petropavlovsk-Kamchatsky, Yuzhno-Sakhalinsk, Vladivostok, Blagoveshchensk, Krasnoyarsk, Chita, Irkutsk to Barnaul, Kemerovo, Novokuznetsk, and etc.; in the north from Mirny, Yakutsk, Neryungri, Magadan, Nizhnevartovsk, Khanty-Mansiysk, Nefteugansk to Tyumen, Ulan-Ude, Syktyvkar, and etc.; in the west from Minsk, Kiev, Moscow, St. Petersburg, Yekaterinburg to Samara, Kazan, Ufa, Perm, Novosibirsk;

in the south from the cities of the Central Asian republics to Ivano-Frankovsk, Odessa, Novocherkassk, Simferopol, Novorossiysk, Vladikavkaz, Voronezh, Stavropol, Astrakhan, and etc. A great number of young people from Urals, Western and Eastern Siberia took an active part in the Symposium. CIS countries were presented by participants from Uzbekistan (Tashkent), Tajikistan (Dushanbe), Azerbaijan (Baku), Kazakhstan (Almaty, Semipalatinsk, Karaganda, Pavlodar), Belarus (Minsk, Gomel), Armenia (Yerevan, Gyumri), Ukraine (Kiev, Odessa, Ivano-Frankovsk, Dnepropetrovsk, Donetsk, etc.), Kyrgyzstan (Bishkek), Moldova (Chisinau). The students and young scientists from Tomsk representing Tomsk Polytechnic University, Tomsk State University, the Institute of Petroleum Geology and Geophysics (SB RAS) and other organizations and institutions took an active part in the symposium.

The scientific results of the symposium were reflected in a special edition consisting of two volumes and available at `www.portal.tpu.ru/science/konf/pgon`.

The editorial board of the symposium and this volume of *IOP Conference Series: Earth and Environmental Science* consider the materials of the symposium to be interesting for researchers and young scientists of universities, research and academic institutes, academies of sciences and their branches, engineering and technical staff of ministries and government departments – for anyone who explores and develops the Earth subsurface.

The editors of this volume acknowledge the administration of the Institute of Physics and its publishing house for the publication of the issue and administration of National Research Tomsk Polytechnic University, represented by the rector, professor P.S. Chubik.

Executive Editor
XVIII International Symposium
"Problems of Geology and Subsurface Development" - 2014,
PhD in Geology and Mineralogy,
Associate Professor G.M. Ivanova

IOP Conf. Series: Earth and Environmental Science **21** (2014) 011002 doi:10.1088/1755-1315/21/1/011002

# Peer review statement

All papers published in this volume of *Journal of Physics: Conference Series* have been peer reviewed through processes administered by the proceedings Editors. Reviews were conducted by expert referees to the professional and scientific standards expected of a proceedings journal published by IOP Publishing.

IOP Conf. Series: Earth and Environmental Science **21** (2014) 012001 doi:10.1088/1755-1315/21/1/012001

# Lithologo-facial, geochemical and sequence-stratigraphic sedimentation in Naunak suite (south-east Western Siberia)

**M Shaminova, I Rychkova[a], U Sterzhanova[b] and T Dolgaya**
Tomsk Polytechnic University, Tomsk, Russia

E-mail: [a]irina.rychkova@mail.ru, [b]usterzhanova@yandex.ru

**Abstract.** The paper describes lithofacial, geochemical, sequence-stratigraphical research results of facial environment reconstruction and oil and evaluation gas – bearing capacity potential for Naunak suite in south-east Western Siberia.

## 1. Introduction

Nowadays, the deficiency of oil\gas productive formations in West Siberian central territories is quite significant, so more and more researchers are turning to the remote territories of this region. The main problem in oil\gas potential area identification in the south-east Western Siberia is considered to be the complicated reservoir distribution pattern due to the areal sedimentation genesis.

## 2. Materials and research methods

The target of this research is Naunak suite of Late Bathonian-Callovian-Oxfordian, formed during the intermediate sedimentogenesis in south-east Western Siberia [5]. The research is based on detailed core analysis of newly drilled wells in Snezhnaya area (figure).

Backstripping and effective forecasting of natural oil\gas reservoirs are based on the detailed comprehensive core analysis with the use of lithologo-petrographic, geochemical, luminescent-microscopic, X-ray-fluorescent, biostratigraphic, sequence – stratigraphy methods were applied, as well as geophysical survey interpretation [4]. The sequence – stratigraphy method has been applied due to the fact that Naunak suite was formed during the intermediate sedimentogenesis. Lithologo-facial depositional changes in vertically studied wells are due to the adjacent sea level fluctuations and sediments entry into accumulation area [7].

## 3. Research results

As a result of comprehensive study of deposits, the following factors were defined: lithofacial rock groups of silt-sandstones, sandstones, silt-mudstones; genetic features of these sediments; backstripping for this period of time and oil\gas-bearing potential have been determined [1, 2].

Silt-sandstones were formed in lacustrine-alluvial environments. The structure is uniform, sub-oriented due to the orientation of flattened fragments parallel to the bedding plane. There are rare clayey intraclasts (fragment size from 0.2-0.5 to 1.0-3.0 cm) and carboniferous plant detritus is plentiful. Fragments of ferns, Czekanowskia torn leaves and coniferous indicate their burial history in rough hydrodynamic environment. The texture is aleuro-psammitic. Fragment size varies from 0.05-0.12 mm, but the predominate size is 0.08-0.01 mm. Fragments make up 90% of the rock volume. Clastic fragments sorting are medium. Fragment shape is subangular, subrounded, and rarely angular. The mineral composition of the fragments is the following: quartz, feldspars (mostly potassium –

orthoclase and microline, rarely acid plagioclase), while rock fragments are presented by microquartzite and quartz – sericite schists. Cement comprises 10% of the rock volume where its structure is porous-pedicular. Hydromica dominates in the cement, however, chlorite, sericite, muscovite, kaolinite, bituminous substances occur rarely.

The bituminous substance is dark-brown and fills interstitials, penetrating into the cementing mass as skins around the grains.

Under the fluorescence microscope the rock has a cemented and porous bituminous structure, where bitumen content is 0.04-0.06%. Bitumen composition varies from resins and asphaltine components to oily whereas bitumen of oily and oily resinous composition predominate (light-yellow and light-brown luminescence). Hydrocarbons have an odor.

Rock have been subjected to epigenetic changes defined in quartz fragments corrosion and regeneration. Plastic deformation pronounced by the bending of chlorite, muscovite and sericite flakes, as well catalysis and feldspar grain decomposition forming silicon-kaolinite aggregates. These sediments could be potentially oil-gas productive.

The second lithofacial group includes sandstones and feldspar greywackes formed in the river bed environment (meander bars, river bed shoals, levees). The rock texture is poorly oriented due to the orientation of flattened fragments parallel to the bedding plane. Thin cracks less than 0.01 mm filled by dark-brown bituminous substance occur along these lines. The rocks have low angle-wave-like, horizontal, lenticular bedding due to micaceous-coaly-clayey material inwash. Coal lenticules (0.1-0.2 cm) occur rarely. In places the bedding is deformed by elements of slumping. Sandstone texture is medium-fine-grained. The fragments content is 85% of the rock volume. The fragment sorting is medium where the fragment size varies from 0.05-0.04 mm with perdominate sizes of 0.2-0.3 mm. Fragment mineral composition is the following: quartz (60%), feldspars (20%) (orthoclase, microcline, plagioclase rarely) while the remaining mass is characterized by microquartzite, zircon (individual grains). Quartz occurs in sub-angular and sub-rounded fragments. The extinction is cloudy and rarely mosaic. Feldspar fragments are sub-rounded, often tabular in shape. Microquartzite fragments are sub-rounded composed of coarse-crystalline quartz aggregates. The cementing mass is 15% of the rock volume. The cement structure is porous-pedicular. The mineral cement composition is mainly hydromica with sericite, chlorite, and carbonates inclusions of dark-brown bituminous substance. The bituminous substance fills the interstitials penetrating into the cementing mass.

Under the luminescence the sandstones have a porous, cement, and fissure- bituminous texture. The bitumen composition varies from resins and asphaltene components to oily ones, where bitumen of oily and oily resinous composition predominate (light-yellow and light-brown luminescence). The bitumen in fissures are of resinous - asphaltene composition where bitumen migration from the fissures into the rock is observed. Epibitumen content is 0.08%. As the rock has a hydrocarbon odour it could be considered as a reservoir.

Epigenetic changes in the rock are more intensive than those of the first lithofacial group, which, in its turn, shows significant eroded and regenerated quartz fragments. Feldspars are disintegrated with further silicon-clayey aggregate formation. There is an increased erosion and corrosion and regeneration level of silicon material in the horizons saturated by bituminous substance.

Silt-mudstones of the third lithofacial group were formed in lacustrine - swamp environments. The structure is oriented due to the orientation of flattened quartz fragments having siltstone size and parallel to the bedding plane. Orientation of clayey minerals and micro fissures filled by bituminous substance is in the same direction. Rock structures are aleurite and pelite. Stratification in the sediments is poorly identified – sub-horizontal, lenticular, wave-like, deformed by turbidity and embodied with light-grey siltstone inclusions. There are pyrite intraclasts from "loose" to lenticular of 0.6-0.8 cm. The rock is comprised of clayey minerals (mainly hydromicas) with detrital impurities (up to15%) and individual zircon grains. Aleurite material is mostly comprised of unrounded and sub-rounded quartz fragments (size of 0.01-0.03 mm), rarely feldspars and microquartzites. Quartz fragments are poorly eroded and regenerated. Clayey materials are mainly hydromicas and sericite as mixed-layered formations. The sections rich in clayey minerals are saturated by the bituminous substance providing the brown color. There are individual grains and pyrite globus (0.11-1.5 mm).

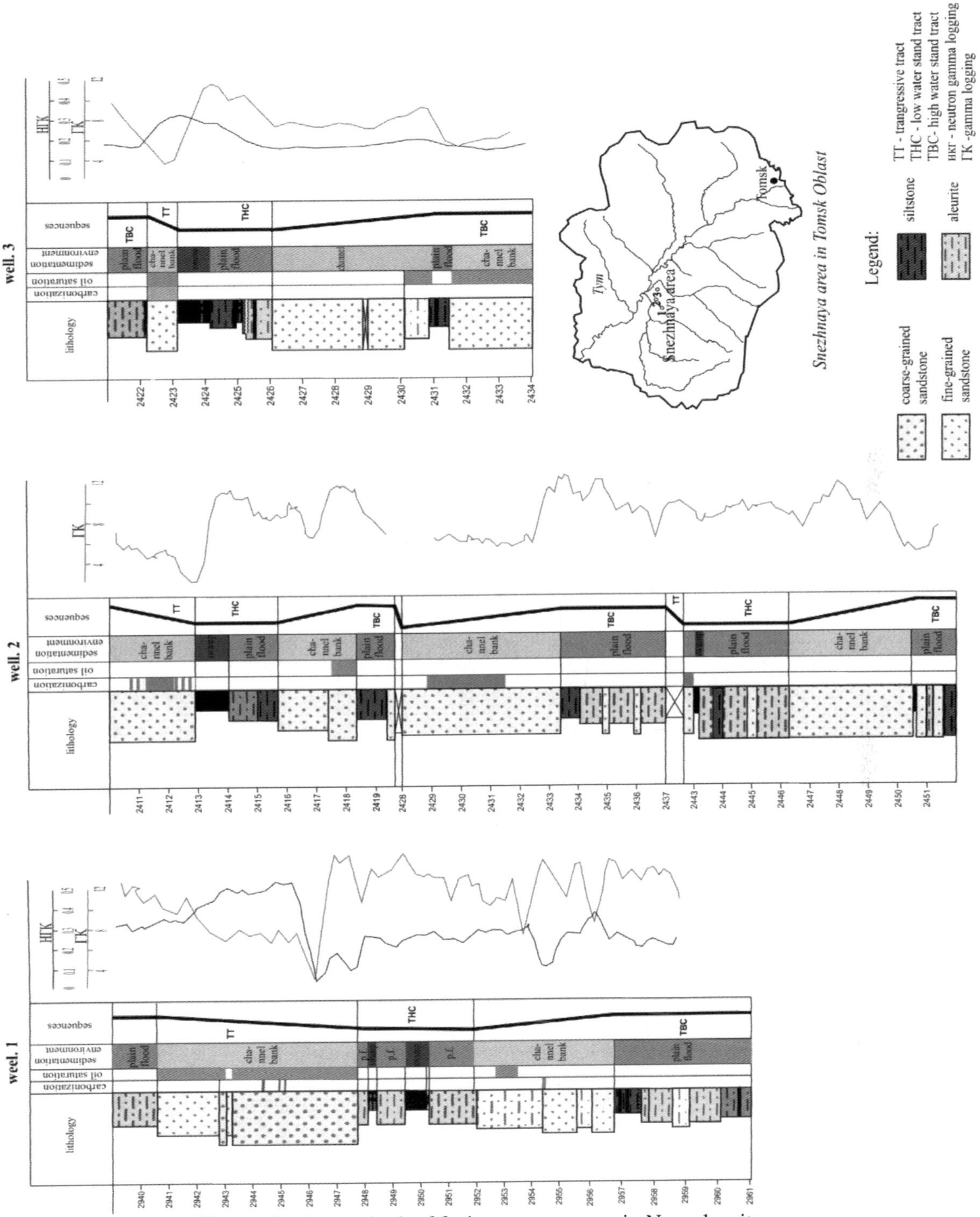


**Figure.** Baskstrippingb of facies enveronmeny in Naunak suite.

IOP Conf. Series: Earth and Environmental Science **21** (2014) 012001 doi:10.1088/1755-1315/21/1/012001

Under the fluorescence microscope the rocks have an irregular dispersed, spotted and fractured bituminous structures. Bitumen composition varies from resins and asphaltene components (dark-brown luminescence) to light ones (bluish luminescence). The bulk rock is dominated by bitumen of resinous composition (brownish luminescence). In the thin fissures (0.01-0.02 mm) predominate resinous asphaltene bitumens, rarely of oily composition (yellowish luminescence). The fissures are oriented sub-parallel to the stratification surface. According to the relationships, the bitumen entry in rock occurred continuously. The bitumen content is 0.06%. By the genesis the bitumens are mixed. In the bulk they are syngenetic, in the fissures - epibitumens. These sediments are considered to be oil\gas generating.

The above-described rocks of lithofacial groups are comprised of a rich complex of plant ichnofossils, which can be related to Naunak phytohorizon. The vegetation community includes horsetails (Equisetites lateralis), ferns (Coniopteris latilobus, C. burejensis), gymnosperms (Phoenicopsis sibirica, Czekanowskia tomskiensis, Nilssonia urmanica) and also numerous seeds (Carpolites cinctus). The coal bands and coaly rocks are determined in the rocks. There is pyrite which is an indicator of sedimentation condition reduction. There is thin carboniferous detritus inwash in the rocks which characterizes stratification.

Application of geochemical methods for studied wells provides more accurate classification, sedimentation reconstruction features, physico-chemical and geodynamic sedimentation environment. Lithochemical calculation of the modules according to Y Yudovich [3, 6] furthered the reconstruction of sedimentation conditions (table).

| Lithofacial group | | $Al_2O_3$ | $SiO_2$ | $K_2O$ | $TiO_2$ | $Fe_2O_3$ | TM | PM | AM | FM |
|---|---|---|---|---|---|---|---|---|---|---|
| Siltstone-mudstone | Siltstone | 13.67 | 78.6 | 3.29 | 0.89 | 3.55 | 0.017 | 0.24 | 0.17 | 0.05 |
| | Mudstone | 19.48 | 74.21 | 1.48 | 0.57 | 4.25 | 0.029 | 008 | 0.26 | 0.06 |
| | | 18.78 | 70.31 | 4.73 | 0.59 | 5.58 | 0.030 | 0.25 | 0.27 | 0.08 |
| | | 21.71 | 69.07 | 3.94 | 0.49 | 4.8 | 0.020 | 0.18 | 0.31 | 0.07 |
| | | 17.2 | 71.32 | 5.2 | 1.16 | 5.3 | 0.068 | 0.31 | 0.24 | 0.07 |
| | Mudstone | 23.88 | 67.69 | 4.2 | 0.67 | 3.56 | 0.030 | 0.18 | 0.35 | 0.05 |
| Sandstone | | | 71.54 | | | 2.63 | | | | 0.04 |
| | Sandstone | 28.97 | 64.16 | 4.51 | | 2.36 | 0.038 | 0.16 | 0.45 | 0.04 |
| | Sandstone | 17.17 | 75.68 | 3.84 | 0.91 | 2.4 | 0.050 | 0.22 | 0.23 | 0.03 |
| Siltstone-Sandstone | | 20.01 | 79.99 | | | | | | 0.25 | |
| | | 13.17 | 86.66 | | | 0.17 | | | 0.15 | 0,002 |
| | | 13.91 | 23.6 | | 59.23 | 2.54 | 0.042 | | 0.59 | 0.11 |
| | | 16.74 | 75.32 | 2.45 | 0.81 | 4.68 | 0.048 | 0.14 | 0.22 | 0.06 |
| | Siltstone | 16.91 | 75.73 | 2.82 | 0.12 | 4.41 | 0.070 | 0.17 | 0.22 | 0.06 |

**Table**. Lithochemical data analysis.

Titanium module (TM = $TiO_2/Al_2O_3$) of clayey rocks proved to be higher than that of arenaceous ones and comprises 0.17-0.38 to 0.038-0.050. Besides, the increased module TM with values of 0.068-0.070 is associated to humid climates of sedimentation while the values 0.017-0.050 – to the semiarid climate. Against the background of the dominated semiarid environment in the studied area, humidization and crust weathering prevailed, where there was an increased iron content. The crust weathering fostered the formation of terrigenous rocks accumulating through time. The humid climate is related to transgressive sedimentation stages and is found in the sediments of wells close to the sea.

Increased humidity of the sea coast resulted in the growth of hydrophilous vegetation: horsetails, ferns and Czekanowskia. If the humidization is analyzed vertically then it refers to the early Naunak suite formation, i.e. Late Bathonian. Aridization is a characteristic feature of Callovian-Oxfordian sedimentation.

Based on the calculations of the alumo-silicon module ($AM=SiO_2/Al_2O_3$), showing the rate of chemical weathering, the rocks formed due to the crust weathering destruction were identified. High values of AM (0.35-0.59) indicate the humid crust weathering washout.

The femic module ($FM=Fe_2O_3/SiO_2$) was used to identify the products of igneous rocks redeposition. The calculation shows the absence of pyroclastic material in rock formation.

The analysis of elemental composition of studied sediments characterized the paliosalinity of the depositional basin. The absence of B, Cl, Na, Rb, Ca elements in the spectra and high content of Fe characterize its desalting, which was proved by the lithologo-facial and biostratigraphical analyses.

The potassium module ($KM=K_2O/Al_2O_3$) [4] was calculated for oil\gas strata forecasting. The rocks with the coefficient 0.10-0.30 (sandstones of the river bed facies) may be considered to be oil\gas-bearing potential.

Geophysical data interpretation defined the reservoir properties and productive hydrocarbon formations.

Decreased values (on gamma-ray logging) and increased values (on neutron-gamma logging) indicated possible reservoirs.

Sequence-stratigraphic sectional layering method was applied to identify the structure and formation relationships in Naunak suite within Snezhnaya cross-section are related to the changeable migrating coast - line. Three main tracts were defined tracts of sedimentation systems conditioned by transgression and regression:

(1) tract of low water stand involves siltstones-mudstones with prograded bedding. During its formation the following were formed- flood plain facies, oxbow-lake facies and swamp facies. Singenetic bituminous rocks were found in the sediments of low water stand. The roof of this tract coincides with the transgressive surface, which is formed during the sea level rise and general transgression towards the continent.

(2) transgressive tract includes sandstones having retrograded bedding formed during fast sea level rise. Deposits of this tract are thought to be potentially oil-gas-bearing ones.

(3) tract of high water stand includes siltstones – sandstones, sandstones and siltstones. In calm hydrodynamic sea basin these deposits are flood plane facies and levees, and characterized by aggraded bedding.

## 4. Conclusions

Based on the analysis of lithologo-facial, geochemical and sequence-stratigraphic structure features of Naunak suite in well cross-sections of Snezhnaya area the polyfacial conditions of their formation was defined. Three lithofacial groups of deposits have been defined. They characterize various conditions of continental basin sedimentation which were influenced by transgressive-regressive marine activity.

Sand bodies formed in the intermediate sedimentation conditions during the transgressive tract formation could be potential hydrocarbon accumulations in Naunak suite. At this time channel facies formed which was proved by geophysical research.

The horizons of the syngenetic bituminous siltstones-mudstones of the lacustrine-swamp facies formed during the low sea level conditions can be related to oil\gas source rocks.

## References

[1] Alekseyev V P 2003 *Lithologo-facial analysis. Workbook in Lithology* (Ekaterinburgh: UGGA) p 147

[2] Baraboshkin E U 2011 *Core sedimentology. Terrigenous reservoirs* (Moscow) p 157

[3] Sklyarov E V 2001 Geochemical data interpretation (Moscow: Intermet Engineering) **1** p 288

[4] Maslov A V 2005 *Sedimentary rocks: Methods of study and data interpretation. Workbook* (Ekaterinburgh: UGGU) p 289
[5] Resolutions of the 6$^{th}$ Interdepartmental Conference on Mesozoic stratigraphy in West Siberia 2004 (Novosibirsk: SNIIGGIMS) p 113
[6] Yudovich Y E and Ketris M P 2011 *Geochemical indicators of lithogenesis (Lithological geochemistry)* (Syktyvkar: "Geoprint") p 740
[7] Galloway W E 1989 Genetic stratigraphic sequences in basin analysis I; architecture and genesis of flooding-surface bounded depositional units *A.A.P.G. Bull* **73** 125-142

IOP Conf. Series: Earth and Environmental Science **21** (2014) 012002 doi:10.1088/1755-1315/21/1/012002

# Application of the solid solution model for the description of the mineral composition of lake Baikal bottom settings

**A V Oshchepkova [1, 2, a], V A Bychinskiy [2, b] and E A Luk'yantseva [1, c]**
[1]Irkutsk State University, Irkutsk, Russia
[2]Institute of Geochemistry SB RAS, Irkutsk, Russia

E-mail: [a]oshepkova-anasta@mail.ru, [b]val@igc.irk.ru, [c]ekaterina-lukyanceva@mail.ru

**Abstract.** The study of clay minerals, accumulated in the Baikal bottom settings, provides valuable information about climate changes in Asia during the Cenozoic Era. However, the depth of a sedimentary section does not allow conducting mineralogical studies at a pitch providing a reconstruction of the paleoclimate record with high resolution. Earlier, the mineral composition of the Baikal sediments has been calculated on the basis of the results of a chemical analysis [1]. At a preliminary stage the simplex-method, used in this approach, requires the determination of a chemical composition of the clay sedimentary components and the calculation of conditional stoichiometric formulas of mixed-layer aluminosilicates, being the markers of the most important paleoclimate episodes. In this article the improved method for calculating the mineral composition of sediments is used to determine the mixed-layer minerals of a solid solution model, is represented. It allowed eliminating the preliminary stage of clay matter chemical composition calculation and computerizing the determination of minerals stoichiometric formulas, corresponding to warm and cold climate episodes in the reconstruction of paleoclimate changes.

## Introduction

The bottom settings of Lake Baikal store information about paleoclimate changes in Asia during the Cenozoic Era. A successful implementation of the project "Baikal Drilling" made it possible to get core samples of Baikal settings up to 600 meters deep [2]. Settings of the Academic ridge, exposed by wells BDP-96 and BDP-98 are of the great interest. The ridge is separated from the lake shores by two deep-water basins. Therefore its settings are formed with the substance, coming from the water layer. It allowed to determine the periods of "warming-cooling", cohering with the marine isotope curve, by the content of biogenic silica in the settings [3].

The study of material composition of Lake Baikal bottom settings, including their clay component, is an important part of paleoclimate reconstructions. It has been established that during the cooling period physical erosion predominated. It was characterized by coarse-grained settings, containing minor amounts of clay material. In warm climate conditions, chemical weathering processes become more active, and, respectively, soil formation processes, related to clay formation, intensify [4, 5]. Low salinity of Baikal water and low temperature prevent a mineral formation in the water layer, and as a result, the setting is formed of fine grain material, brought from the catchment basin. The great thickness of sedimentary layer makes a detailed X-ray phase analysis long-lasting and expensive. Therefore, before starting a detailed mineralogical analysis, it is advisable to use computational methods based on data of settings chemical composition [1, 6].

**Calculation of the mineral composition of Baikal bottom settings, based on the data of a chemical analysis**

In the works of E.P. Solotchina [5], G. Muller [7], the results of studies of a mineral composition of Baikal settings are presented. Due to the X-ray phase analysis, layered aluminum silicates, quartz and plagioclase (mainly, albite) were found in the deep-water settings of the Academic ridge. Potash feldspar and amphibole were found only in trace amounts. Clay minerals are presented as illite, chlorite, smectite, muscovite, kaolinite, being the main components of hemipelagic settings of the lake during the interglacial and glacial periods [5, 8]. Muscovite and chlorite are considered to be detrital minerals, brought out from the bed rock of the catchment area, the great amount of which indicates the climate cooling. Mixed-layer minerals – illite and smectite, formed in soil, increase in number during warm climate periods. It allows identifying the warm and cold climate periods based on the proportion of these mineral groups.

Previously, an algorithm has been developed to calculate the mineral composition based on chemical analysis data, and it applies the simplex-method, a significant limitation of which is the application of classical crystallochemical formulas of minerals. But mixed-layer minerals are interlaid minerals that are two- or three-layered aluminum silicates with adsorbed elements in the interlayer space. Due to the erratic chemical composition, the formalism of a mineral description is based on the application of variable coefficients in the stoichiometric formula that excludes the use of simplex-method in these calculations. In works [1, 6] it has been shown that if a total setting mineral composition is known, it is possible to calculate the total chemical composition of clay minerals, and on the basis of the latter to determine conventional stoichiometric formulas of the main groups of mixed-layered aluminum silicates: smectite, chlorite and illite. In this approach, the correlation of the total setting chemical composition with the data of the X-ray phase analysis is an obligatory stage of the calculation. As the X-ray phase analysis was performed for a restricted number of samples from the key horizons, it is hard to apply the simplex-method for the whole sedimentary section. The conventional stoichiometric formulas of clay minerals, calculated for samples corresponding to warm and cold periods, may add considerable error to results of mineral composition calculation for the same Holocene-Pleistocene settings from other cores taken in basins with different conditions of sedimentary accumulation.

Using the software complex “Selector” [9], the mineral composition calculation can be solved without a preliminary determination of the clay stoichiometry. The physicochemical model calculates a mineral paragenesis depending on the chemical composition and independent variables of the system state (temperature, pressure, volume, activity factors) [10]. Three models of solid solutions are used for a complex mineral phase description: subregular, Darken's quadratic formalism model and model of perfect disordered solid solution [10, 11, 12]. In the work the model of disordered solid solution was taken as the basis to calculate the equilibrium of the following mineral phases: illite, illite-smectite, chlorite. At the final stage, the equilibrium solid solution is calculated into a common conventional formula according to the mole quantity of minals presented in a solid solution based on methods proposed by A.G. Bulakh [13].

For Baikal bottom settings, the mineral composition is reconstructed; the setting is presented as a mixture of clay minerals (chemical weathering products) and minerals formed under completely different conditions: quartz, feldspar, muscovite, chlorite. The appropriate use of methods of the thermodynamic modeling of non-equilibrium mineral composition requires calculating the metastable Gibb’s potential for xenogenic minerals (muscovite, feldspar). This allows determining the metastable equilibriums and provides a high reproducibility of the physicochemical modeling with data of the X-ray phase analysis.

The principle of the thermodynamic model is a list of potential mineral in equilibrium, composed according to the data of the X-ray phase analysis. The greatest attention has been paid to the formation of the solid solutions of mixed-layered aluminum silicates – illite, smectite, chlorite. The actual isomorphic substitutions (replacements) occurring during the formation of mixed-layer mineral phase, can be described by mathematical methods using the list of minals, including all variations of its chemical composition changes. For illite and smectite 16 minals were selected, and 7 for chlorite. Temperature and pressure parameters were set close to standard: temperature – 20°C, pressure – 1 bar.

**Results of the mineral composition reconstruction based on solid solution model application.** The presented method for calculating mineral composition is used in the study of bottom settings from the well BDP-96 (the Academic ridge). Samples were taken from 18.45 and 32.59 m deep. The sample from the depth of 18.45 m is characterized by a high content of biogenous silica (39.6 – 46.2 %), and corresponds to the 11th warm interglacial stage according to the marine isotope-oxygen curve; its age is 410 thousand years. The setting from the depth of 32.59 m was accumulated during the 18th cold climate episode of the marine isotope-oxygen curve; its age is 760 thousand years. The biogenic silica content in this sample forms the first percentages [5].

These samples shall be considered as standard samples that allow assessing the calculation accuracy (table 1).

**Table 1.** Mineral composition of Baikal bottom settings from well BDP-96, mas. %.

| Depth, m | | Terrigene Sediment component (100%) | | | | | | | | |
|---|---|---|---|---|---|---|---|---|---|---|
| | | Non-layered minerals | | | Layered minerals | | | | | |
| | | Quartz | Feldspar | Total | Illite | Illite-smectite | Chlorite-smectite | Chlorite | Kaolinite | Muscovite |
| 18.45 | XRD | 12.60 | 12.60 | 25.20 | 23.00 | 34.00 | 5.20 | 4.00 | 1.90 | 6,70 |
| | Calculation | 12.75 | 20.91 | 33.66 | 23.93 | 30.96 | | 4.40 | 0.00 | 5.59 |
| 32.59 | XRD | 17.70 | 16.80 | 34.50 | 0.00 | 31.80 | 4.10 | 6.40 | 2.60 | 20,80 |
| | Calculation | 26.86 | 19.44 | 46.31 | 13.22 | 22.46 | | 5.81 | 0.00 | 11.76 |

Note. XRD – data of X-ray phase analysis; calculation – data received using the software "Selector".

It should be noted that it is not easy to coordinate the results of X-ray phase analysis with the calculation data and results of chemical analysis. As in any algorithm for a mineral composition calculation minals with the stoichiometric formula and strictly-defined coefficients are used, while in X-ray phase analysis the phase of mixed-layered minerals are defined as a set of aluminum silicate layers different in structure. Such phases possess an inconstant chemical composition.

However, in spite of different magnitudes of crystal structures content, evaluated by different methods, the proportions of mixed-layered silicates and non-layered minerals are preserved. The main purpose is not to develop a method of mineral composition precise calculating, but to develop an express method to assess the clay minerals content, which allows identifying cold and warm episodes and decoding climate changes within the region during the studied period. Both methods distinguish invariably of warm and cold climate episodes. The change of illite / illite-smectite proportion turned out to be a reliable indicator of paleoclimate changes; the X-ray phase analysis describes this proportion as the content of illite layers, while the physicochemical models describe it as the increasing of illite content [6].

In simplex-method calculations by a chemical composition of clay part setting the conditional formulas have been calculated, describing the illite-smectite group in aggregate [1]. It has been shown on its base that the content of non-layed minerals along with muscovite and chlorite increases in settings accumulated in cold climate episodes, and illite and illite-smectite prevail in warm climate periods. The conditional formulas of mixed-layered minerals were calculated out of the total clay chemical composition at the first stages of the study, which deprives this approach of universality.

The application of the solid solution model allows determining the conventional formulas of mixed-layered aluminum silicates, related to glacial and interglacial episodes. They were calculated automatically without preliminary calculations. As it was noted above, it is hard to show the structure and composition of mixed-layered mineral, such as illite, with a strict stoichiometric formula. Usually, it is described as $K_{0.75}(Al_{1.75})Re_{0.25}(Al_{0.5}Si_{3.5})O_{10}(OH)_2$, where Re – is divalent cations of Mg or Fe, interlayer cations $K^+$ can be partly replaced with $Ca^{2+}$, $Mg^{2+}$, $H^+$, sometimes – with $Na^+$ [5]. The following conventional formulas of mixed-layered minerals are automatically calculated in the solid solution model:

Depth 18.45 m:

illite $K_{0.5}Fe\ Al\ [Al_{0.5}Si_{3.5}]_4\ O_{10}(OH)_2$; (23.93 % weight)

illite-smectite $K_{0.013}Na_{0.053}Ca_{0.106}Fe_{0.218}Mg_{0.43}Al_{1.439}[Al_{0.107}Si_{3.893}]_4\ O_{10}(OH)_2$. (30.96 % weight)

Depth 32.59 m:

illite $K_{0.5}Fe\ Al\ [Al_{0.5}Si_{3.5}]_4\ O_{10}(OH)_2$; (13.22 % weight)

illite-smectite $K_{0.013}Na_{0.074}Ca_{0.094}Fe_{0.226}Mg_{0.426}Al_{1.434}[Al_{0.109}Si_{3.891}]_4O_{10}(OH)_2$. (22.46 % weight)

It is essential to determine the general stoichiometric formula because the layers of various types, separated by different interlayered intervals, alternate in mixed-layered minerals [5].

These peculiarities of mixed-layered minerals structure and composition reflect changes in physical and chemical conditions of weathering process, thus, the mixed-layered phases, related to both warm (illite), and cold (illite-smectite, smectite) episodes, may be formed from the same solid solution.

**Conclusions**

Thus, the data calculated using the physicochemical model do not give accurate quantitative information about the mineral composition, but these data contain an expressive evaluation of the clay to non-layered minerals proportion, which allows reconstructing the catchment weathering conditions based on chemical composition only.

The mixed-layered minerals were included in the model as solid solutions (a set of minals), enabling to describe a wide range of aluminum silicate layers proportions. The calculated results were compared with the data obtained by XRD-method (see Table 1). It should be noted that modeling of mineral composition is the way to identify warm and cold climate episodes. Despite the variations of calculated amounts of mineral phases from the XRD-analysis data, high accuracy in identifying the main climate episodes by proportions of clay and detrital phases is preserved. In warm periods the illite content and total amount of clay minerals increase (the amount of illite, illite-smectite, chlorite-smectite), in cold periods, the non-layered minerals (quartz, feldspar) and detritak muscovite and chlorite prevail in settings.

The evaluation of the developed method will be tested on key sections of Baikal deep-water, settings of the Holocene-Pleistocene period. A wider application of the developed method requires a specified list of minals of illite and montmorillonite solid solutions, which is the main target of further studies.

The work has been supported by Irkutsk State University; Individual Research Grant No.111-13-218.

**References**

[1] Oschepkova A V, Kuz'min M I and Bychinskii V A 2013 Mineral composition reconstruction of Baikal's deep-water deposit based on chemical composition *Bulletin of Irkutsk State University Series "Earth sciences"* **6**(1) 122-132

[2] Kuz'min M I 2001 Deep drilling on Lake Baikal: main results *Geology and Geophysics* **42** (1-2) 8 – 34

[3] 2008 *Global and Regional Climatic and Environmental Changes in Siberia in the Late Cenozoik,* ed A P Derevjanko (Novosibirsk: Press SB RAS) p 511
[4] Millo J 1968 *Clay's Geology* (Leningrad: Nedra) p 359
[5] Solotchina E P 2009 *Structural typomorphism of Clay Minerals in Sedimentary Sections and Crusts of Weathering* (Novosibirsk: Geo) p 234
[6] Kuz'min M I 2014 Chemical composition of sediments in Baikal deep-water boreholes as a basis for reconstructions of climatic and environmental changes *Geology and Geophysics* **55** **(1)** 3 – 22
[7] Müller J 1998 TEM analysis of smektite-illite mixed-layer minerals of core BDP-96 Hole 1: Preliminary Results *The 1998 BBD Baikal Symposium of the Japanese Association for Baikal International Research Program (JABIRP) November 5-8, Yokohama, Japan* pp 90–100
[8] Kuz'min M I 2000 Clay minerals in bottom sediments in the Lake Baikal as indicator of paleoclimate *Geology and Geophysics* **41** (10) 1347 – 1359
[9] Karpov I K 1981 *Physicochemical modeling in Geochemistry* (Novosibirsk: Nauka) p 247
[10] Avchenko O V, Chudnenko K V and I A Alexandrov 2009 *Fundamentals of physical and chemical modeling of mineral systems* (Moscow: Nauka) p 229
[11] Aranovich L Ya 1991 *Mineral equilibria of multicomponent solid solutions* (Moscow: Nauka) p 256
[12] Kuperin V A 1981 *Thermodynamics of minerals of variable composition and geological thermobarometry* (Kiev: Naukova Dumka) p 160
[13] Bulakh A G 1964 *Guide and Tables for Calculation of Mineral Formulas* (Moscow: Nedra) p 131

IOP Conf. Series: Earth and Environmental Science **21** (2014) 012003 doi:10.1088/1755-1315/21/1/012003

# Mineral and Elemental Composition Features of "Loose" Oolitic Ores in Bakchar Iron Ore Cluster (Tomsk Oblast)

**M Rudmin[1], A Mazurov[2] and L Bolsunovskaya**
Tomsk Polytechnic University, Tomsk, Russia

E-mail:[1]rudminma@tpu.ru, [2]akm@tpu.ru

**Abstract.** Geo-technological investigation considerations of iron ore deposits within the Bakchar ore cluster are being carried out. The mineral and elemental composition of "loose" ores have been studied, embracing such important aspects as the distribution pattern of valuable and harmful impurities, the determination of element concentrators (such as vanadium, phosphate and sulphur) in basic minerals and the analysis of ore composition varaiation in volume ore cluster.
Based on investigation results the mineral and elemental composition characteristic features of "loose" ores were defined. Although hydrogoethite was the basic identified ore mineral, such minerals as goethite, lepidocrocite, leptochlorite, siderite and hisingerite were also found. The deportment of calcium phosphate (anapaite) and phosphates of rare-earth elements (monazite, killarite), which are associated with the harmful impurity- phosphorous, are described. It has been defined that the ore constituent composition contains such persistent impurities as vanadium and manganese, the content of which is 0.35% and 0.03%, respectively. The "loose" ores are continuous in mineral composition, both in area and cross-section throughout the Bakchar ore cluster. Based on the sample element composition analysis the most perspective areas for further mineral processing could be: western with the fraction of 1....0.2mm. and eastern- fraction of 1...0.1mm.

## 1. Introduction

Bakchar ore cluster is located in south-eastern West-Siberian iron-ore basin, about 200km. north-westward from Tomsk (Figure 1). The total cluster area is 1200km$^2$ and the ultimate inferred resources involve 28bln. tonnes, classifying this formation as unique. These iron-producing deposits are typical coastal-marine facies sediments- gridstone, oolitic ores (minette), sandstone, siltstone (aleurolite) and clays. The ores are generally oolites of gothite-hydrogeothite and chlorite-hydrogeothite composition and are located in three formations (upward): Narimsk, Kolpashevsk and Bakchar [1, 2]. "Loose" ores formed under conditions of oolitic sand facies and are confined to the Bakchar bottom formation [3].

Since the ore finding and up to the present day numerous researchers, such as A. Babin, I. Nikolaev, N. Belous, A. Kondakov, U. Kazanski, A. Mazurov, V. Domarenko, A. Pshenichkin, L. Rikhvanov, E. Asochakova and others, have been exploring and surveying the iron ore deposits in the West-Siberian basin [1, 2, 4, 5]. However, despite the significant contribution of the above-mentioned researchers, the outstanding Bakchar ore valuation still remains an open problem.

Based on mineral composition and ore specifications the following ore types were distinguished by these researchers [1, 2, 4, 5]: hydrogoethite oolitic loose ore; gothite-hydrogoethite with siderite cement; leptochlorite-hydrogoethite oolitic ore with chlorite-argillaceous cement; goethite-hydrogoethite oolitic ore with hydromica cement; glauconite-hydrogoethite oolitic ore with siderite-chlorite cement. The ore components of these types are gothite-hydrogoethite and chlorite-hydrogoethite oolites, while the difference includes the nature and composition of the cement itself, which, in its turn, is determined by the facies environment. The classification of the above-mentioned ore types can be identified only in laboratory conditions by applying the detailed optical microscopic method. In this case, it is impossible to perform ore geometrization by type during geological prospecting and exploration, thus the identification of only 5-7 ore types is impractical and of little importance.

As a result of the detailed investigation of material composition and ore texture-structure specifications three natural ore types were identified: “loose”, medium and poorly cemented (or continuous) and closely cemented (sideritic). These ore types macroscopically contour ore sequence cross-sections as observed in drill-hole core samples.

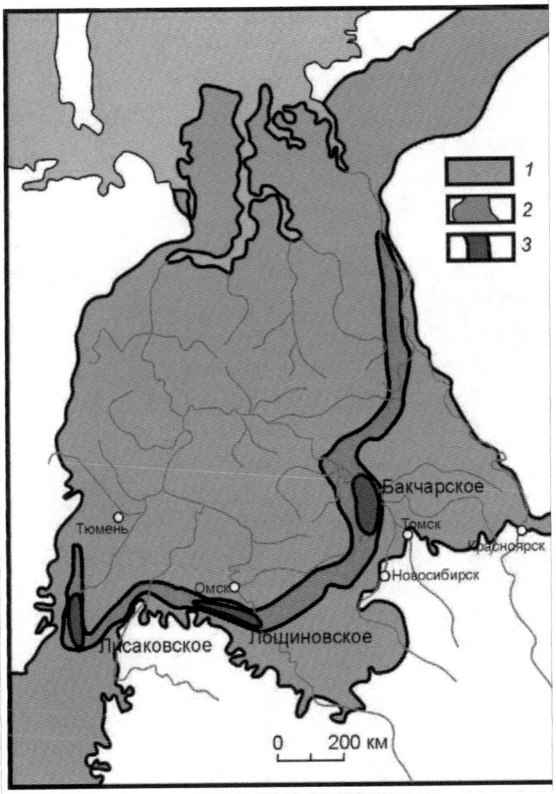


**Figure 1.** Map of West-Siberian iron-ore basin [2, 4]: 1 – Cretaceous-Paleogene sea area; 2 – occurrence zone of coastal-marine iron ores; 3 – area of vast iron ore clusters (deposits).

## 2. Data Sources and Methodology

Based on the results of optical observations, X-ray fluorescene and electron microscopy, infrared spectroscopy (IR – spectroscopy) and mass-spectroscopy with inductive-coupled plasma (ICP-MS) the mineral and elemental composition characteristic features of “loose” oolitic ores have been considered. Cores were retrieved from the lower Bakchar formation (“loose” ore deposits) of the western and eastern ore cluster areas. The core samples were divided into the following granulometric grades: more than 1mm; -1 - +0.5 mm; -0.5 - +0.2 mm; -0.2 - +0.1mm and less than 0.1mm. The testing of these core sample groups was performed by the "wet" screening method and further investigated under ore microscopy. Next, polished pressed blocks were made from above-mentioned fractions, then investigated in reflected X-ray fluorescene (HORIBA XGT-7200) and electron (Hitachi S-3400N) microscopy. Fraction mineral composition of less than 0.1mm was investigated on the infrared spectroscopy (IR-Fourier IR Prestige-21). The sample elemental composition was determined by ICP-MS method at the premises of Tomsk Polytechnic University (TPU).

## 3. Results and Discussion

In view of the results the granulometric composition of "loose" ores within the ore cluster area was determined as consistent. The following grade yields were determined: fraction more than 1mm-6.3%; -1 - +0.5 mm – 13.6%; -0.5 - +0.2 mm – 60.7%; -0.2 - +0.1mm – 12% and less than 0.1mm-7.3%. The highest grade yield was for fraction (-0.5 - +0.2 mm), which indicates that the bulk of ore components is concentrated here.

Based on optical and microscopic investigations the Bakchar cluster oolites can be classified as chlorite-hydrogoethite and hydrogoethite grains being of relatively consistent mineral composition. The average elemental composition of chlorite-hydrogoethite and hydrogoethite grains was determined on the X-ray fluorescene microscope. Due to the alumino-silicate mineral predominance, the first group is distinguished by reduced iron and phosphate content, but by increased silica, aluminuim and potassium content. Hydrogoethite grains are characterized by the following elemental composition (based on X-ray fluorescene analysis-XFA data): Fe – 58,59 %, Si – 2,73 %, P – 1,17 %, Al – 3,17 %,, K – 0,15 %, Mg – 0,15 %, Ca – 0, 16 %, Ti – 0,09 %, V – 0,16 %, O – 33,23 % and others.

Oolites and chlorite-hydrogoethite peas have a rather diverse mineral composition, being of 0.5mm. Such grains have consistent ilmenite impurities (figure 2), phosphate minerals and, rarely, zircon. The impurity size of titanian minerals is from 1 to 80-100 microns, including ilmenites and, presumably, pseudobrookites ($TiO_2$ – 20-30 %; $Fe_2O_3$ – 65-70 %). In some cases, tens of titanian impregnations can be found in one ore grain. Besides, there is zircon impregnation of 30-100microns in chlorite-hydrogoethite peas. According to the elemental composition, titanian and zircon mineral impregnations in oolites and peas are similar to those of ore minerals in Tugan ilmenite-zircon placer deposit (40km north-eastward from Tomsk). This assumes the fact that the source area for these components is common.

Consistent phosphate minerals of 1-2 to 50 microns are found in ferrous oolites and peas. There are different phosphates in composition: kularite, anapaite and monazite, whereas only anapaite impregnation is of 20-50 microns, while the remaining phosphates are abundant in ore components as micro-inclusions, concentric zones (figure 3A) and dentritic aggregates. However, the latter can only be found in chlorite-hydrogoethite pea composition. Rare-earth phosphates (kularite) which form either micro-inclusions or single interrupted circular zones (figure 3) are predominant in hydrogoethite oolitic with regularly zonal structure. In some cases, significantly disintegrated grains of prismatic habitus can be found, which, in composition correspond to monazitet with an increased content of La, Ce (18-20 % for each). Not only quartz, feldspar, glauconite, leptochlorite, magnetite, sphalerite and rutile (figure 3C), but also zircon, brazilite and galenite (figure 3B) can be found in oolite ores. Different terrigenous fragments ( quartz, feldspar, goethite, hydrogoethite, magnetite, glauconite and clay mineral concentrations) act as "seed" components of oolite kernels.

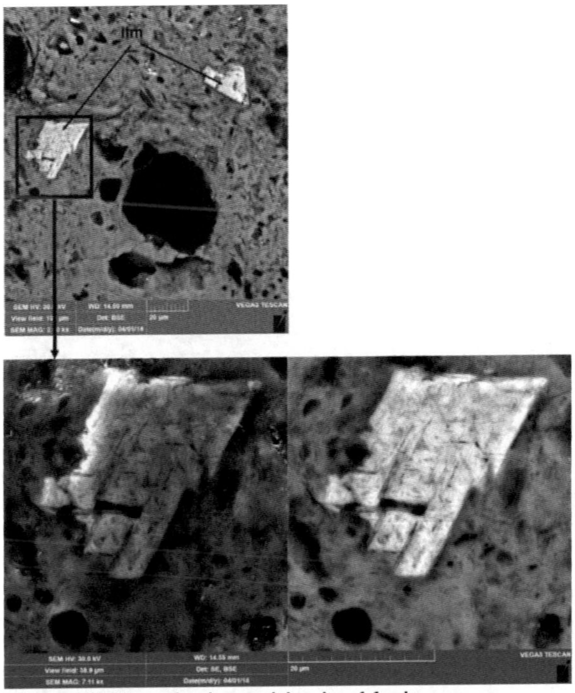


**Figure 2.** Ilmenite impurities in chlorite-hydrogoethite grain fraction of 1…0.5 mm [ backscattered electron image ( bottom left) and secondary electrons (top and bottom right)]

According to the IF-spectroscopy results, characteristic sample wavelength frequency of less than 0.1mm was obtained. The mineral composition fraction has a continuous behavior within the ore cluster area. Visible divergence could be observed only in those samples which are confined to the Kolpashevsk formation. This divergence can be explained by the occurrence of carbonate mineral in the sample itself and this fact is based on the specific IF spectrum frequency peak of 1420 and 877 $cm^{-1}$. Specifying the fraction less than 0.1mm of "loose" Bakchar formation ores, it should be noted that these samples are of heterogeneous composition, which, in its turn, significantly impedes the precise and detailed diagnostics of the mineral phases. However, all the samples are approximately of one and the same composition.

Based on the IF-spectroscopy results, the following minerals can be found in the samples: ferrum oxides (hydrogoethite, lepidocrocite), clay minerals of the nontronite-beidellite group, purpurite and hisingerite. Ferrum oxides are determined by the specific frequency peaks of 912, 800..796, 488 $cm^{-1}$; clay minerals -maximum double peaks of 1032, 1008 $cm^{-1}$, as well as 912 $cm^{-1}$ The double peak of 1032, 1008 $cm^{-1}$ indicates the presence of aluminosilicate anion in the sample; however, this could superpose anion peak $PO^{4-}$ which impedes the identification of phosphate minerals. Purpurite is identified in some spectrums of double peak 796, 777 $cm^{-1}$. Carbonate minerals with wave frequency of 1420 и 877 $cm^{-1}$ ,which are probably ankerite, can be found in Kolpashevsk formation samples [6]. Carbonate minerals were also identified on the basis of X-ray structural analysis.

Elemental composition of "loose" ores are characterized by an average content of the following: iron - 35 and 38 %, phosphate - 0.41 and 0,32 %; vanadium – 0.16 and 0.18 % within the western and eastern areas, respectively. In this case, the iron content increases up to 5…8 % in a fraction of maximum yield percentage (fraction 0.5…0.2 mm).

The granulometric grade 1…0.2 mm within the western area includes a total yield of about 80 %, whereas the average content of iron, phosphate and vanadium would be 37.5, 0.49 and 0.17 % respectively, which, in its turn, simplifies the further treatment process of these ores. The granulometric grade 1…0.1 mm within the eastern area includes a total yield of about 86,2 %, whereas the average content of iron , phosphate and vanadium would be 39.4, 0.35 and 0.19 %, respectively, thus, regarding these ores as high grade ones.

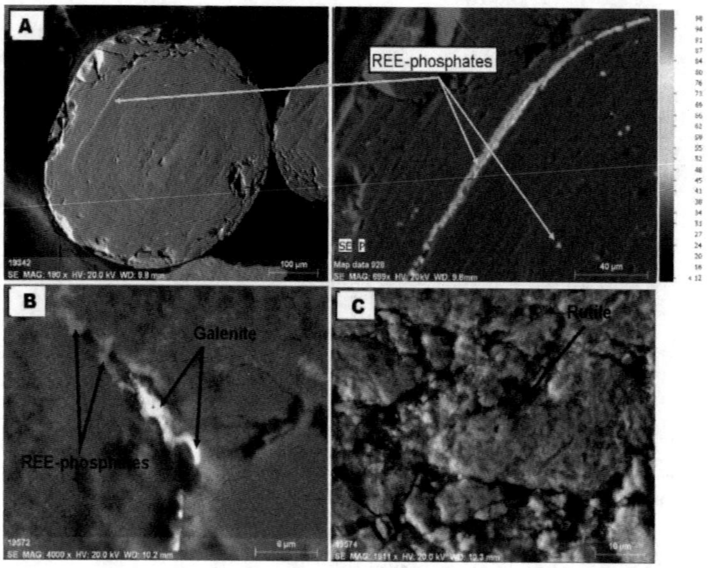


**Figure 3.** Micro-inclusions in hydrogoethite oolites of fraction 0.5…0.2 mm ( backscattered electron image): A- concnetric zones and micro-inclusion of rare-earth phosphates (kularite); B- galenite in rare-earth phosphate within micro-cracks; C- singe rutile impregnation.

## 4. Conclusions

In this paper the Bakchar cluster "loose" ores have been investigated. The results suggest that:

1. "Loose" oolitic ores are of medium-grained structure. The average yield of fraction more than 1mm includes 6.3 %; fraction of 1…0.5mm – 13,6 %; fraction of 0.5…0.2 mm – 60,7 %; fraction of 0.2…0.1 mm – 12 %; fraction of less than 0.1mm – 7,3%. The highest grade yield was for fraction (-0.5 - +0.2 mm), which indicates that the bulk of ore components is concentrated here.
2. "Loose"ore-types include hydrogoethite and chlorite-hydrogoethite oolites, lepidochlorite-clay grains, lepidocrocite, hisingerite, clay minerals, siderite, glauconite, quartz fragments, opal and feldspar. Inter-mineral impregnations in oolites and peas include quartz, feldspar (orthoclase), sphalerite, rutile, zircon, beidellite, magnetite, calcium phosphates (anapaite), phosphates of rare-earth elements ( monazite, kularite). Different terrigenous fragments - quartz, feldspar, goethite, magnetite, glauconite and clay mineral concentrations- act as "seed" components. The major hydrogoethite oolite bulk is concentrated in fraction 0.5...0.2mm with characteristic impregnation- minerals- kularite, ilmenite, rutile and galenite. The most significant group (more than 0.5mm) involves an equal distribution of chlorite-hydrogoethite and hydrogoethite oolites and a rather impregnation diversity. Fraction less than 0.1mm includes clay minerals (nontronite-beidellite group), hydrosilicate minerals (hisingerite) with subordinate quantity of lepidochlorite (cronstedtite), purpurite, siderite, oolite ore

IOP Conf. Series: Earth and Environmental Science **21** (2014) 012003 doi:10.1088/1755-1315/21/1/012003

fragments and glauconite.

3.The basic "loose" ore mineral is hydrogoethite, where the iron content is 70%, while the remaining ferrous minerals include goethite, lepidocrocite, lepidochlorite, siderite and hisingerite. The harmful phosphate impurities can be consistently found in the ores. The minerals associated with phosphate impurities are calcium phosphates ( anapaite) of 5…60 microns and phosphates of rare-earth elements (monazite, kularite) of nor more than 10 microns. Calcium phosphates are impregnations in fractions of more than 0.5mm, the average yield of which is not more than 20%. In the productive fraction 0.5…0.2 mm the phosphate impurities is associated with kularite which is observed as concentric zones and microscopic impregnations of 5microns in oolite structure. Microscopic phosphate inclusions in the productive fraction significantly affect processing results due to the fact that it is impossible to extract such inclusions mechanically. The remaining quantity of harmful impurities such as arsenic, copper, zinc, lead and tin, does not exceed allowable value for oolite ores. Besides phosphate there are also such impurities as vanadium and manganese, the content of which is up to 0.35 % and 0.3 %, respectively, in ore components. The content of titanium in ores is 0.8% and is associated with the following mineral impregnations: ilmenite, rutile and their fragments. The size of ilmenite and rutile impregnations is up to 80…100 microns, which means that it is possible to reveal them in association with quartz and feldspar impurities, if the coarseness is less than 0.2…0.1 mm.

4. Based on the sample analysis, the "loose" ore-types, retrieved from different areas and formations, are continuous in mineral composition, both in area and cross-section throughout the Bakchar ore cluster.

5. The elemental composition of "loose" ores within the western area includes the following average content: iron-35 %, phosphate- 0.41 % and vanadium-0.16%, whereas within the eastern area, the average content includes: iron- 38 %, phosphate- 0.32 % and vanadium-0.18 %. The most productive granulometric grade 1…0.2 mm, within the western area, under conditions of 80% yield includes the following average content: iron- 37.5%, phosphate-0.49% and vanadium- 0.17%. The most productive granulometric grade 1…0.1 mm, within the eastern area, is about 86.2% yield. The average content of iron, phosphate and vanadium in this group is 39.4 %, 0.35 % and 0.19 % , respectively. Thus, the fraction 1…0.2 mm (in the western area) and the fraction 1…0.2 mm (in the eastern area) are the most promising ones in future processing.

6.It can be concluded that the basic process in oolite ore formation is oxidation of previously formed chloritic and hydromicaceous minerals with a high iron content, as well as microbiological processes and colloidal exsolution. In general, the formation of iron-ore oolites can be characterized as a multi-phase and heterogenous process.

## References


[1] Babin A 1969 Bakchar iron ore deposit (geology, occurrence characteristics and genesis of iron ores) *Ph.D. Dissertation*

[2] West Siberian iron ore basin 1964 *SO RAN SSS*

[3] Rudmin M 2014 Facial features and magnetic susceptibility of ore-hosting sediments of Bakchar iron ore occurrence (Tomsk region). *Bulletin of Tomsk Polytechnic University* **324**(1) 48–55

[4] Mazurov A, Bojarko G, Emeshev V and Komarov A 2006 Prospects of Bakchar ore field development *J. Ores and metals* **2** 64–70

[5] Nikolaeva I 1967 Bakchar oolitic iron ore deposit *SO AN SSSR* p 129

[6] Barbara Stuart 2004 *Infrared spectroscopy: fundamentals and applications*

IOP Conf. Series: Earth and Environmental Science **21** (2014) 012004 doi:10.1088/1755-1315/21/1/012004

# Volumetric geochemical model of Chortovo Koryto gold-ore deposit (Patom range)

**R Gavrilov[1], M Khmelev and R Abramova**
Tomsk Polytechnic University, Tomsk, Russia

E-mail: [1]GavrilovRY@ignd.tpu.ru

**Abstract.** Core sampling analysis from vertical exploration wells was conducted by the statistical methods. Assayable gold samples were analyzed by the atomic absorption method, while trace elements- emission-spectral semi-quantitative method. It was determined that the grade distribution of gold and trace elements within the deposit itself does not match one of the partition laws. Applied complex mode of processing of geochemical data provided the interpretation of anomaly geochemical field structure. The analysis of geochemical association distribution showed that their structure is concentric-zonal within bulk-metasomatic halo. The geochemical association distribution, in general, is related to hydrothermal minerogenesis stages in a deposit. Based on the mining-geological Datamine studio 3 program 3-D geochemical deposit model was plotted.

## 1. Introduction

An important investigation stage in forecasting and prospecting ore deposits includes exploration and modeling by applying sophisticated mining-geological programs. Not only are familiar methods up-dated, but also new advanced ones are introduced in processing geochemical data.

Future ore forecasting is based on previously developed prognostic exploration models: geological-geochemical, geological-geophysical, geological-structural and others. By applying developed models comparative analysis of these models to perspective areas and bodies of industrial mineralization can be used.

Previous geochemical studies within Chortovo Koryto ore field involved not only the geological history remodeling of chemical elements in ore-hosting metasomatic haloes, but also the identification of bulk-geochemical halo structure. It was determined that the ore-hosting geochemical gold-silver-arsenic deposit halo has a flat cubic configuration oriented along the ore-controlling and solution-feeding fault and dipping westward.

The research aim was to plot a 3D geochemical deposit model to develop forecasting-exploration criteria for gold mineralization.This model was plotted on the basis of obtained interpretation results of anomaly geochemical field structure (AGFS) [1].

## 2. Regional setting

The geological characteristics of the deposit are particularized in [3].

The deposit is in the north of Irkutsk oblast in Patom range (figure 1) and is confined to the ore-grade gold of black-shale formation. The deposit is located in the brachyfold syncline, composed of early Proterozoic carbonaceous rocks of Mikhailovskoe suite (figure2). The dip angle of formation rocks is from 0 to10-20°. There are high-order step folds within the fold itself. The formation is composed of rocks metamorphosed to epidot-amphibolitic facies: poorly sorted arkose, fine-grained sandstones, siltstones and mudstones. Carbon content in rocks of mineralized zones vaeries from hundred to 2 % volume. Metasoamntic rock alterations are superimposed by propyltization and beresitization forming gold-sulphide-carbonate-quartz veins and veinlets.

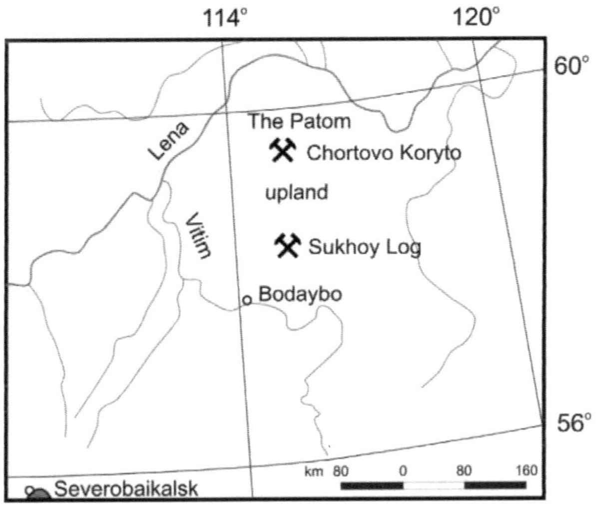


**Figure 1.** Map of Chortovo Kuryto deposit.

Ore deposit, with contoured cut-off grade of 0.5 gr/tonne and westward dipping at an angle of 0 – 20°, extents 1800m in length and about 500m in width. Its maximum thickness of 140m is in the central part of the ore body. Gold concentration varies from n×0.01 to n×100 gr/tonne. The ore deposit size and its reserves can change depending on cut-off grades.

## 3. Research methodology

Chortovo Koryto was investigated by applying core sampling boring. Based on the exploration results the relevant reserve categories were estimated. Assaying included: section-type core- to determine gold grade and section-type chip to determine trace elements in gold –ore grade and their behavior within the ore zone. Core sample length is in average 1m., section-type chip sample- 3m. 3D plotted geochemical model is based on detailed study of 6 exploration well profiles.

Gold samples were analyzed by assaying and atomic-absorbent methods in laboratory environment in Lensk gold-ore company. Method sensitivity is 0.1 gr /tonne and 0.01 gr/tonne, respectively.

The trace-element content in mineralization was analyzed by the emission-spectral semi-quantitative method in the laboratory of Bodaibinsk geo-exploration expedition. The following elements were determined Ag, As, Pb, Co, Cu, Ni, Zn. While the remaining elements were found in limited amount and could not be used in statistic calculations.

Integrated indicator calculations were based on analytical data obtained from section-type chip samples (n>4000). Background is determined by the analysis results of 170 samples, retrieved from those wells located outside the ore-metasomatic zones.

To plot 3D geochemical model and interpret anomaly geochemical field structure (AGFS) calculations of grade correlation, mineralization energy, dispersion and geochemical spectrum variations were conducted, as well as cluster and factor analyses.

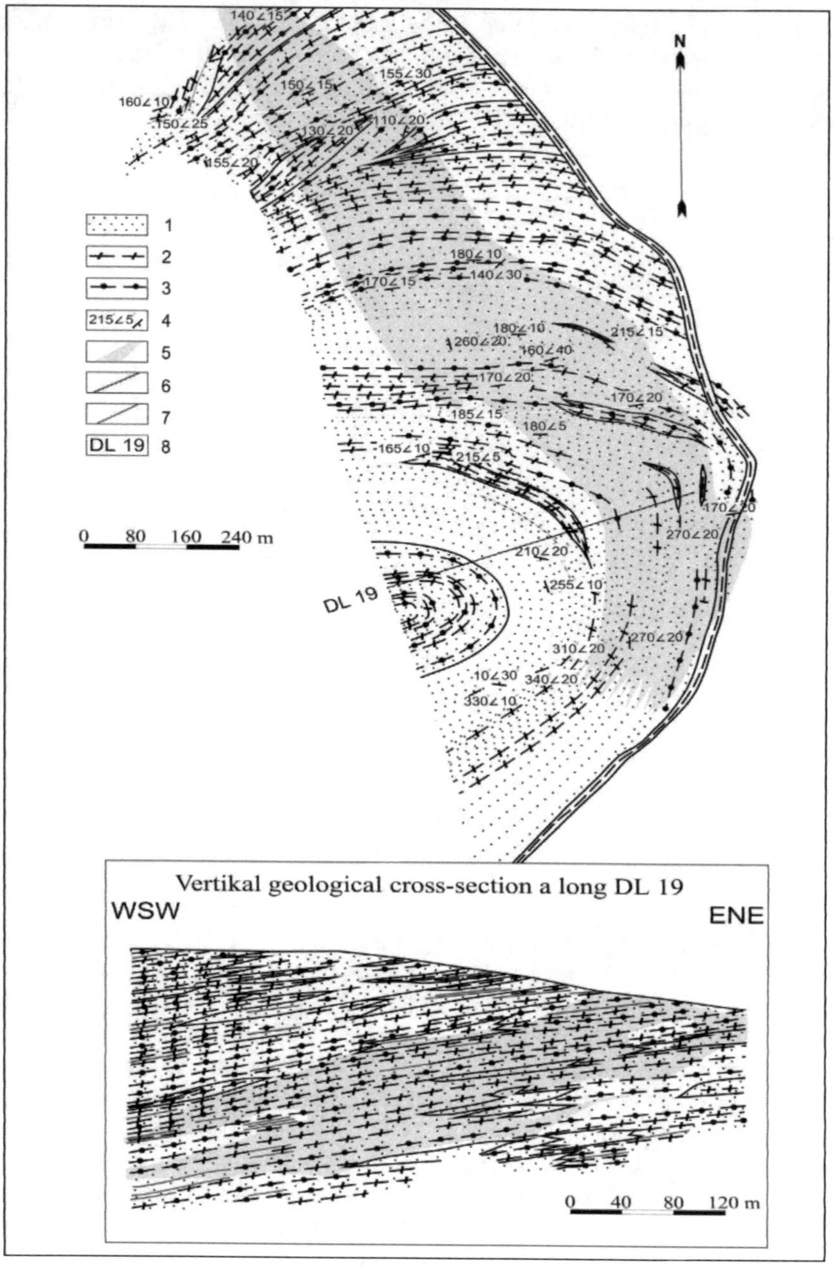


**Figure 2.** Geological map of Chortovo Koryto deposit structure: 1) generally, poorly- sorted grey to dark grey, from fine-grained to medium-grained, indistinct layered to coarsely stratified feldspar-quartz sandstones, silty sandstones, carbon and carbon free; 2)generally, poorly sorted dark- grey anisometric, from fine-grained to coarse-grained finely-stratified to coarse-stratified feldspar-quartz siltstones, sandstone siltstones, carbon and carbon free; 3) sericitic mudstones with insignificant content ( up to 25 volume %) and / or lack of fragmental fraction of sand and / or silt size, carbon and carbon free; 4) azimuth and layer angle of penetration; 5) sulphide mineralization zone; 6) zones of thin cleavage and rock breaking; 7) back zones (chlorite, albite and beresite) of ore-hosting metasomatic halo [3].

## 4. Spatial distribution of gold and trace elements

Grade distribution of gold and trace elements within the deposit itself does not match one of the partition laws.

High-contrast gold halo is presented by a large-volume body extending N- S. The internal halo structure is characterized by pipe-like, acicular and tabular bodies. Within the ore zone the element concentration incomparably exceeds the clarke (percentage abundance) level. In the south direction there is a significantly decrease of halo areas and ore zones.

Within the deposit sub-clarke silver concentrations are predominate. Silver halo is characterized by a nest-like structure. Most of the halo volume includes moderate-contrast concentrations. Moderate and high-contrast haloes are pipe-like and acicular bodies.

Field allowable of the high-contrast arsenic halo shows an oval -shaped elongated body which covers practically all the ore zone.The inner halo includes pipe-like, acicular and flame-shaped bodies. Most of the ore zone embraces haloes with elements incomparably exceeding high clarke concentration. Increased arsenic concentration is due to its mineral -carrier – arsenic pyrite and insignificant amount of other sulphides.

The basic deposit bulk is sub-clarke lead concentration. Moderate contrast haloes have nest-like distribution which are pipe-like and acicular. Increased lead concentration is due to basic mineral-carrier- galenite found in the ore zone.

The highest volume in ore-hosting metasomatic halo embraces sub-clarke zinc concentrations. Poor contrast haloes are oval-shaped. Ther halo structure is characterized by oval-shaped, pipe-like and isometric nests, rarely acicular shaped. Most likely, increased element content is associated with sphalerite.

In deposit volume the contrast cobalt halo is oval-shaped. The inner halo structure is characterized by pipe-like, tabular and isometric bodies. In most of the ore zone the element concentration is significantly higher than the clarke level itself. Such contrast haloes are conditioned by cobalt arsenic and other sulphide minerals within the large-volume metasomatic haloes.

Within the metasomatic halo the presence of contrast nickel halo is conditioned by the presence of its mineral carrier- pyrrhotine, as well as cobalt arsenic and some other minerals, containing it as an impurity. Haloes with different intensity are located within the ore zone contour. In cases of the coupling of the most contrastive haloes with high-contrast gold haloes is rather rare.

Moderate-contrast copper concentrations occupy most of the ore-hosting metasomatic halo. The inner halo structure includes a pipe-like, isometric and acicular body system.

Thus, the most contrast anomalies in large-volume metasomatic halo are produced by Au, As, Ag and Pb.

## 5. Discussion

The interpretation of anomaly geochemical field structure (AGFS) was performed by applying the mathematical statistics method.

According to the grade correlation method it was defined that relevant positive correlation is distinctive for all oregenetic elements. However, in this case, it was practically impossible to classify these elements into separate groups.

Intensive hydrothermal-metasomatic ore formation is described by the integrated indicators of ore-formation energy [4] and dispersion and the variations of geochemical spectrum [2].

Ore formation energy is calculated on the basis of the total product of element concentration clarkes by their logarithms. Background value changes from –1 to 2 units within the mineralization zone are rarely observed. Indicator values could be up to $10^4$ units and even more. Ore formation energy values demonstrate a direct dependence from As content value.

IOP Conf. Series: Earth and Environmental Science **21** (2014) 012006 doi:10.1088/1755-1315/21/1/012006

# Calculation of quarzite crystallinity index by infrared absorption spectrum

**O Razva, A Anufrienkova, M Korovkin, L Ananieva and R Abramova**
Tomsk Polytechnic University, Tomsk, Russia

**Abstract**. Quarzites retrieved from the Sopks-248 deposit and Belokamenka deposit, Antonovsk field cluster (Russia, West Siberia) were studied by the infrared spectroscopy method. The silicate thickness alteration rate was determined and the most crystalline quartzite variations were identified by defining the crystallinity index $K_i$ according infrared absorption spectrum. To calculate "crystallinity index" the ratio of quintet peaks of 778 $cm^{-1}$ as to peaks of 695 $cm^{-1}$ were applied. It should be noted that the most crystalline quartzite variations are characterized by the relative value of crystallinity index.

## 1. Introduction

Due to the recently increased production of new materials and products from quartz, the demand for high-quality quartz raw materials has increased. Its application is rather diverse, including fiber optic communication systems, special glass, ceramic materials and unique products from silicon carbide (SiC) and silicon nitride ($Si_3N_4$).However, of great commercial importance is the silicon "solar" quality in manufacturing photovoltaic cells – solar panels.

In the context of increasing scarcity of rock crystal which was and is traditionally used as a high-purity quartz, the more and more attention is being paid to quartzites, which could be a potential high quality raw material source in high technology [1].

In this regard we have investigated the quartzites of Antonov field cluster (Russia, Western Siberia) [2]. According to its origin quartzites are sedimentary-metamorphic sediments being a lithification product formed under conditions of early metagenesis of quartz- sericite hydromica series [3]. As a result of biogenic siliceous strata metamorphism amorphous silica crystallization occurred and the highly pure crystalline alpha-quartz phase emerged.

To evaluate the quartzite alteration degree, the quartz crystallinity index $k_i$ proposed by Murata & Norman [5] to determine the X-ray diffraction peak intensity (212) corresponding to $2\theta = 67,74°$ was used. At the same time [4] it was proposed to determine the quartz crystallinity index by infrared (IR) spectroscopy. It was observed that the change in the crystallinity degree of chalcedony and occurrence of the crystalline alpha -quartz phase, the absorption double peak within the range of 800...778 $cm^{-1}$ in infrared absorption spectra was modified. For calculations the ratio of the weak peak value 778 $cm^{-1}$ was proposed to be applied to its short wavelength shoulder. Another methodological approach [4, 5] was proposed to use the ratio of intensities of the IR absorption peaks at 778 $cm^{-1}$ and 695 $cm^{-1}$, which belong to the vibrations of different symmetry types [6]. In [9] to calculate the crystallinity index the use of the peak intensity change of infrared absorption at 1145 $cm^{-1}$ was proposed. However, in the IR spectra of studied fine-grained quartzite this peak was slightly distinguished (figure 1) and the proposed method, in this case, was inconclusive.

The method to estimate the substance separation degree according to geochemical spectrum of all specific chemical elements in a spot sample was firstly used by I Biborov and S Bistrov [2]. Dispersion spectrum indicators and geochemical spectrum variations (DSI and CSV respectively) are calculated on the basis of the standard formula of variance and variations of use element concentrations.

In the volume mineralized zone the DSI value increases significantly. Maximum concentration indicators are $10^5$ units or more. Anomaly aresnic content of insignificant amount affects the integrated indicator value. It is this fact that explains the spatial overlapping of the haloes.

The DSI and CSV values throughout the deposit are distributed differently. Mineralization zone contours are distinguished by the high-contrast CSV halo, which, in its turn, is conditioned by the significant difference in element concentrations in the mineralized zone itself and its borders. It has been determined that the irregular distribution pattern of ore-genetic element content is plotted as indicator values CSV>140 %.

It is obvious that these indicators characterizing intensive ore formation processes could determine the zone borderlines of hydrothermal-metasomatic alterations, but, in this case, impede the CSV interpretation of ore features.

Based on the specific features of the geochemical spectrum and cluster analysis 5 sample groups were identified.

In the first and fourth sample clusters there were no elements that could have significantly influenced their geochemical spectrum. The second cluster is characterized by Cu, Zn, Ag, and Pb associated with the galenie-chalcopyrite-sphalerite mineral association. The third sample cluster shows high Co and Ni content found as a result of pyrite-pyrrhotite association accumulation. The last cluster (fifth) is charactertized by the dominating As which actively interacts in the formation of arsenic pyrite-pyrite-prryholite association. Identified geochemical associations correlate to mineral formation stages determined on ther basis of previous geological surveys [3]. The limited amount of analyzed elements made it impossible to define a concise concept of the zonal structure of geochemical associations.

To determine the oregenetic element associations with identical behavior during the geochemical process, the factor analysis was applied. Three associations were distinguished, i.e. 64% of total dispersion.

The first factor, embracing 30% of total dispersion, is characterized by significant Pb, Zn and Ag load. The second factor is maximal influenced by Co, Cu and Ni, where the total dispersion is 20%. The third factor is characterized by the significant load of only one element- As. Thus, this fact indicates the oregenetic element input process during hydrothermal metasomatic ore formation. Theidentified geochemical associations correlate with the mineral formation stages [3]. It is obvious that the formation of all three described associations indicate the different conditions and time period in the formation of ore solutions within the mineralized zone.

The processed geochemical data by statistics methods allowed plotting a 3D geochemical model of the ore-gold deposit in black shale formation made it possible to (Figure 3).

Later Ag-Pb-Zn-Cu associations are also spatially overlapping the borderlines of mineralized zones, however, Co-Ni geochemical association is found within these zones.

Based on the results of previous investigations it was determined that the basic metal mass accumulates during the galenite- chalcopyrite-sphalerite stage of mineral formation.

Mineralized zone is located in the centre of the bulk metasomatic halo, spatially matching with the high-contrast Co-Ni, Ag-Pb-Zn-Cu and As geochemical associations. At the same time, the mineralized zone contours are plotted on Co-Ni and Ag-Pb-Zn-Cu geochemical associations.

Contrast association halo Co-Ni corresponds spatially with the contours of ore bodies, fragmentarily with Ag-Pb-Zn-Cu associations, occupying the farmost peripheral areas of the mineralized zone. Co and Ni supply of early hydrothermal solutions and deposited with pyrite-prryholite mineralization. The second group is characterized by the presence of only one element –

As. Its accumulation, basically as arsenopyrite, occurs during the arsenopyrite-pyrite-prryholite mineralization.

It has been determined that the concentric-zonal structure of geochemical associations within the bulk metasomatic halo (from central to periphery) is the following: As → Ag-Pb-Zn-Cu → Co-Ni. Identified behavior in the spatial distribution of geochemical associations could be applied not only insurface (areal lithogeochemivcal survey) but also depth (wells, mining) investigations of perspective areas.

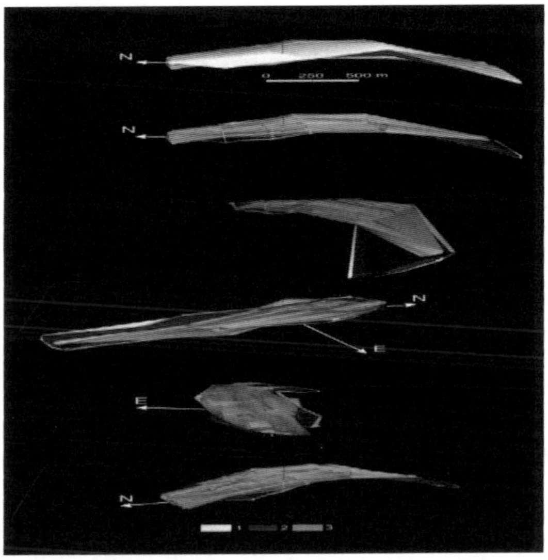


**Figure 3.** 3D geochemical model ofChortovo Koryto gold-ore deposit 1 – Co-Ni association; 2 – Ag-Pb-Zn-Cu association; 3 – As association.

## 6. Summary and conclusions

Applied set of processing methods for geochemical data made it possible to interpret the anomaly geochemical field structure (AGFS) and to determine its concentric-zonal structure. The most important results in processing geochemical data from Chortovo Koryto deposit was obtained through such methods as nonparametric statistics, i.e. cluster and factor analysis.

Based on the processed geochemical data obtained by the mathematical statistical method the AGFS in Chortovo Koryto deposit was interpreted and a 3D geochemical model was plotted. This localized the border contours of gold-ore mineralization within investigated mineral resources.

Designed 3D geochemical gold-ore deposit model would stipulate the potential targets in licensed areas.

## References

[1] Voroshilov V 2009 Anomaly structures of geochemical fields of hydrothermal gold-ore deposits: formation mechanisms, geometrization, model types, forecasting large-scale mineralization *J. Ore Geology* **51** 3–19

[2] Viborov S and Bistrov I 1990 Application of integrated indicator of fracturing geochemical field in predicting mineralization *J. Izvestaya vuzov* **4** 102–110

[3] Kucherenko I, Gavrilov R, Martinenko V amd Verkhozin 2008 Petrographic-geochemical features of ore-hosting metasomatic halo in Chortovo Koryto gold-ore deposit (Patom range) *Bulletin of the Tomsk Polytechnic University* **312** ( 1) 11–20

[4] Safronov N, Mesheryakov S and Ivanov N 1978 *Ore mineralization energy and mineral exploration* (Leningrad: Nedra) p 215

IOP Conf. Series: Earth and Environmental Science **21** (2014) 012005 doi:10.1088/1755-1315/21/1/012005

# Gold in Pyrite within Altai – Sayan Folded Belt Gold Deposits

**A Pshenichkin[1], Y Oskina, O Savinova and T Dolgaya**
Tomsk Polytechnic University, Tomsk, Russia

E-mail: [1]paya@tpu.ru

**Abstract**. This article describes the result analysis of gold in pyrite for a number of some Altai – Sayan folded belt gold deposits. The analyzed pyrite samples showed gold grade from 0.04 to 20.0 g/t., which proves the fact that the gold concentrations in pyrites is irregular. Pyrites in productive mineral associations include significant metal enrichment, while pyrites of early poor gold-bearing associations and wallrock metasomatites are less prolific.

## 1. Introduction

Pyrite is one of the most abundant sulfide minerals in gold-ore deposits of various geneses. The accurate interpretation of decoded information on crystal morphology, physical properties and mineral element impurities provides insight into the nature of pyrite genesis, as well as genesis of ore deposits [3, 4, 7, 8]. Based on the identified typomorphic pyrite properties, the following factors have been defined: mineralogical and geochemical zoning of the deposits, level of their erosion truncation, and location of perspective ore bodies to the depth and at the flanks and "blind" ore bodies. The gold grade in pyrite and high concentrations of gold in pyrite (either in ore bodies or wall rock metasomatites) are examples of potential ore bodies and deposits [1, 5, 9, 10]. This information is of scientific importance. In most deposits there exists a positive correlation between the pyrite crystals with pyritohedron habit and the gold grade both in pyrites and in ore bodies [2, 4, 6, 7].

## 2. Data Sources and Methodology

Pyrites in ore bodies and wallrock metasomatites were analyzed by neutron-activation and polarographic methods. The samples were retrieved from two areas: Kuznetsk Ala Tau which included Kommunarovsk ore field, Balakhchino ore field and Saralinsk ore field and Sayan Mountains, mainly the eastern part which included Konstantinovo deposit, Medvezhye deposit, Lysogorsk deposit, Tardansk deposit and Oktyabrsk deposit.

| Mineral associations | n | Au g/t grade | | | V |
|---|---|---|---|---|---|
| | | from | to | x | % |
| **Kommunarovsk ore field (Kuznetsk Ala Tau)** | | | | | |
| Quartz-gold –actinolite | 14 | 0.22 | 10.0 | 5.45 | 80 |
| Quartz-gold –scheelite | 5 | 0.22 | 5.0 | 2.92 | 76 |
| Quartz-gold –tellurobismuthite | 9 | 0.10 | 20.0 | 5.56 | 164 |
| Pyrite in wallrock metasomatites | 10 | 0.04 | 10.0 | 2.42 | 165 |
| **Balakhchino ore field (Kuznetsk Ala Tau)** | | | | | |

| | | | | | |
|---|---|---|---|---|---|
| Quartz-gold polymetallic sulfide | 56 | 0.10 | 20.0 | 3.95 | 124 |
| Beresite pyrite | 28 | 0.10 | 5.0 | 0.94 | 126 |
| **Saralinsk ore field (Kuznetsk Ala Tau)** | | | | | |
| Quartz-gold – sulfide | 9 | 0.4 | 20.0 | 5.6 | 159 |
| Pyrite in wallrock metasomatites | 7 | 0.1 | 5.3 | 3.7 | 95 |
| Metamorphogenetic pyrite | 3 | 0.04 | 0.1 | 0.08 | |
| **Olkhovsk –Chibizhek ore field (eastern Sayan Mountains) Konstantinovo ore deposit** | | | | | |
| Carbonate-pyrite-pyrrhotite | 75 | 0.10 | 20.0 | 4.77 | 89 |
| Quartz-gold - sulfide | 91 | 0.10 | 20.0 | 4.73 | 102 |
| Pyrite in wallrock metasomatites | 8 | 1.00 | 9.0 | 4.90 | 60 |
| **Medvezhye ore deposit** | | | | | |
| Carbonate-pyrite-pyrrhotite | 18 | 0.10 | 3.0 | 1.05 | 135 |
| Quartz-gold - sulfide | 59 | 0.08 | 9.9 | 2.76 | 104 |
| Pyrite in wallrock metasomatites | 10 | 0.10 | 0.5 | 0.30 | 142 |
| **Lysogorsk ore deposit** | | | | | |
| Quartz-gold –pyrite-bismutite | 9 | 0.10 | 9.0 | 3.64 | 80 |
| Pyrite in wall rock metasomatites | 4 | 0.07 | 2.7 | 1.25 | 98 |
| **Tardansk ore deposit (eastern Sayan Mountains)** | | | | | |
| Skarn-gold-ore-sulfide | 14 | 0.05 | 5.0 | 1.57 | 147 |
| Quartz-gold - sulfide | 5 | 0.05 | 7.5 | 2.45 | 132 |
| Pyrite in wallrock beresites | 3 | 0.05 | 3.6 | 1. 35 | |
| **Oktyabrsk ore deposit (eastern Sayan Mountains)** | | | | | |
| Quartz-gold - sulfide | 8 | 0.10 | 10.0 | 3.2 | 87 |
| Pyrite of carbon-bearing shale | 24 | 0.05 | 0.6 | 0.27 | 24 |

**Table 1.** Gold content in pyrites of gold deposits in Altai – Sayan folded belt

## 3. Results and Discussion

Gold was found in all analyzed pyrite samples from 0.04 to 20.0 g/t. In this case, the gold grade in pyrites within various ore deposits and associations was irregular. High gold grade was observed in pyrites from productive mineral associations, while low gold grade- older weak mineral associations in pyrites in wallrock metasomatites (table 1).

Pyrites of Kommunarovsk ore field are more prolific. For example, the average gold grade is approximately 5.45 and 5.56 g/t. in pyrites of quartz-gold –actinolite (Maslovsk, Yanvarsk, Kaliostrovsk) and quartz-gold–tellurobismuthite (Znamenitinsk, Kuznetsovsk, Yanvarsk) associations. The gold grade is twice lower in pyrites of quartz-gold –scheelite associations (Spasskoe), and pyrites in wall rock metasomatites from all ore field deposits.

The same average gold grade – 5.6 g/t. was identified in pyrites of quartz-gold-sulfide associations from Saralinsk ore field.The highest gold grade in pyrites in wallrock metasomatites were found in this region, i.e. - 3.7 g/t (table 1). The deeper the Kaskodniy ore deposits and Andreyevsk ore fields, the lower the gold grade. These concentrations are wave-like with the amplitude of 260-280 m where enrichment occurs in the most prolific zones of ore bodies. Diagenetic pyrite in enclosing shales contains the minimal gold grade– 0.08 g/t. According to the high gold grade in pyrites and the predominate development of pyrite crystal with pyritohedron habit in ore field quartz viens, the following potential ore areas were identified: Central, Kuznetsovsk, Vstrechni, Verkinski.

Increasing erosion truncation level was observed in Olkhovsk – Chibizhek ore field from Lysogorsk (drift-in horizons) to Medvezhye (depleted in 11 horizons), from Konstantinovo (depleted in 5 horizons) to north-east Tarchinski ore field (explored by core drilling), where there are very low metal concentrations in ores and pyrites. The highest gold grade could be observed in pyrites of all Konstantinovo ore deposit mineral associations, i.e. 4.73-4.90 g/t. in the Olkhovsk – Chibizhek ore field. The pyrites of the productive associations from Lysogorsk (3.64 g/t) and Medvezhye (2.75 g/t) ore deposits have lower gold grade. Pyrites in wallrock metasomatites have 3-9 times less gold grade than ore pyrites (except Konstantinovo ore deposit, where the average gold grade is a bit higher than that of in ore pyrites) (table 1). The deeper the Medvezhye ore deposit the lower the gold grade is in pyrite. These decreasing concentrations are wave-like with the amplitude of 200-240 m. Similar wave-like alterations are observed in most element-impurities in pyrite deposits [1, 5].The deeper the Konstantinovo ore deposit, the higher the average gold grade in pyrites - from 3.1 g/t in the horizon of 417 m to 5.9 g/t in the horizon of 297 m. Such ore deposits could be potential.

Pyrites of quartz-gold polymetallic sulfide associations in Balakhchino quartz veins exhibit a rather high metal concentration – 3.95 g/t (ranging from 0.1 to 20.0 g/t). There are both horizontal and vertical zoning of gold distribution in pyrites of ore bodies. Horizontal zoning in pyrites shows the gold grade decrease from Maiskaya (highly temperature vein – 6.9 g/t),) to Oktyabrskaya pyrites (medium temperature vein – 3.2 g/t), then to pyrites of Chapayevskaya and Izotovskaya (low temperature veins – 3.07 g/t). This demonstrates the decrease of vein erosional truncation level towards this direction. A significant gold grade increase in all quartz veins is observed at depth. High gold grade in accessory pyrites of Balakhchino diorite – monzonite massif – 0.11 g/t. is also observed.

Pyrites of Tardansk skarn-ore deposit and Oktyabrsk quartz –vein deposit of Tyva ore field are characterized by a reduced gold grade in comparison to other deposits of this region (table 1). The gold grade in pyrites is up to 10 g/t (the average is 3,2 g/t) in Oktyabrsk deposit (Amylo-Sistighemsk region) in Stepanovsk and Grigoryevsk quartz veins. The gold grade is significantly lower in other ore deposit veins (Nikolskaya, Neizvestnaya, Poperechnaya, Baba). Metamorphogenetic pyrites in carbonic shales contain only 0.27 g/t. gold.

Medium gold grade in pyrites from Tardan skarn-ore bodies is 1.57 g/t. Fine-grained gold can be found not only in pyrite but also in ores. Downward the gold grade increases gradually in pyrites. In quartz veins of Kopto (6 km to the north-east of Tardan) ore formation the gold grade in pyrites of quartz veins and wallrock beresites is higher – 2.45 and 1.35 g/t respectively, in comparison to those in Tardan skarn-ore bodies.

Gold concentration in pyrites from various temperature mineral associations and ore deposit geneses is left- asymmetrical, sometimes 2-mode and often extremely irregular. This can be explained by the fact that there are large-sized, fine-dispersed and submicroscopic gold concentrations in pyrite. Large-size gold concentrations are typical for rather low-temperature mineral associations, while fine-dispersed – for high -temperature ones. There is often no stable relationship between the gold grade and other impurity elements in pyrites of deposits. For example, only in Lysogorsk ore deposit there is a positive bonding of gold with Zn (0.50), Bi (0.69), Ag (0.89), Cu (0.55), Co (0.88), Ni (0.70) and a negative one with As (0.75). Sometimes there is a positive bonding between Au and Cu – Konstantinovo ore deposit (0.41), Balakhchino ore deposit (0.53), Kommunarovsk ore deposit (0.51), Au and Mn – Balakhchino ore deposit (0.47), Au and Co (0.52) and Ni (0.62) – Kommunarovsk ore deposit; a negative bonding – Au and As in Balakhchino ore deposit (0.85).

## 4. Conclusions

The research determined that pyrites in gold ore deposits of Altai – Sayan folded belt contain increased gold grade, including pyrites in wallrock metasomatites of these ore deposits, where, in some cases, the metal content is higher than in the pyrites of non –gold deposits. The findings show the possible approach in evaluating ore bodies and ore deposits, level of their erosional truncation, depth prospects and flanks.

**Acknowledgements**

The research has been conducted with financial support from Tomsk Polytechnic University. Project: INR_114_2014. Analyses were performed in the nuclear reactor of the Institute of Nuclear Physics TPU (by) by the neutron-activation method and in the laboratory "Geology of Gold", TPU Department of Geology by the polarographic method. We would like to thank V Chesnokov and N Novikova who conducted these analyses.

**References**

[1] Korobeinikov A, Pshenichkin A 1985 Gold deposits -geochemical peculiarities of the pyrite. *J. Geochemistry* **1** 93-104

[2] Kouzmanov K, Bailly L, Ramboz O and Rouer J 2002 Morphology, origin and infrared microthermometry of fluid inclusions in pyrite from the Radka epithermal copper deposit, Srednogorie zone, Bulgaria, *Mineralium Deposita* **37** 599–613

[3] Pshenichkin A 1977 Thermoelectric properties of the pyrite in gold deposits of various types. *J. Izvestia TPI* **247** 74-78

[4] Pshenichkin A 1980 Pyrite as a prospecting characteristic for gold. *J. Mineralogical articles* **7** (Genetic information in minerals) 150-151

[5] Pshenichkin A 1983 Elements-impurities in pyrites as typomorphic features of gold deposits. Geological and natural resources exploration problems in Siberia *Theses of Usov M.A Conference* (Tomsk: TPU) 86-187

[6] Pshenichkin A 1989 Kristallomorphology of pyrite and its application in search for gold. *J. Geology and Geophysics* **11** 65-75

[7] Pshenichkin A and Ananyev Yu 2013 Exploration and local forecast technologies for gold deposits by the pyrite typomorphic features. Mineral resources base development: from Obruchev V A, Usov M A, Urvantsev N N to our days (Tomsk: TPU) 409-423

[8] Simon G, Kesler S E and Chryssoulis S 1999 Geochemistry and textures of gold-bearing arsenian pyrite, Twin Creeks, Nevada; implications for deposition of gold in carlin-type deposits *Economic Geology* **94** (3) 405-421

[9] Tauson V, Kravtsova R, Smagunov N, Spiridonov A, Grebenshchikova V and Budyak A 2014 Structurally and superficially bound gold in pyrite from deposits of different genetic types *Russian Geology and Geophysics* **55** (2) 273-289

[10] Tauson V L and Lipko S V 2013 Pyrite as a concentrator of gold in laboratory and natural systems: A surface-related effec, book chapter, ISBN: 978-162257851-1 *Pyrite: Synthesis, Characterization and Uses,* Nova Science Publishers, Inc.

The topic of this research is to investigate the quartzite crystallinity index changes of different technological brands selected from ore deposits «Sopka-248» and Belokamenka, Antonov field cluster (Russia, Western Siberia) by infrared spectroscopy via a various calculation methods in determining this or that crystallinity index.

## 2. Research method and results

IR absorption spectra were recorded on the spectrophotometer Specord M40 within the range of 400...4000 $cm^{-1}$ with a resolution of 0.01 $cm^{-1}$ , and a spectrophotometer 21 IRPrestige ( Shimadzu Co.) with Fourier conversion (FTIR-8400S) in the interval of 300 ... 4000 $cm^{-1}$ with a resolution of 0.001 $cm^{-1}$ (FT-IR) via software IRsolution. The test samples were prepared from selected quartzite samples under similar conditions. The test sample is grinded in an agate mortar to a fraction of less than 2 microns and mixed with both exclusively pure and purified for analysis KBr powders and then placed in a mold and compressed under high pressure into transparent thin disks. When measuring the IR spectra with spectrophotometer FTIR-8400S the weighed sample is 0.1...0.5 mg. Fragments of the infrared absorption spectra of quartzite are shown in figure 1.

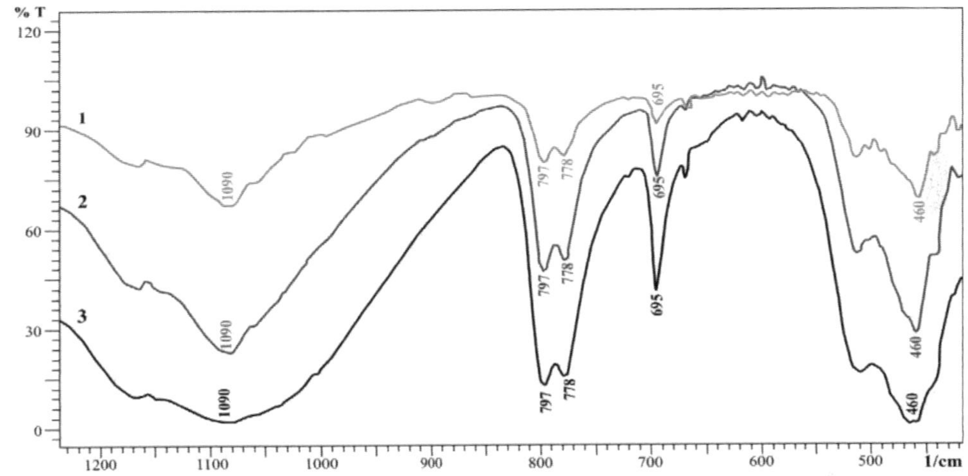


**Figure 1.** Fragments of infrared absorption spectra of samples on different quartz content basis.

With a further content increase of investigated material in prepared sample, a regular intensity increase of infrared absorption peaks was observed, and, in cases of high concentrations, more than 1.5 % their shape distorted and their magnitudes exceed the measurement limit range. In this case, the crystallinity index value calculated by various methods, also changes, including the IR intensity ratio of peaks at 778 and 695 $cm^{-1}$ (table 1). Therefore, equal test sample mass was provided for an accurate comparative measurement analysis.

**Table 1.** Crystallinity index value ($k_i$) of samples on different quartz content basis.

| №№ (measurement) | Weight of quartz on basis, % | $k_i$, (calculation by the method [5]) | $k_i$, (calculation by the method [6,7]) | Sample description |
|---|---|---|---|---|
| 1 | 0.50 | 1.30 | 2.57 | Belokamenka field. Quartzite of white colour, saccharoidal. Samples in binocular microscope |
| 2 | 0.25 | 0.90 | 2.20 | |
| 3 | 0.75 | 0.50 | 1.74 | |
| 4 | 1.00 | 0.077 | 1.1 | |
| 5 | 1.25 | 0.047 | 0.95 | |

The calculation the quartzite crystallinity index was based on the methodology of Indian scientists [6, 7], who proposed the infrared absorption peak ratio of 778 and 695 $cm^{-1}$.

Crystallinity index calculation was performed according to the following formula:

$$k_i = a/b,$$

where, a/b – ratio peak intensity of 778 $cm^{-1}$ to peak of 695 $cm^{-1}$, which were determined by the baseline method (figure 2).

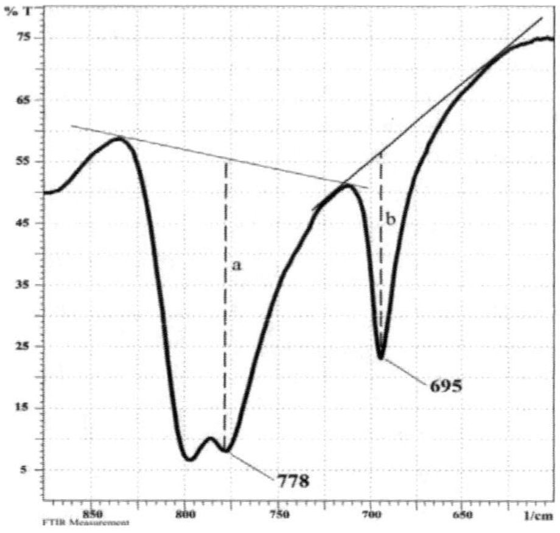


**Figure 2.** Calculation method of crystallinity index in relation to changes of absorption peaks 778/695 $cm^{-1}$ in the infrared absorption spectrum.

Absolute crystallinity index values obtained by various methods differing to the absolute value are presented in table 2 for comparison. In all proposed methods crystallinity index value has a numeric value and is calculated directly from the measurements of the infrared absorption peak intensity and the magnitude of the IR absorption. These averaged values of crystallinity index obtained from three measured samples are presented in table 2. In all cases, a regular increasing trend of calculated crystallinity index values with quartzite conversion degree increase was observed, despite the fact that the calculated values of $k_i$ are different.

Obtained by different methods, the purest white fine-grained quartzite diversity from «Sopka-248» [9] deposit is characterized by the lowest calculated crystallinity index value of – 2.21. Different quartzites from various local ore deposit areas, especially in high fragmentation areas, as result of supergene processes, undergo the following changes: quality deterioration, chemical composition and color alteration and the crystallinity index value also increases to 2.42...2.50.

**Table 2.** Quartzite crystallinity index.

| Deposit | Sample | $k_i$, (calculation method [5]) | $k_i$, (calculation method [6,7]) |
|---|---|---|---|
| Sopka-248 | White quartzite | 2.21 | 2.21 |
| | Gray quartzite with iron oxides adhesion | 2.52 | 2.42 |
| | Grayish quartzite with clay adhesion | 2.66 | 2.49 |
| | Black quartzite from periphery | 2.75 | 2.50 |
| | Jasper-like quartzite of brownish-cherry color with black veins | 5.6 | 2.55 |
| Belokamenka | White, saccharoidal quartzite | 3.27 | 2.47 |
| | White, transparent quartzite | 3.8 | 2.67 |
| Bural-Sardag | White coarse-grained quartzite | 4.6 | 3.8 |

Fine-grained pure white quartzites of Belokamenka deposit are characterized by even higher crystallinity index values. To compare the quartzite alteration degree in Antonov field cluster, crystallinity index value of super-quartzites from Bural-Sardag deposit were examined. Super-quartzites from Bural-Sardag deposit are composed of larger grains with elongated porphyritic inclusions, subparallel orientation indicating that they have formed (recrystallized) in long-term dynamic stress conditions [1].

Super quartzites from Bural-Sardag deposit are characterized by high crystallinity index values in comparison to those of the Antonov field cluster (ki = 3,8…4,6).

## 3. Conclusion

Quartzite crystallinity index values calculated by different methods to infrared absorption spectra represent their conversion regularity despite the different values. Using these methods makes it possible to define and estimate quartzite recrystallization degree in different metamorphism stages including such as Antonov field fine-grained quartzites where traditional mineralogical and petrographic methods seem to be ineffective. Due to the fact that the purest samples are characterized by the lowest estimated crystallinity index values, such an evaluation could be applied in further mineralogical-technological mapping.

## References

[1] Nepomnyashchikh A I, Krasin B A, Vasilyeva I E, Yeliseyev I A, Yeremin V P, Popov S I, Sinitskiy V V, Fedoseyenko V A, Spiridonov A M, Vorobyev Ye I and Gnilusha V A 2000 Kremniy dlya solnechnoy energetiki *Bulletin of Tomsk Polytechnic University* **303** (1)176-190 [in Russian]

[2] Zakharov V B 2000 Geologicheskoe stroenie Antonovskoi gruppi mestorogdeniy kvarcitov *Mining Journal* 7 7 [in Russian]

[3] Ananyeva L G and Korovkin M V 2003 Mineralogo-geokhimicheskoye izucheniye kvartsitov Antonovskoy gruppy mestorozhdeniy *Bulletin of Tomsk Polytechnic University* 306 (3) 50-55 [in Russian]

[4] Barsanov G P, Plyusnina I I and Yakovleva M E 1979 Osobennosti sostava, nekotorykh fizicheskikh svoystv i struktury khaltsedona (Moskov: Nauka) **28** 3-33

[5] Murata K J and Norman II M B 1976 An index of crystallinity for quartz *American Journal of Science* **276** 1120-1130 [in Russian]

[6] Ramasamy V and Suresh G 2009 Mineral Characterization and Crystalline Nature of Quartz in Ponnaiyar River Sediments, Tamilnadu, India *American-Eurasian Journal of Scientific Research* **4** (2) 103-107

[7] Lazarev A N, Mirgorodskiy A P and Ignatyev I S 1975 *Kolebatelnyye spektry slozhnykh okislov* (Leningrad: Nauka) s 296 [in Russian]

[8] Shoval S, Ginott Y and Nathan Y 1991 A new method for measuring the crystallinity index of quartz by infrared spectroscopy *Mineralogical Magazine* **55** 579-582

[9] Korovkin M V, Ananyeva L G and Antsiferova A A 2012 Otsenka stepeni preobrazovaniya kvartsitov metodom infrakrasnoy spektrometrii *Bulletin of Tomsk Polytechnic University* **320** (1)16-18 [in Russian]

# Microstructural analysis of quartz grains in Vasyugan suite sandstones of layer $U_1^{1-2}$ in Kazanskoe deposit

**D Cherdantseva [1] and L Krasnoshchekova[2]**
Tomsk Polytechnic University, Tomsk, Russia

E-mail: [1]dannia@tpu.ru [2]krasnl@yandex.ru

**Abstract.** Microstructural analysis of quartz grains in sandstones revealed preferred directions which define and influence porosity and permeability anisotropy in oil and gas reservoirs In this research, we investigated the Upper Jurassic sandstone reservoir sediments from 14 wells in Kazanskoe field. The authors studied: the orientation of elongated quartz grains, and intergranular fracture within grains, as well as the pore space in oriented thin sections of sandstones. The analysis of elongated quartz grains in the bedding plane showed three main types of preferred directions in quartz grain orientation along different axes in sandstone reservoirs. Obtained results allow identifying a variability of facies and dynamic depositional environment for Upper Jurassic sandstone formation. Subsequently, these results can be used in field modeling, as well as pattern optimization of injection and production wells.

## 1.Introduction

Kazanskoe oil and gas field is located in the southeastern part of Tomsk region, were the commercial hydrocarbon potential is associated with Upper Jurassic terrigenous sediments. It is necessary to solve a number of issues for the successful development of this field. First of all, the characteristic features of sandstone bodies - reservoir distribution, as well as, their formation conditions and areas including the reservoir properties should be identified. And secondly, to study the reservoir anisotropy. It should be noted that the anisotropy factor is fundamental in updating the essential geological structural alteration of oil and gas field reservoir rocks. Numerous publications of the last few years have highlighted such a fact as the intrinsic heterogeneity of horizontal permeability [1,2,3,4].

In recent years, an increasing number of geologists use microstructural analysis of rock forming minerals as an optical method for studying sediments, which allows obtaining information about the dynamic mode of their formation [5,6,7].

## 2.Data Setting

The investigations of the sandstone layer was based on the sediment material from 14 wells of the Kazanskoe oil field. More than 70 meters of core samples were described, 16 rock samples of thin reservoir sandstone sections from $U_1^{1-2}$ layer were studied in details. The core sampling depth embraced 2481.3 –2511.2m. Paleomagnetic oriented core samples and oriented thin sections were used The results of the sampling examination can be used in field modeling, existing anisotropy in reservoirs benefits the pattern optimization of injection and production wells, as well as, monitoring enhanced oil recovery operations. The sampling examination involved several stages which are described below.

## 3.Morphological analysis of quartz grains in thin sections

*3.1 Instrument and materials.* Morphological analysis was performed using north-oriented thin sections. Measurements included MWD dimension of elongated quartz grains in sandstones. These measurements embraced elongated grains to the northern direction. The number of measurements required to describe the charactertics of orientation distribution involved 100-120 grains in thin sections. Besides, these investigations, obtained on the microscope, we measured the orientation distribution was observed on thin section photos, where the minimum number of measurements included 200-300 grains and further data processing in ArcGIS and Grapher 9 programs and charting. This suggested method of measuring and interpretating grain orientation allows to simplify the exceedingly enormous amount of obtained numerical data.

The entire process of measuring and calculation was divided into several stages:

1. taking potos of oriented thin sections in an interval, processing and grouping these photos in CorelDRAWX5 program.
2. plotting preferred directions of elongated quartz grain orientation by ArcMap program.
3.recalculating azimuth directions by a special module for ArcInfo program.
4. plotting rose-diagrams based on azimuth data and further visualization with “StereoSN” and Grapher 9 programs.

*3.2 Data processing* Statistically processed measurement results are shown on a pie chart. Based on calculations of the linear orientation of quartz grains semicircular arc data was obtained (from 0 to 180°), which, in its turn, proved the fact that the second semicircular is symmetrical to the first one. For data visualization and comparison of elongated quartz grain orientation with other data, the petrostructural program Grapher 9 was applied on the basis of which rose diagrams were plotted and lines for preferred elongated quartz grains $L_1$, $L_2$, $L_3$ and were drawn (figure 1).

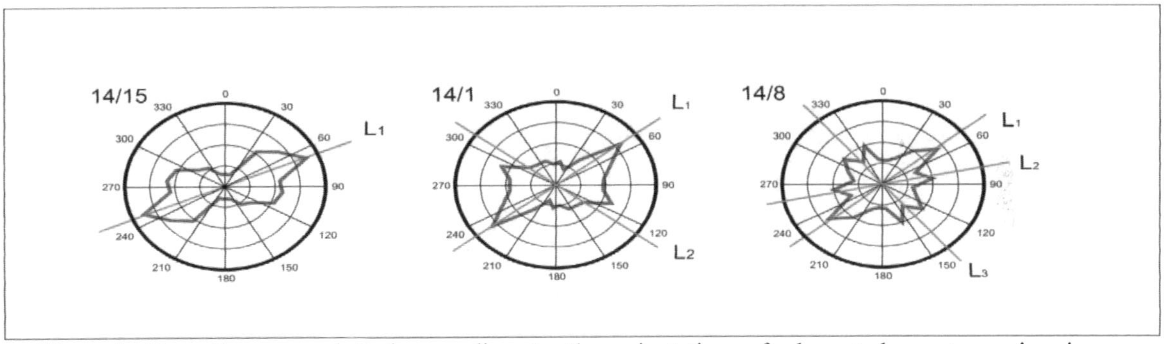


**Figure1.** Rose-diagrams plotted according to the orientation of elongated quartz grains in sandstones from $U_1^{1-2}$ layer in well № 14 Kazanskoe field. $L_1$, $L_2$, $L_3$ □ preferred direction of elongated quartz grains in thin sections.

*3.3 Types of orientation for elongated quartz grains*

Analysis of quartz grains according to their elongation in the bedding plane has revealed three main types of preferred directions for major axis orientation of quartz grains in investigated sandstones. The first orientation type is predominant and is characterized two directions ($L_1$, $L_2$) in elongated quartz grains orientations, the first of which □ $L_1$ is dominant. The second type can be identified by the presence of one maximum longitudinal orientation of the grains □ $L_1$ with slightly developed transverse orientations. The third type of orientation is much rarer the first two and is characterized by polymodal distribution ($L_1$, $L_2$, $L_3$).

*The first type with* bimodal orientation □ $L_1$ and $L_2$ is typical for most of the investigated thin sections. In sample 14 / 1 (figure 1) there are two preferred directions of elongation for grains, one of which is in the subordinate position. The intensity of statistical maximum is 13-16% or even more.

IOP Conf. Series: Earth and Environmental Science **21** (2014) 012007 doi:10.1088/1755-1315/21/1/012007

The angle between the peaks varies from 40° to 80°. Such orientation may be related to the coastal (internal) zone of a water basin, which is characterized by constant sea wave action. Waves lapping on the shore at an angle transport sandy material obliquely to the shoreline, while the reverse flushing movement is perpendicular to the coast. Longitudinal orientation is typical for grains with flat and elongated shapes, while cylindrical and spindle grains due to their mobile dislocation by rolling occupy cross position in the water flow. This is reflected in the diagrams as a minor peak (figure 1).

A distinctive feature of the second type is one or two single-order elongation directions - $L_1$ which can be seen in sample 14/15 (Figure 1). The intensity of statistical maximum is quite high - 15-17%. Angle between the principal maximum is fixed within the limits of 40°. In this case we can conclude that preferred arrangement of quartz grains is consistent to the direction of fast water flow □ the sharp grain end is oriented to the flow direction. A similar pattern of distribution is possible when moderate or strong driving environment influences the sediments, for example, longshore currents. It should be noted that an important factor is that the orientation path has different flow heterogeneity. In case of superimposed factors there could be a more complicated orientation image. However, the grain arrangement is still regular.

The third type of grain orientations in sandstones is characterized by spindly located elongated quartz grains in thin sections - $L_1$, $L_2$, $L_3$ , which can be seen in sample 14/8 (figure 1). It is impossible to identify the predominant direction. Accordingly, we can assume that sediment resuspension has influenced the formation of such a grain distribution pattern. Such processes form turbidites. In this case, it is possible after transportation and transformation of clastics into sediments have not been lithified yet, because of a large number of plastic clay material.

*3.4 Subderivations R*ose-diagram analysis of preferred distribution for elongated quartz grains in investigated rocks showed the predominance of two distinct sub-orthogonal directions in quartz grain orientation, i.e. north-east and north-west. Each type of sediment environment can be characterized by a specific type of quartz grain orientation in sandstones. Obtained data in combination with other characteristics of sedimentary rocks (textural and structural features, particle size distribution) can be used to determine depositional facies. As a result of our further investigation – study of microstructural quartz patterns - previously supposed and above -described mechanisms in the formation of linear orientation in elongated grains was proved.

## 4.Microstructural analysis of optical axes in quartz grains

*4.1 Methodology and methods* Microstructural analysis is based on the measurement of the optical axes in quartz grains, which describes the internal structure of the rocks in geometrical terms. The research method included determining the optical axis coordinates in quartz grains. To find the spatial orientation of the optical axis it is necessary to make two measurements in Stolik Fedorova (special attachment on the microscope determining the position of the crystallographic axes in minerals): strike azimuth and angle of the optical axis. The selection included 100-150 measurements of quartz grains in thin sections. After statistical processing by applying "StereoSN" orientation elements had the form of a pie chart, by which the geological interpretation was given. By analyzing the pattern diagram of the optical axis we can indicate the presence of orientation in the grains or its absence. Quartz optical axis are located at a higher position in the grain elongation. During the investigation it was noted that a significant part of the optical axes projections in quartz grains are located in the rock bedding plane with deviations in the range of 5-25°.

*Types of optical axes orientation.* Petrostructural analysis of quartz grains in sandstones allows identifying three types of optical axes orientations (figure 2).

*Type 1* is characterized by the presence of two concentration zones: most of the optical axes are scattered conically in the zone around linearity □ $L_1$ with a radius of 35-40°; the maximum of density in the belt up to 8%. This type of optical axes orientation is typical for tidal zone at the seaside (figure 2 – sample 14/10).

*Type 2* is characterized by two maximums in the outputs of the optical axes, which clearly has one circular zone concentration of a small circle and linearity □ $L_2$, located subnormal to $L_1$. Maximum of density reaches 11%. A similar pattern orientation can be formed in the case of coeffect of tidal and longshore currents (figure 2 – sample 14/1).

*Type 3* is difficult for interpretation; however, we can note the pattern of a small circle for optical axes - belt type chart. Such a pattern appears in a mobile depositional environment with typical currents, which have alternating direction, whereby there is no preferred direction in grains orientation. Perhaps, postsedimentary (diagenesis, katagenesis) processes, which occur in sediments after their lithification, had a significant influence on the formation of this pattern (figure 2 - sample14/9).

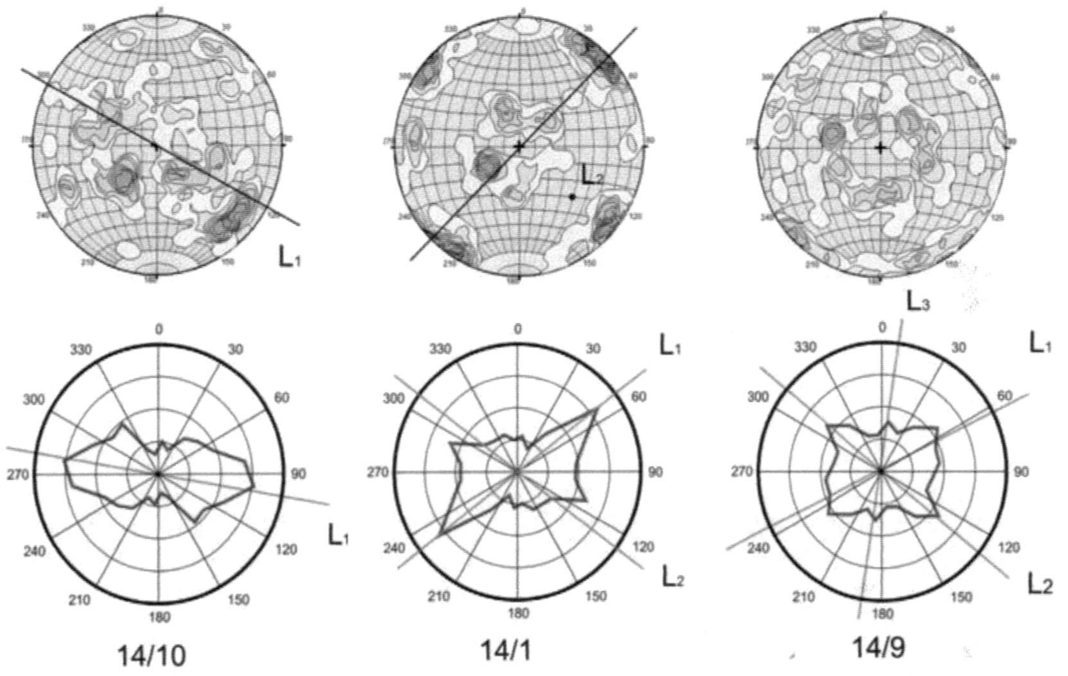


**Figure 2.** Diagrams with quartz optical emergences of axes (top row) and rose-diagrams with orientation of elongated quartz grains (bottom row) in sandstones from Kazanskoe deposit.

## 5. Accompanying investigations

*5.1 The spatial orientation of the fractures and pores in sandstones* Fractures in quartz grains and location (orientation) of pores in sandstones have also been studied in oriented thin sections in which orientation of elongated quartz grains and the optical emergences of axes in quartz have already been defined.

*5.2 Comparison of petrostructural and petrophysical data* Directions of preferred orientation for elongated grain material coincide with the directions of cataclase fractures in quartz grains in combination with the main axis of the magnetic anisotropy ellipsoid. The main direction of the orientation elements in the rocks is defined as north-east with a subordinate north-western direction (figure 3).

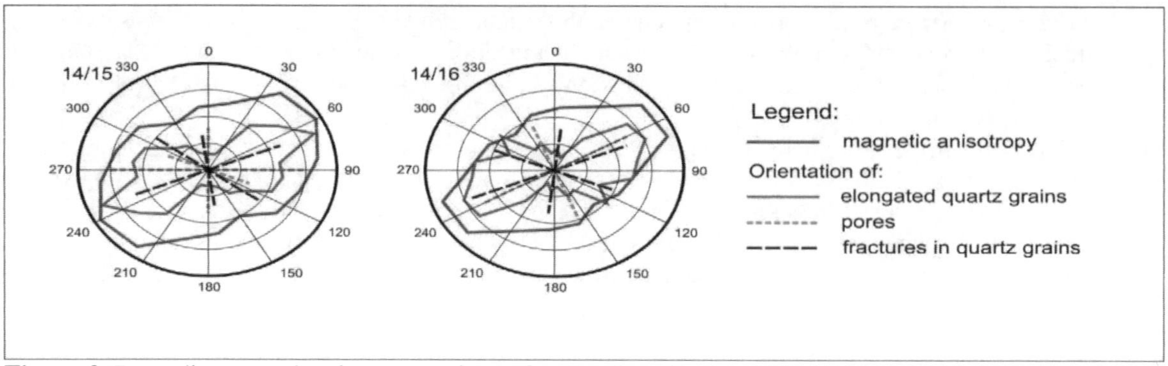


**Figure 3.** Rose-diagrams showing comparison of magnetic anisotropy ellipsoid and main directions of orientation elements in sandstones.

## 6.Interpretation of the results

Dynamic and facies conditions of reservoir $U_1^{1-2}$ formation in Kazanskoe field depend on sand□clastic sediment with numerous pore spaces, which was formed from medium-grained clastic material at the early stage of sediment accumulation process.

Shelf aqueous medium promoted the preferred orientations pattern process which is defined wave-cut character and longshore currents. At the time of sedimentation wave flow velocity can be defined as moderately strong and according to the morphological and microstructural analysis tracked north-easterly to the shoreline. Extension of pores and pore channels in sandstones coincides with the direction of elongation in quartz grains. In the same direction the largest number of fractures in the terrigenous cataclase material of sandstones was recorded,showing simultaneous development of additional directions of grains fragmentation.

## 7.Conclusion

Integrated petrophysical and petrostructural studies revealed that sand layer has a spatial anisotropy of reservoir properties. Direction of spatial orientation in elongated quartz is well- matched with the anisotropy magnetic properties of rocks and orientation of cracks and pores in the sandstones.

Spatial orientation of grains associated with the hydrodynamic features of the clastic reservoirs formation determines the structure of the porous matrix and the anisotropy of its filtration characteristics.

According to the microstructural analysisthey have a north-east direction to the coastline, which is consistent with the proposed model of Kazanskoe deposit where there are facies with bar sedimentary structure.

Methodology developed by the authors included lithological and petrophysical investigations to predict: anisotropy in sediment rocks with high permeability and reservoir properties. It can also be used to study sedimentary rocks and for the modeling of hydrocarbon deposits.

## References

[1] Du J and Wong R 2002 Stress-induced Permeability Anisotropy in Fractured Reservoir *paper SPE 79019 presented at the SPE International Thermal Operations and Heavy Oil Symposium and International Horizontal Well Technology Conference, Calgary, Alberta, Canada, 4-7 November*

[2] Sahin A, Menouar , Ali A and Saner S 2003 Patterns of Variation of Permeability Anisotropy in a Carbonate Reservoir *paper SPE 81472 presented at the Middle East Oil Show* (Bahrain 9-12 June) p 12

[3] Hansen C E and Fanchi J R 2002 Producer/Injector Ratio: The Key to Understanding Pattern Flow Performance and Optimizing Waterflood Design *SPE 75140 presented at the SPE/DOE Improved Oil Recovery Symposium, Tulsa, Oklahoma,* 13-17 April, p 12

[4] Schön J H, Georgi D T and Fanini O 2003 Imparting Directional Dependence on Log-Derived Permeability *SPE Reservoir Evaluation & Engineering* **6 /1** 48 – 86

[5] Baas J H, Hailwood E A, McCaffrey W D, Kay M and Jones R 2007 Directional petrological characterisation of deep-marine sandstones using grain fabric and permeability anisotropy: Methodologies, theory, application and suggestions for integration *Earth-Science Reviews* **82** (1–2) 101–142

[6] Ketcham R 2005 Three-dimensional grain fabric measurementsusing high-resolution X-ray computed tomography *Journal of Structural Geology* **27** 1217–1228

[7] Merkulov V and Krasnoshchekova L 2014 Estimation of influence oil and gas reservoirs filtration anisotropy in the modeling field *J. Gas industry* **3/703** 22–27

IOP Conf. Series: Earth and Environmental Science **21** (2014) 012008 doi:10.1088/1755-1315/21/1/012008

# Noble-metal mineralization in olivine clinopyroxenite within idgimskiy gabbro-peridotite-pyroxenite complex (West Sayan Mountains)

**T Y Cherkasova[a] and T V Korotchenko[b]**
Tomsk Polytechnic University, Tomsk, Russia

E-mail: [a]tomasstik@yandex.ru, [b]tatyana1003@mail.ru

**Abstract**
The study of noble-metal mineralization in olivine clinopyroxenite within idgimsky gabbro-peridotite-pyroxenite complex was carried out. Palladium minerals were identified: palladium telluride and palladium and platinum telluride with mercury impurity. The discovered noble-metal mineralization in olivine clinopyroxenites suggests the presence of perspective low sulfidation platinemetal mineralization in mafic within Verkhne-Amyl ore field.

## Introduction

The study is concerned with noble-metal mineralization samples collected in idgimsky gabbro-peridotite-pyroxenite complex. The provenance area is supposed to be a small massif located 7..8 km south-east from the well-studied Kalininsk massif which contains rock associations similar to the composition of the samples being studied. The samples containing platinum group minerals (PGM) were studied at Innovative Scientific-Education Center "Uranium Geology", National Research Tomsk Polytechnic University using electronic scanning microscope Hitachi S3400N (operator S.S. Ilyenok).

## Methodology

Macroscopically, olivine clinopyroxerites are young, earthy green colour, with small dark veinlets of partially replaced olivine. Grain size ranges from 2 to 4 mm Rock structure is compact, heterogeneous and massive. The texture is panidiomorphic and medium-coarse grained. Olivine comprises 10% of the rock bulk. Clinopyroxene is represented by diopside making up the bulk of the rock, with grain size ranging from 3 to 6 mm. The same weakly altered diopsidites comprising late dykes in dunite-harzburgite rocks of the ophiolitic series were identified within Kalininsk massif by A. Ekhanin [1]. However, the presence of noble-metals in these dykes has not been identified.

Sulfide impregnation is visually defined and is 8-10% of the rock bulk. Sulfide grains are randomly dispersed within the rock and can be identified both in clynopyroxene and at the boundary of olivine. The sulfide grain composition is not homogemeous and basically includes copper pyrite, pyrrhotite and pentlandite, but less in bornite. Along with the identified minerals, fibrous pyrite and copper mica are also found. False galena and galena are identified as single ambiguous grains. Palladium mineralization was identified as intergrowth with platinum phase in copper pyrite grain enclosed in pentlandite (Fig.) and independent phase of palladium telluride. This type of mineralization includes: palladium telluride with mercury impurity, compositionally similar to temagamite $Pd_3HgTe_3$;

palladium and platinum telluride; palladium and platinum sulfide, compositionally similar to cooperite (Pt,Pd)S or braggite (Pt,Pd)S.

**Results and discussion**

Similar ore mineralization was identified in olivine clynopyroxenites within Urals Dzelyatyshor massif [2] and interpreted by the authors as a new sulfidation gold platinemetal type of ore mineralization. According to A. Pystin [2], Dzelyatyshor massif, based on its geological features, chemical composition of rock-forming minerals and geochemical characteristics, cannot be associated with ophiolites where it is embedded. That's why, authors consider Dzelyatyshor massif to be either a fragment of dismembered intrusive, or a part of platinum-bearing dunite-pyroxenite-gabbro complex in Ural association. The composition of noble-metal mineralization in Dzelyatyshor massif is diverse: native gold (Au), native silver (Ag), braggite ((Pt,Pd)S, kotulskite PdTe , vysotskite (PdS). In the studied olivine clinopyroxenites, the following minerals were identified: native gold (Au), native arsenic As with impurities (Co, Ni), native silver (Ag), awaruite (Ni, Fe), zinc yellow metal (Cu, Zn).

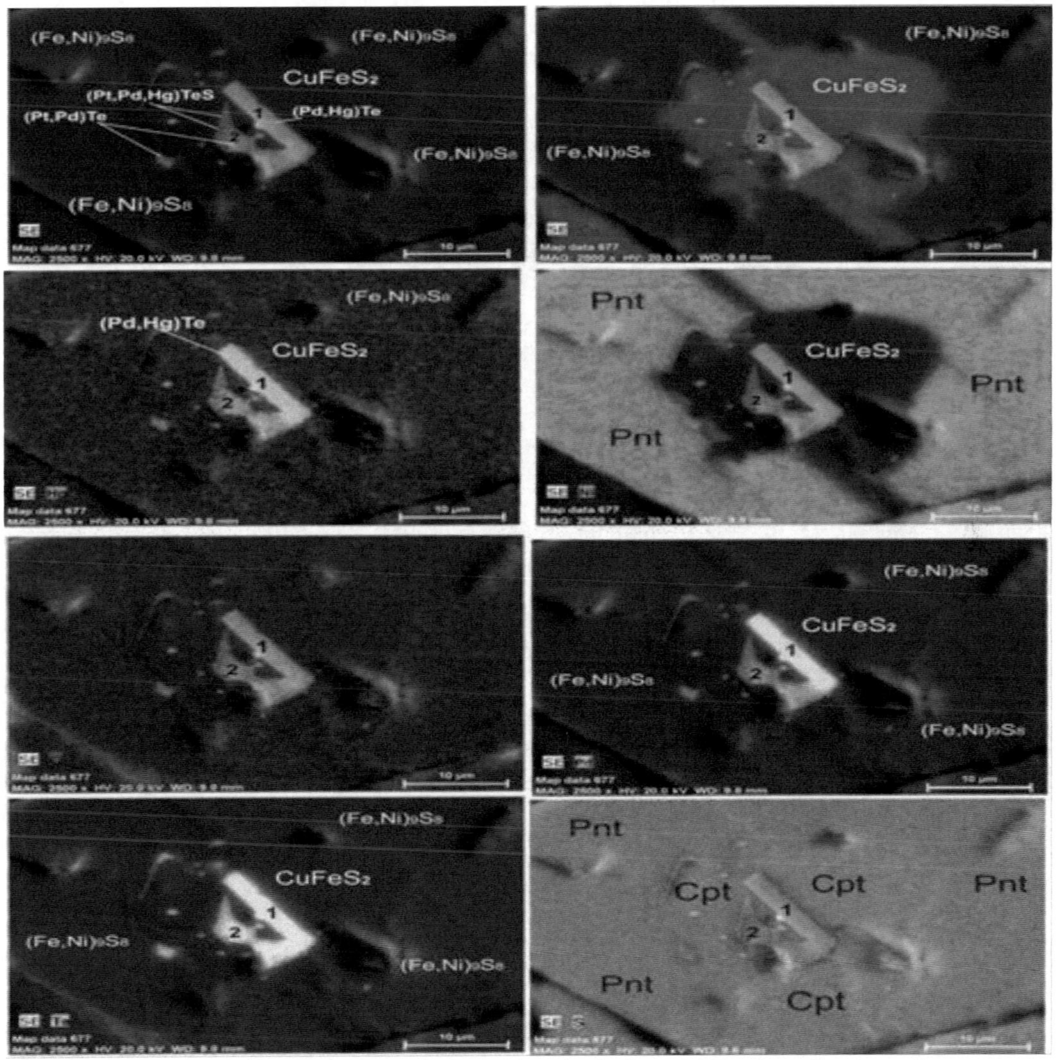


**Figure.** Ore grain fragment element composition including PGE mineralization. Cpt – copper pyrite, Pnt – pentlandite; 1 – palladium telluride with mercury impurity, 2– palladium and platinum telluride.

Such a diversity of native metals and their alloys, as well as the presence of minerals which contain platinum, palladium and mercury impurities prove the high degree of ore solution fluid saturation and their renewability. For example, in the Owendale Complex from New South Wales (Australia), noble-metal mineralization was identified within pegmatoid, biotite- and magnetite-poor clinopyroxenites ("P-units") which form irregular lenses and vein-like bodies within biotite- and magnetite-rich, fine grained clinopyroxenites. The concentration of PGM is probably related to the presence of a reducing fluid phase during final evolution stages of "P-units", themselves emplaced in the most oxidized pyroxenites of the Owendale Complex [3].

The similar point of view concerning the origin of noble-metals in mafic-utlramafic complexes of different genesis is expressed by Shengrong Li, Zhenmin Gao [4]. According to their opinion, the abnormal enrichment of noble-metal element is closely associated with hydrothermal fluid that flew out on the sea floor through deep cycling and reaction with ultramafic-mafic igneous rocks forming noble-metal rich fluid.

There is a variety of research authors of which share the above-mentioned point of view, as magmatic ore-bearing fluids are considered to be basic condition for commercial Cu-Ni PGE ore deposit. The sulfides became PGE-rich due to interaction with a large volume of ore-bearing fluids that were introduced into different types of mafic-ultramafic bodies during a juvenile stage of their development as a conduit through which magma has passed to the main chamber [5,6,7]

Thus, the prospects of the studied area concerning the presence of noble metal mineralization have been discussed since 1950s. At this time, A. Krukov [1] identified the following platinum-group metals in placers of the Kalna river: osmiridium (OsIr), rutheniridosmine (RuIrOs), isoferroplatinum ($Pt_3Fe$), tetraferroplatinum (PtFe), chalybeate, tulameenite ($Pt_2FeCu$), ferronickelplatinum ($Pt_2FeNi$), high nickel platinum ($Pt(FeNi)_2$), sperrylite ($PtAs_2$), allopalladium (PtHg), as well as sulfides, sulfedenes, tellurides, arsenide ore minerals, palladium and rhodium antimonides. Then, in 2003, A.Krivenko, M.Polipskiy [1] identified sperrylite ($PtAs_2$), various palladium alloys Pt-Fe-Ni, palladium containing antimonide ($Pd_2CuSb$) and osarsite ((Os,Ru)AsS) in impregnated chromitites within Kalninsk massif during chrome exploration surveying.

With contemporary element composition analysis methods there appear new opportunities to obtain new data about noble-metal accumulation, redistribution, and concentration in mafic-ultramafic complexes of different genesis. Based on the conducted study, it is possible to state that not only chromitites, but also olivine clynopyroxenites within idgimsky gabbro-peridotite-pyroxenite (West Sayan Mountains) are the source of platinum group noble-metal mineralization. The main consumers of such research data are Noril'sknickel LtD, Krasnoyarsk Research Institute of Geology and Mineral Resources, one of the perspective development trends of which is prospecting and exploration of noble metal deposits located in Krasnoyarsk Krai [8].

*The investigations were supported by National Research Tomsk Polytechnic University, project №: LRU INR_114_2014.*

## References

[1] Ekhanin D A 2010 Geologiya i rudonosnost' Kalninskogo ul'trabazitovogo massiva (Zapadnyj Sayan) *Avtoreferat na soiskanie uchyonoj stepeni kandidata geologo-minerologicheskikh nauk* (Krasnoyarsk) p 23 [in Russian]

[2] Pystin A M, Pystina Yu I, Generalov V I and Potapov I L 2010 Low-sulphide gold-platinum mineralization in the Polar Urals *Journal "Proceedings" of the Komi Science Centre of the Ural Division of the Russian Academy of Sciences* **2**(2) 48-53

[3] Johan Z, Ohnenstetter M, Slansky E, Barron L M and Suppel D 1989 Platinum Mineralization in the Alaskan-type Intrusive Complexes near Fifield, New South Wales, Australia Part 1. Platinum-group Minerals in Clinopyroxenites of the Kelvin Grove Prospect, Owendale Intrusion *Mineralogy and Petrology* **40** 289-309

[4] Shengrong Li and Zhenmin Gao 2000 Source tracing of noble metal elements in Lower Cambrian black rock series of Guizhou-Hunan Provinces *China Science in China Serias D Earth Sciences* **43** (6) 625-632

[5] Latypov R M, Chistyakova S Yu and Alapieti T T 2008 PGE reefs as an iin situ crystallization phenomenon: The Nadezhda gabbronorite body, Lukkulaisvaara layered intrusion, Fennoscandian Shield, Russia *Mineralogy and Petrology* **92** (1-2) 211-242

[6] Tuba G, Molnár F, Ames D E, Péntek A, Watkinson D H and Jones P C 2014 Multi-stage hydrothermal processes involved in "low-sulfide" Cu(-Ni)-PGE mineralization in the footwall of the Sudbury Igneous Complex (Canada): Amy Lake PGE zone, East Range *Mineralium Deposita* **49** (1) 7-47

[7] Péntek A, Molnár F, Tuba G, Watkinson D H and Jones P C 2013 The significance of partial melting processes in hydrothermal low sulfide Cu-Ni-PGE mineralization within the footwall of the Sudbury Igneous Complex, Ontario, Canada *Economic Geology* **108** (1) 59-784

[8] Cherkasova T Yu 2010 Geologo-geneticheskaya model' obrazovaniya i potentsial'naya rudonosnost' mafit-ul'tramafitovykh massivov nizhnederbinskogo kompleksa (Vostochnyj Sayan) *Avtoreferat na soiskanie uchyonoj stepeni kandidata geologo-minerologicheskikh nauk* (Tomsk) p 21 [in Russian]

# Anomaly Geochemical Fields in Siberian Hydrothermal Gold Deposits

**V Voroshilov, O Savinova, Yu Ananev and R Abramova**
Tomsk Polytechnic University, Tomsk, Russia

E-mail: 1v_g_v@tpu.ru

**Abstract.** The composition and internal structure of geochemical fields associated to hydrothermal gold deposits within the Siberian territory were investigated. The concentric zonal structure of ore-forming geochemical fields embracing accumulations of Au, Ag, Bi, Pb, Zn, Cu, Te, As in ore bodies and their adjacent locations, and Ni, Co, V, Cr, Mn, Ba, Ti – within the external margin of gold ore formations were determined. The thermometric properties of gas-fluid inclusions in minerals of hydrothermal gold deposits were described. The results specified not only high-mineralized but also weakly-salted fluids are involved in the formation of the deposit. The latter is subjected to both retrogressive boiling and the mechanisms of direct and reverse osmosis. In this case, ascending and descending fluids produce eddy fluxes during Earth rotation, where minerals of different composition are formed in this flow path. Produced spiral mineral-geochemical fields can be observed on satellite images as hierarchy circular pattern system. Both the analysis of anomalous geochemical field structure and satellite images make it possible to predict the ore bodies of different grades in complicated landscape geological conditions.

## 1. Introduction

The target problem in modern applied geochemistry involves such a fact as the low efficiency of existing geochemical methods used in mineral exploration. The reason is the depletion of easy-to-discover deposits, outcropping as primary and secondary dispersion haloes of tracer-element mineralization. Today, the prospecting of new deposits is being conducted in complicated landscape geological conditions where ore bodies are detectable only at great depths, and associated anomaly geochemical fields outcrop only as fragments, which, in its turn, is difficult to structure without additional information. Personal understanding of a geologist plays the major role in this case. This, however, does not foster reliable results and forecast. Current research is governed by the new geochemical data interpretation methods for mineralization forecasting in complicated landscape geological conditions. Anomaly geochemical fields develop during the formation of thermal fluid ore-generating systems, that is why, additional information on the development of hierarchic geochemical field system is based on the result investigation of gas-fluid inclusions preserved in hydrothermal minerals [1-3]. Temperature, composition and fluid salinity varies consistently in time and space, due to such phenomena as retrogressive boiling, direct and reverse osmosis and the blending of juvenile and vadose waters. Fluid migration scale during ore emplacement causes not only significant chemical rock composition alterations, but also changes its physico-mechanical properties. These alterations can be observed in the geochemical field structure and are recorded as line and circular patterns on spectro-zonal satellite images. Circular pattern hierarchy is easily identified within any landscape geological conditions, which, in its turn, makes it possible to use these patterns in ranking ore-forming geochemical fields and predicting future mineral deposits.

## 2. Methodology

The investigation of anomaly geochemical field structure accompanying gold - ore deposits was based on V. Voroshilov's method [4]. Geochemical field ranking was performed within the framework of the existing Russian gradation: ore body- deposit- ore field- ore cluster- ore district.

In microthermometrical investigation of gas-fluid inclusions in minerals, the following research tools were used: computer measurement package designed on the basis of microthermochamber MDSG600 ( produced by "Lincam, England), microscope "Axio Imager" (Germany), with a set of long-focus lens (maximum amplification 50x), video-camera and administrative computer ( all investigations were conducted in Laboratory of Natural Research Methods of Rocks and Ores at Department of Geology and Mineral exploration, Tomsk Polytechnic University). These facilities are used to measure the temperatures of the transition phases in real time within the interval of -180 - +500° C, to observe corresponding increases and to obtain digital micro photos. Measurement error is not more than 0.5° C.

Geological interpretation of spectrozonal satellite images was conducted by the geoinformation systems Erdas Imagine and ArcGIS. Interpreted data was retrieved from geological, depth-geophysical and geochemical information. Within the framework of the investigation, spectrozonal resolution images of high and low resolution Modis, Landsat ETM+ were used, as well as radar data- SRTM and Aster Global DEM. Processing, interpretation, analysis of satellite data and the modeling of geological and ore-geochemical systems have been conducted in accordance with the recommended and approved guidelines [5].

## 3. Results and discussion

Within the investigated deposit there are three basic developed gold-ore formations: (1) gold-polysulphide-quartz, (2) gold-skarn and (3) gold -(arsenic)- sulphide. The geochemical field of gold-bearing pyrite-polymetallic Rudni Altai deposits, traditionally associated with gold-sulphide formation, was investigated.

Quartz-veined gold-ore deposits are widely found in the Altai-Sajansk fold and for quite a perceptible time they have been the major geological-industrial type of lode mineralization. These deposits are spatially associated to magmatic massifs of diorite-granodiorite composition, accompanied by beresite-type wallrock alteration having similar mineralogical composition and close geochemical properties. Anomaly geochemical fields of gold-polysulphide-quartz formation have concentric zonal structure with Au, Ag, Bi, Pb, Zn, Cu, Te, As accumulations located within the interior zones. There are also three spatial isolated associations within this interior zone: (1) Au, Ag, Bi, Te; (2) Au, Pb, Zn, Cu; and (3) Au, As. These three industrial gold- bearing associations can be found in major field deposits. In small ore occurrences the high gold concentrations are associated with Ag, Bi, and Te in geochemical fields, due to poor polymetallic and arsenic mineralization.

Ore bodies' periphery is characterized by the formation of Cr, Ni, Co associations in beresite, which, in its turn, is conditioned by poor gold-bearing pyrite impregnations. In small occurrences with poorly-developed zoning this association combines with polymetallic and arsenic associations.

Anomaly geochemical fields of gold-skarn formation deposits of different mineral types, in general, are similar and can be characterized as concentric structures with centripetal zoning. Au and other trace elements accumulate in ore bodies of which associations slightly vary depending on the mineralization type. External ore body boundaries, involving slight pyrite and pyrite-magnetite mineralization, can be found in all deposit types of Co, Ni, Cr, V associations, whereas, Ba, Mn, and sometimes Ti accumulate longwise ore-controlling faults.

Gold sequentially accumulates in pyrite with arsenic pyrite, then with galena and chalcopyrite and finally with tellurides and sulphosalts in association with late copper sulfide. Every subsequent process involves the gold redistribution and relative gold accumulation in late associations.

A typical gold-(arsenic) sulphide ore formation is Olimpiad deposit in Yenesei ridge. The anomaly geochemical field has a pronounced concentric structure typical of hydrothermal deposits: ore-channel structure, where external ore bodies' periphery is distinguished by Ni, Cr, V associations, while gold and groups of associated trace elements concentrate within the internal zone. Arsenic and antimony are found in the Olimpiad deposit, whereas Ag, Cu, and W, to a far smaller degree. Wallrock micaceous-carbonate-quartz metasomatites involve Mn and Ba accumulations.

The more increasingly the hydrothermal process the more distinct the spatial isolation of geochemical associations. In areas of intensive mineralization reduction there are overburden associations.

Similar zoning is observed in other large-volume gold-vein deposits in black-shale formations. In particular, in the Bakirchiksk ore district deposit (Kazakhstan) an intimate gold to As, Sb, Ag and W correlation was established, which is governed by massive gold associations with early pyrite-arsenic-pyrite mineralization [6]. Polymetallic association is a less gold-bearing lode, while Co, Ni, V and Ba have a negative correlation to gold and is spatially confined to the ore body periphery.

Gold-sulphide deposits in Patom crater black-shale formation have a structurally similar anomaly geochemical field [7].

The gold-bearing deposit geochemical field of the sulphide-polymetallic formation has not only similar features of all hydrothermal deposits, but also its own specific characteristic features. In the Rudni Altai sublateral ore-controlling irregularities are detected by the anomaly concentrations of Ti, V and Mn. Areal volcanism which is observed in a geochemical field as associations of H Ni, Cr, Co (basaltoid volcanism) and Mo, Sn (acid volcanics) was formed in the intersection with tectonic structures of NW strike slip. According to the metallogenic behavior these structures could be classified as ore clusters. Anomaly Cu, Pb, Zn, Ag concentrations confined to these structures highlights the location of individual ore fields and deposits. In particular, gold-silver deposits are peripherally located in geochemical fields of ore clusters. In this case, rather intensive geochemical haloes are associated with basaltoid volcanism, while gold-silver mineralization is associated with acid volcanics, well-defined in volcanic depressions. Industrial gold and silver concentration deposits expose a rather high concentration of Ba, Ag, As, Au, Sb, and Te.

Thus, oregenetic geochemical fields of hydrothermal gold-ore deposits have a concentric zoning structure with gold accumulations in the central section of this anomaly structure. The chalcophilic trace element groups of gold mineralization and their quantitative correlations significantly vary, however, for each individual deposit type, typical geochemical associations can be defined reflecting the gold-bearing mineral assemblage compositions. For example, Kuznetsk Altai gold-polysulphide-quartz formation, these are Au-Cu-Pb-Zn-Ag-As; Yenesei Ridge gold-(arsenic) sulphide formation – Au-As-Sb-W; and Gorny Altai gold-copper skarn deposits - Au-Cu-Ag-Bi.

The spectrum of siderophilic and lithophylic elements forming the anomaly concentrations in the geochemical fields of hydrothermal gold depositsare, generally, single-type and include Ni, Co, V, Cr, Mn, Ba, but rarely Ti. In most cases, these elements have a distinct spatial antipathy to gold and its associated chalcophilic elements.

On scale of ore clusters and districts those elements indicating ore-generating magmatism: Mo, Be, Y, Sn are of scientific importance. The anomalies of these elements often exhibit a spiral ring structure and are confined to magmagene ring structures with a diameter of tens kilometers. Spiral structures of anomaly geochemical fields are often plotted on a deposit scale, as well as individual ore bodies [8]. Self-similar variable ranking geochemical fields indicate their versatile formation mechanism and the possibility of applying the structures of geochemical fields in predicting ore bodies.

There are two basic problems accompanying this prediction technique:

(1) geometrizing zonal plotted poly-elemental geochemical fields; and (2) ranking multi-phase oregenetic geochemical fields in complicated landscape geological conditions.

IOP Conf. Series: Earth and Environmental Science **21** (2014) 012009 doi:10.1088/1755-1315/21/1/012009

Mathematical procedure in geometrizing anomaly geochemical fields has been developed for many years and at present there are 4 methods to implement different approaches in objectively solving this problem [4]:

1) defining quasi-homogeneous spatial regions (cluster analysis observation, artificial neural networks);

2) identifying stable element associations and analyzing their positional application (factorial, discriminate, regressive methods);

3) calculating indexes of total intensive re-distribution of chemical elements ( mineralization intensity, geochemical spectrum dispersion, ranking dispersion);

4) calculating zoning indexes based on the universal vertical geochemical zoning concept or cetrifugal-centripetal differentiation of elements within the hydrothermal process.

Integrated application of all these methods minimizes the human factor in geometrizing the anomaly geochemical fields and advancing objective prediction of mineralization according to the geochemical data.

In complicated landscape geological conditions where the anomaly geochemical fields are only fragments, to structure them by using above-mentioned methods is sometimes unsuccessful. In this case, an independent ranking criterion for geochemical fields is necessary which could be a geological feature closely associated with the deposit formation itself and would be identified in any landscape geological environment.

Such features could be detected ring structures (those interpreted from spectral-zoning satellite images) with a natural hierarchy. To understand what the interrelation between anomaly geochemical fields and ring structures is, the temperature evolution pattern and fluid salinity during hydrothermal deposit formation was examined.

The examination of preserved gas-fluid inclusions in minerals show that after the separation of aqua-saline fluid from magma, its salinity increases from 45 to 60% (NaCl equivalent) due to periodic boiling, and then gradually decreases in proportion to solution temperature drop, correlating to NaCl saturation curve [2,3,9–11].

During boiling, the separated vapor-gas mixture condensates into a high-temperature (350–400 °C) low-mineralized solution. Its salinity initially increases, but at the temperature of 270–300 °C it decreases. This can be explained by the fact that at the initial mineralization stage the high fluid pressure promotes the so-called phenomenon- reverse osmosis, i.e. solution dewatering in low permeable country rocks. The retrogressive boiling processes also further the entrance of low-mineralized waters into the environment. During penetration these non-equilibrium hot waters ( in balance to rocks) being enriched by country rock components (Co, Ni, Cr, Fe –from mafic minerals; Mn – from carbonates) migrate further alongwise weakened zones, forming in the surrounding rocks upward flows. In time, these solutions, receding from the main channel, then cool, blend together in the downward flows and escaping from above-mentioned elements form their geochemical anomalies, being typical of the hydrothermal system frontal zones.

Ore formation during the initial mineralization stage could not be associated with solution dilution, but possibly be associated with the sharp pressure variations as a result of tectonic movements. In the process of fluid pressure drop as a result of solution intensity decrease, the reverse osmosis inevitably becomes a direct one. Critical osmosis pressure boundary depends on the concentration gradient and correlation between fluid pressure and lithostatic load. As seen in Figure 1 fluid pressure inversion occurrences begin at solution temperature decrease to 270-300° C and solution salinity of up to 13-18 mass% of NaCl equivalent. At this moment the water inflow from the environment and the salinity reduction (up to 2-5 mass% of NaCl equivalent) start against the mineral formation temperature drop. This process is embodied in the geochemical ore features. Particularly, pyrite and arsenic-pyrite accumulating during solution flow inversion and significantly enriched by Co and Ni imported from the host rocks.

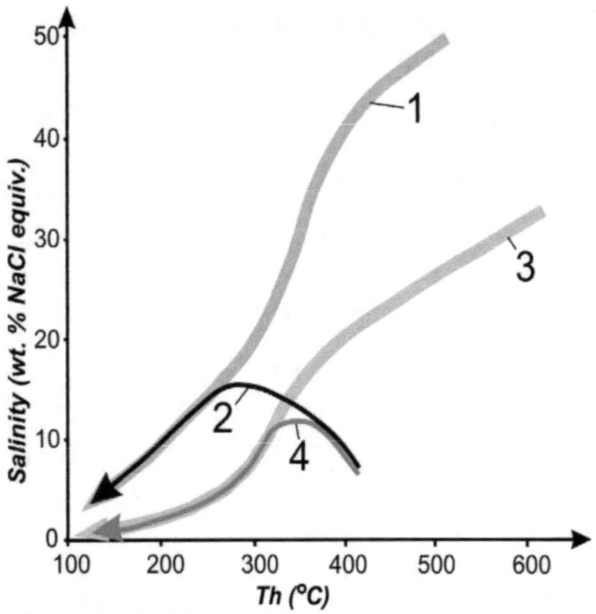


**Figure 1.** Salinity and homogenization temperature evolution of gas-fluid inclusions in quartz hydrothermal deposits (in accordance to data from [1–3, 9–13]): (1) gold-copper-porphyry; (2) quartz-gold-suphides and gold-suphides; (3) in prograding skarns; and (4) in retrograding skarns.

Downward fluid flows affected by Corioli forces form cyclones controlling the hydrothermal product distribution, in which the structure of emerging anomaly geochemical fields is embodied in.

In spite of the deposit mechanism diversity of ore mineralization, the eddy fluid flow would lead to the concentric (ring) distribution pattern of trace element mineralization anomalies. These changes in rock composition affect their physico-mechanical properties and influence the spectral features and vegetation composition, which, in its turn, can be observed as new formations on satellite images. Based on experimental data, the regular spatial correlation between multi-phase anomaly geochemical fields and detected ring structures (observed on satellite images) have been pervasively revealed [8]. It can be highlighted that most detected ring structures, principally magmatogene, are the results of emerging eddy fluid flows.

## 4. Conclusion

Investigation results of gas-fluid inclusions in hydrothermal minerals show the extensive process development of retrogressive boiling, direct and reverse osmosis, blending of juvenile and vadose waters during ore formation. As a result the system of upward and downward fluid flows, the interaction of which promotes the concentric zoning structure of emerging geochemical fields.

Geochemical fields of hydrothermal deposits have a concentric- zonal structure withAu, Ag, Bi, Pb, Zn, Cu, Te, As accumulations in the central anomaly structure zone, while Ni, Co, V, Cr, Mn, Ba, Ti – in the external margin of gold-ore bodies.

Under conditions of Earth rotation the upward and downward fluid flows form rotating-cyclones and anti-cyclones. In this case, oregenetic geochemical fields have a spiral structure. These changes in rock composition affect their physico-mechanical properties and influence the spectral features and vegetation composition, which, in its turn, can be observed as new formations on satellite images. Thus, hydrothermal systems and associated geochemical fields include cause-effect and positional correlation with detected ring structures.

Therefore, the natural hierarchy of ring structures, interpreted from space data, could be used as an independent tool in ranking oregenetic geochemical fields.

## References

[1] Roedder E 1984 Fluid inclusions *J. Rev. Mineral.* **12** 646
[2] Reiff F 2009 *Conditions and formation mechanisms of granitic ore-magmatic systems (based on thermobarogeochemical data)*
[3] Borisenko A, Borovikov A et al. 2006 Composition of magmatogenetic fluids, geochemical and metal-bearing factors *J. Geology and Geophysics.* **47 (1)** 1308-1325
[4] Voroshilov V 2009 Anomalous Structures of Geochemical Fields of Hydrothermal Gold Deposits: Formation Mechanism, Methods of Geometrization, Typical Models, and Forecasting of Ore Mineralization *J. Geology of Ore Deposits.* **51 (1)** 1–16
[5] Posteloev A, Ananev Yu, Zhetkov V, Nazarov V and Kuznestov A 2007 *Remote Methods in Geological Surveying, Forecasting and Mineral Exploration* ( Rudni Altai) (Tomsk: STT ) p 228
[6] Narseev V, Gostev Yu, Zakharov A et al 2001 *Bakirchik (Geology, Geochemistry and Mineralization)* (Moscow: CRD GSI ) p 174
[7] Gavrilov R, Kucherenko I, Martinenko V and Verkhozin A 2009 3-D geo-geochemical model of mesothermal gold-bearing Chertovo-Korito deposit (Patom ridge) *Bulletin of Tomsk Polytechnic University* **315 (1)** 30–43
[8] Voroshilov V 2012 Eddy features of oregenetic geochemical fields *J. Izvestaya TPU* **321 (1)** 46–51
[9] Gonzalez-Partida E, Camprubi A 2006 Evolution of mineralizing fluids in the Zn–Pb–Cu(–Ag ± Au) skarn and epithermal deposits of the world-class San Martín district, Zacatecas, Mexico *Journal of Geochemical Exploration* **89** 138–142
[10] Vagina E 2012 *Mineral complexes and genesis of Chertovo-Korito (Patom ridge) gold deposit* (Tomsk; Dissertation) p 141
[11] Mollai H, Sharma R and Pe-Piper G 2009Copper mineralization around the Ahar batholith, north of Ahar (NW Iran): Evidence for fluid evolution and the origin of the skarn ore deposit J. *Ore Geology Reviews* **35** 401–414
[12] Ettlinger A and Meinert L 1990 Gold in the Soviet Union: A study of the Veseli Mine, Siniukhinskoe District,Sibiria USSR *Mines and Petroleum Resources (*Brit. Colum Ministry of Energy, Prof Paper) 2–13.
[13] Gusev A 2003 *Gold metallogenesis in Gorny Altai and southern Gorny Shopii* (Tomsk: STT) p 350

IOP Conf. Series: Earth and Environmental Science **21** (2014) 012010 doi:10.1088/1755-1315/21/1/012010

# Structure-geochemical zoning of Topolninsk gold-ore field (Gorny Altai)

**T V Timkin[a], D S Lavrov[b], O Y Askanakova[c] and T V Korotchenko[d]**
Tomsk Polytechnic University, Tomsk, Russia

E-mail: [a]timkin@tpu.ru, [b]tooloot@sibmail.com, [c]AsoJlb@rambler.ru, [d] <tatyana1003@mail.ru>

**Abstract.** Geochemical zoning of prospective mineable gold-bearing skarns was carried out. The geochemical field abnormal structures of different hierarchy levels associated with gold-skarn formations were revealed. The interrelation between the structure of ore-geochemical fields and associated ring structures was studied. Complex structure-geochemical criteria for gold mineralization prospecting and evaluation were proposed.

## Introduction

Topolninsk gold-ore field is located on the right bank of the rivers Anui and Karama between the Topolnoye and Stepnoye villages in Soloneshensky region of the Altai Territory (figure 1). Structurally, this gold-ore field is confined to the Anui structure block on the west by the Talitstkiy and on the east by the Katunskiy blocks along the large-scale fault zones - Baschelakskiy and Kuyachinskiy, respectively. According to metallogenic zoning, Topolninsk gold-ore field is related to Anui copper-gold-ore alluvial cluster in Anui ore area within North Altai gold belt.

The investigated area is composed of lower Silurian deposits, basically terrigene-Chinetinsk, clastic-carbonate Polatinsk suite, dismembered clastic-carbonate Gromotukhinsk series, terrigene-Chesnokovsk, lower Devonian clastic-carbonate Kamyshensk and Baragashsk suites.

Intrusive formations occupy one third of Topolninsk gold-ore field. They are represented by two massifs (Topolninsk and Karaminsk) and numerous dykes of different composition within Topolsk gabbro-granodiorite-granite complex, as well as stocks and linear subvolcanic bodies of Kuyagansk rhyolite-dacite-andesine complex.

According to [1], host rocks are represented by hornfels and, in places, by skarns within the vast exocontact massif area. Dispersed skarnification is observed throughout the gold-ore field. According to the study results, propylitic type has altered within the ore deposit. These metasomatites are more local and confined to skarn peripheries, which allows classifying them as post-skarn metasomatites [2,3]. The overlying ore formation is basically concentrated in skarns, post-skarn propylites and partially sericitization. The basic ore mineralization occurs mainly in association with gold-skarns. The content of sulfides varies between 2% and 7%, with pyrite, arsenic pyrite, magnetic pyrites and copper pyrite being well-developed. Sphalerite, molybdenite, erubescite, chalcosine, galena, fahlite, tellurides and Bi, Pb, Ag sulfotellurides are rarer [4].

Ore mineralization is associated with the intense primary and secondary dispersion of a wide range of elements which are typical for this area and type of mineralization. However, the identified chemical composition of element assemblages and their distribution patterns unveil a rather specific composition feature of abnormal geochemical fields [5].

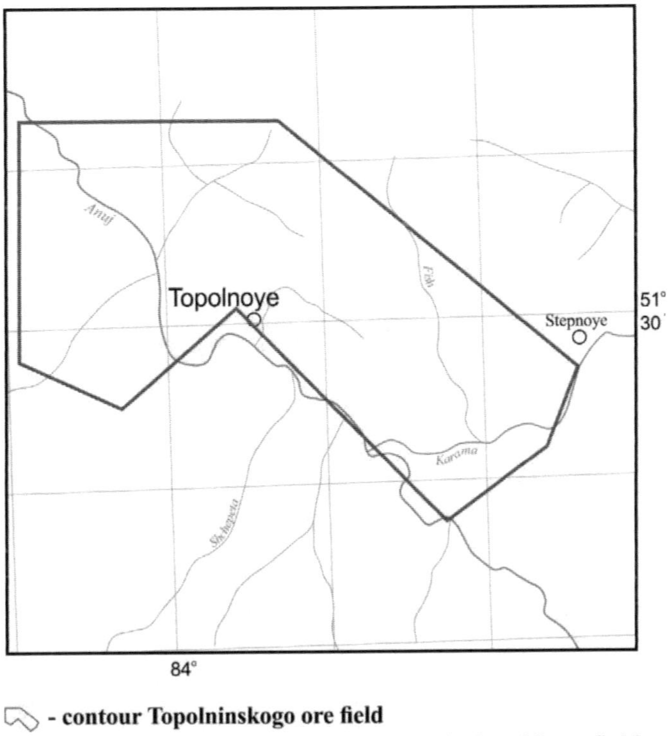


**Figure 1.** Geological setting of Topolninsk gold-ore field.

**Research technique**

The present study is based on the lithogeochemical surveys in scale 1:25000, 1:10000 along the secondary dispersion haloes (more than 8000 samples) within Topolninsk gold-ore field carried out by JSC "Gorno-Altaiskaya Ekspeditsiya" in 2012-2013 and geochemical sampling data. The collected data were processed by standard statistical programs, whereas geometrization was carried out by GIS-technology, in accordance with the developed procedures [6,7].

Besides, multiband satellite images Modis, Landsat ETM+, Aster, IRS and digital terrain models SRTM and Aster DEM were used. The interpretation and analysis of multi-band satellite images, as well as geological and ore-geochemical system modeling were carried out according to the recommended methodology and approaches [8].

As Topolninsk gold-ore field is characterized by complicated landscape conditions and interpretation of geochemical data can hardly be sufficient for complete reconstruction of abnormal geochemical field, the study of the interrelation between the ore-geochemical field structure and associated ring structures was carried out first. Thereafter, it helped to make forecast and accurate identification of the most perspective gold-ore areas.

**Research results**

The results of the study are as follows:

- background and minimum abnormal concentrations of gold and trace-elements in secondary dispersion haloes were identified;
- geochemical associations were revealed and association zoning within the gold-ore field was outlined;
- interrelation between ring structures and the ore-geochemical field structure, as well as the location of gold mineralization within the investigated area was revealed;
- structure-geochemical criteria for gold mineralization prospecting and evaluation were proposed.

Background and minimally abnormal concentrations of elements were calculated in accordance with standard methodology developed by A. Soloviev [9]. As element distribution does not correspond to the normal law ($A/S_A>3$ и $E/S_E>3$) in the investigated area, background and abnormal values were calculated based on the log-normal distribution model (table 1).

**Table 1.** Background and minimum abnormal values

| Element | Background $\tilde{X}$ n*10$^{-3}$ % | Standard multiplier | Minimum abnormal N content to correlate points (gradations of spectral analysis), in n*10$^{-3}$ % (Au – in ppm) | | | | | | | | |
|---|---|---|---|---|---|---|---|---|---|---|---|
| | | | N=1 | N=2 | N=3 | N=4 | N=5 | N=6 | N=7 | N=8 | N=9 |
| Au | 0.003 | 2.05 | 0.024 | 0.013 | 0.01 | 0.008 | 0.007 | 0.007 | 0.006 | 0.006 | 0.006 |
| Ag | 0.005 | 1.14 | 0.008 | 0.007 | 0.006 | 0.006 | 0.006 | 0.006 | 0.006 | 0.006 | 0.006 |
| Bi | 0.05 | 1.07 | 0.063 | 0.059 | 0.058 | 0.057 | 0.056 | 0.056 | 0.055 | 0.055 | 0.05 |
| Pb | 1.4 | 1.46 | 4.4 | 3.2 | 2.7 | 2.5 | 2.3 | 2.2 | 2.2 | 2.1 | 2.1 |
| Cu | 3.04 | 1.3 | 6.9 | 5.4 | 4.9 | 4.6 | 4.4 | 4.24 | 4.14 | 4.06 | 3.99 |
| Zn | 4.03 | 1.7 | 21 | 13 | 10 | 9.2 | 8.4 | 7.9 | 7.5 | 7.2 | 7.0 |
| Co | 1.2 | 1.57 | 4.7 | 3.2 | 2.7 | 2.4 | 2.2 | 2.1 | 2.0 | 1.98 | 1.92 |
| Ni | 3.6 | 1.3 | 8.7 | 6.7 | 6.0 | 5.6 | 5.3 | 5.1 | 5.0 | 4.9 | 4.8 |
| Cr | 3.9 | 1.4 | 11 | 8.4 | 7.3 | 6.7 | 6.3 | 6.1 | 5.9 | 5.7 | 5.6 |
| Mn | 61.5 | 1.18 | 102 | 88 | 82 | 79 | 77 | 76 | 75 | 74 | 73 |
| Sn | 0.2 | 1.4 | 0.5 | 0.4 | 0.3 | 0.29 | 0.3 | 0.3 | 0.3 | 0.3 | 0.2 |
| W | 0.5 | 1.1 | 0.71 | 0.66 | 0.63 | 0.62 | 0.61 | 0.6 | 0.6 | 0.6 | 0.6 |
| Mo | 0.09 | 1.4 | 0.26 | 0.19 | 0.16 | 0.15 | 0.14 | 0.14 | 0.13 | 0.13 | 0.13 |

Based on the factor analysis, 3 resistant associations of basic trace-elements which reflect spatial and temporal evolution of ore-forming fluids were identified throughout the ore field: F1–Au,Ag,Bi,Cu,(As,Sb), F2–Sn,Pb,Zn,W,Mn, F3–Cr,Ni,V. The first factor corresponds to gold-ore mineralization while the number of the second association elements can be related as "rock". Factor 2 is spatially related to the intrusive rocks, which in its turn may identify temperature zoning of ore mineralization. Factor 3 identifies the zones of dispersed sulfide mineralization.

Based on the interpretation of the multiband high-resolution satellite image, the complex of hierarchically organized ring structures was revealed. Figure 2 illustrates spatial distribution of the identified associations, ring structures and faulting within the studied area.

According to figure 2, the following conclusion can be made: within the investigated ore field, ore occurrences are associated with Au,Ag,Bi,Cu,As,Sb anomalies, bounded on the north-west and on the north east by faults, small ring and bogen (up to 9,5 km) structures. The similar regularities are identified in Co, Ni and V distribution. They form the external zone in relation to Au,Ag,Bi,Cu,As,Sb haloes and are thrusted to the frontal zones which is reflected in the contours of ring structure.

It is obvious that most contrasting anomalies Au,Ag,Bi,Cu,As,Sb are identified in the peripheries of the defined ring structures. This can be explained by the fact that the frontal enrichment zone of abnormal geochemical field is represented by a number of identified ore occurrences and dispersed mineralization zones confined to the tectonic structures which are less favorable locations for ore mineralization [10].

Figure 3 depicts the structure of abnormal geochemical field of the perspective area characterized by three main factors defining the total dispersion as70 %. They form contrasting restricted structure, with central (nuclear), intermediate and frontal zones of different size, composition and concentration degree of major and trace-elements being identified. The central zone is characterized by gold-, silver-, copper-, bismuth-, wolframite-, algam associations having the largest concentrations. Within the intermediate zone of abnormal geochemical field structure, such large element concentrations are not observed, however, it separates the central zone from the frontal one. Cobalt, nickel, chromium, and vanadium accumulate in the frontal zone, with some increase in concentrations of other elements being observed (figure 3).

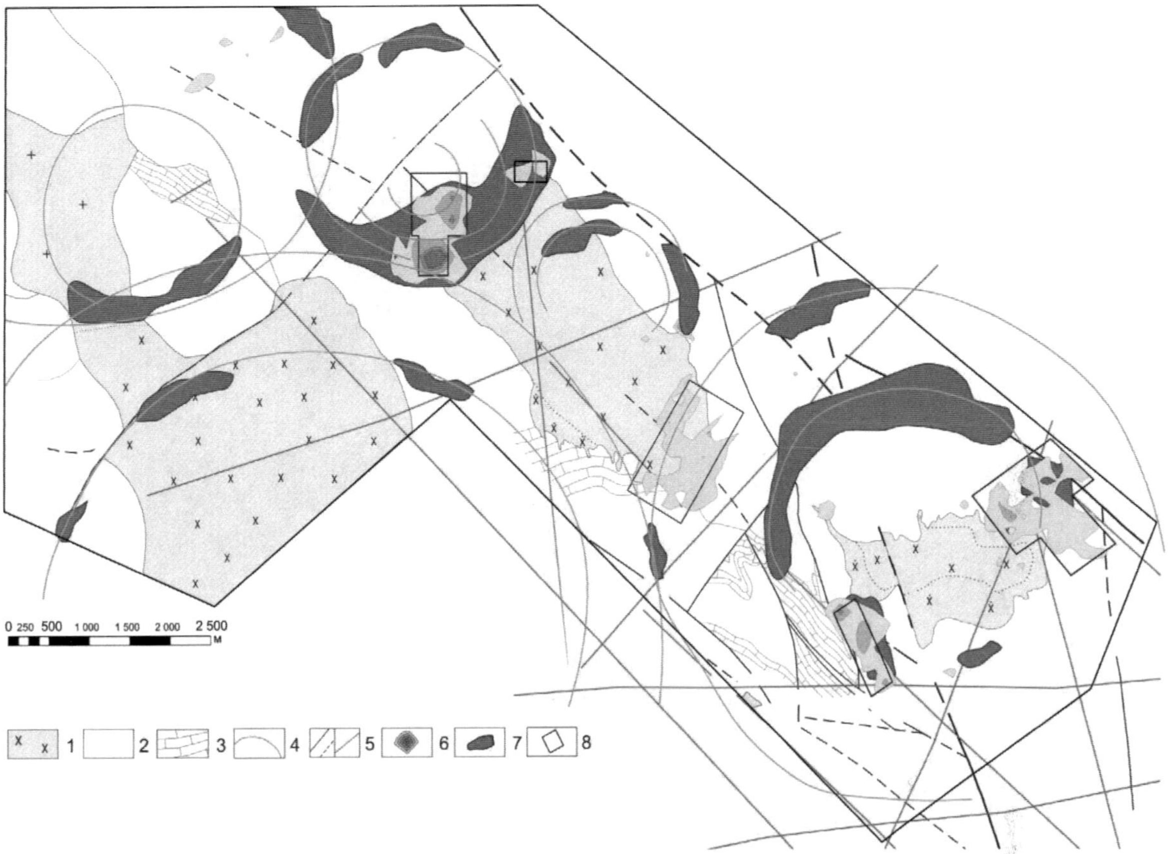


**Figure 2.** Distribution of geochemical associations in ring and linear structures in Topolninsk gold-ore field: 1) granodiorites, granites, diorite; 2) calcareous sandstone, siltstone, mudstone, sandstone; 3) limestone; 4) ring structures; 5 lineation as faults: a) proved, inferred; б) interpreted; anomalies; 6) Au-Ag-Bi-Cu-(As-Sb); 7) Co-Ni-V; 8) perspective zones.

The obtained results have revealed that element association distribution is concentrically zoned within the field. Zoning is defined by the distribution of high-contrast element halos: some in the central zones and some in the periphery.

IOP Conf. Series: Earth and Environmental Science **21** (2014) 012010 doi:10.1088/1755-1315/21/1/012010

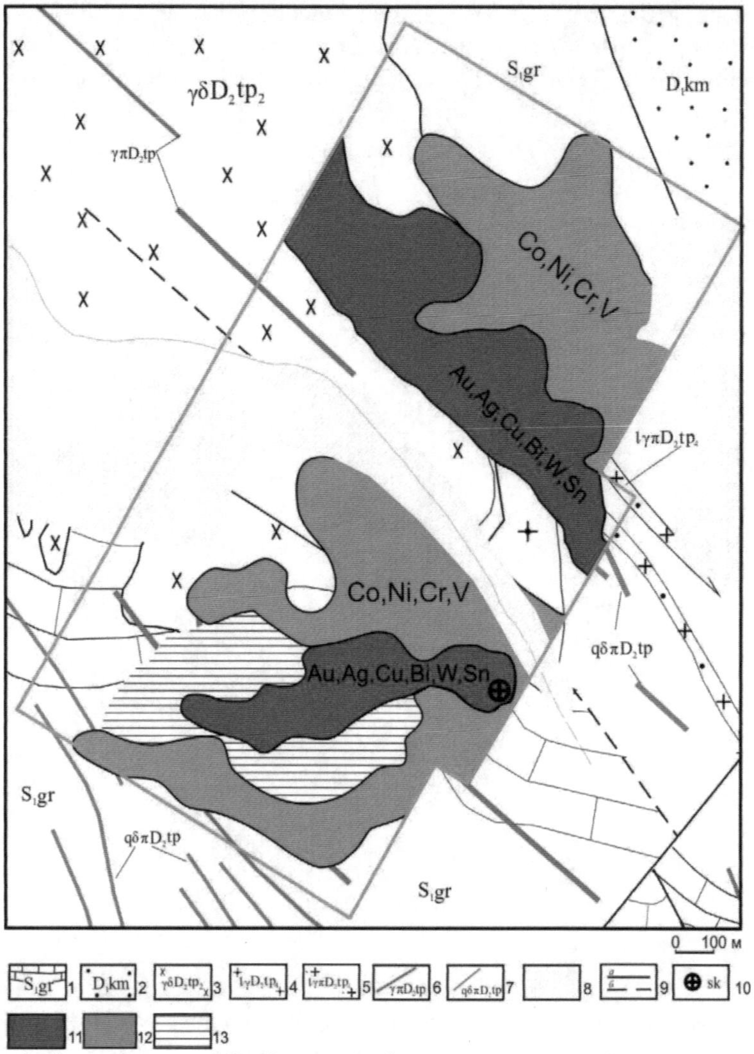


**Figure 3. Geochemical zoning model of perspective area**: 1 – argillaceous slates siltstone, sandstones, limestones; 2 – sandstones, siltstones; 3–6 Topolsk gabbro-granodiorite-granite complex: 3 – granodiorite biotite-hornblend, 4 – leucogranite amphibole biotite, 5 – alkali-type leucogranite. 6 – granite-porphyry dykes, 7 – diorite- porphyry dykes; 8 – quaternary deposits; 9 – contact lines: a – proved, б – inferred; 10 – skarns and skarned rocks; 11 – central concentration zone (Au-Ag-Cu-Bi-W-Sn); 12 –frontal concentration zone (Co-Ni-Cr-V); 13 – intermediate zone

## Conclusion

Based on the lithogeochemical surveys, multiband satellite images Landsat ETM+, IRS and radar mapping SRTM, geochemical zoning and cosmostructure of Topolninsk gold-ore field (Gorny Altai) has been studied.

Due to integrated use of geochemical data and medium/high spatial resolution satellite imagery, it was possible to specify geological structure and obtain absolutely new geochemical information:

- Gold-skarn formations is accompanied by Au,Ag,Bi,Cu,(As,Sb); Sn,Pb,Zn,W,Mn and Co,Ni,V associations in the secondary geochemical field within the investigated area.
- The zones of spatial overlapping of several geochemical associations, which form abnormal geochemical concentric structure in the secondary geochemical field, are the most perspective exploration of this type of mineralization.

- The identified ring structures of ore deposit are associated with intrusive bodies, both blind, and exposed by erosion. The structures of higher level, i.e. field, demonstrate the interrelation of hydrothermal-fluid systems and host rocks.
- The identified levels of ring structures are proved by geochemical data.
  The number of revealed regularities could be regarded as forecasting and prospecting criteria of both future gold-ore deposits and ore fields.

**Acknowledgement**
*The investigations were supported by National Research Tomsk Polytechnic University, project №: ВИУ_ИПР_114_2014.*

## References

[1] N I Gusev and A I 1998 Gusev Gorny Altai gold producing ore-magmatic systems (Moscow: Rudy i Metally) 67-78
[2] Ali Asghar Calagari and Ghader Hosseinzadeh 2000 The mineralogy of copper-bearing skarn to the east of the Sungun-Chay river East-Azarbaidjan Iran *Journal of Asian Earth Sciences* 423-438
[3] M T Einaudi, L D Meinert and R J Newberry 1981 Skarn deposits *Economic geology* **75** 317-391
[4] O V Logvinenko and T V Timkin 2014 Veshchestvennyy sostav i posledovatelnost mineraloobrazovaniya rudoproyavleniya Log-26 Topolninskogo zolotorudnogo polya (Gornyy Altay) *Tomsk State University Journal* **383** 212-220 [in Russian]
[5] Chongmin Liu, Shuqi Hu, Shengming Ma and Liling Tang 2014 Primary geochemical patterns of Donggua Mountain laminar skarn copper deposit in Anhui *Journal of Geochemical Exploration* **139** 152-159
[6] V G Voroshilov 2009 Abnormal structures of geochemical fields of hydrothermal deposits: formation mechanism, geometrization, model types, forecasting large-scale mineralization *Ore Geology* 3-19
[7] John C Davis 1990 *Statistics and Data Analysis in Geology* (Moscow: Nedra) p 472
[8] A A Potseluev, Yu S Anan'ev, V G Zhitkov, V N Nazarov and A S Kuznetsov 2007 Remote methods for geological research, forecasting and mineral exploration (Ore Altai) (Tomsk: STT) p 228
[9] A P Solovov 1985 Geochemical exploration methods for mineral deposits (Moscow: Nedra) p 294
[10] T V Timkin 2011 Localization Criteria of Gold Mineralization in Maisk-Lebedsky Ore Field (Mountain Shoria) *Vestnik of Irkutsk State Technical University* 58-64

IOP Conf. Series: Earth and Environmental Science **21** (2014) 012011 doi:10.1088/1755-1315/21/1/012011

# Petroleum potential of the Neocomian deposit of Nyurolsky megadepression

**E N Osipova[1], F S Prakoyo and L K Kudryashova**
Tomsk Polytechnic University, Tomsk,

E-mail: [1]osipovaen@list.ru

**Abstract.** The distribution of shelf deposits and Achimovsky reservoirs of the Neocomian deposits of Nyurolsky megadepression was studied. Based on paleotemperature simulation, the values of the heat flow, distribution paleotemperatures and generation paleosource of the Bazhenov oil were received, evaluation of resources relative density distribution was performed. The areas for priority hydrocarbons exploration were proposed.

## 1. Introduction

Nyurolsky megadepression, confined to the southeastern part of the West Siberian plate, is the research target. Deposits of hydrocarbons (HC), which were earlier developed here, are associated to the Upper Jurassic oil and gas complexes (OGC). The aim of this work is to study the petroleum potential of underexplored Cretaceous (Neocomian) OGC.

The area under study covers two oil and gas fields - Kaymysovsk and Vasyugansk where 44 fields were discovered (figure 1a).

Source of hydrocarbon traps formation in the deposits of the Upper Jurassic and Cretaceous OGC is the disseminated organic matter (DOM) of Bazhenov formation sediments, but the intensity determinant of hydrocarbon generation at DOM catagenesis is geotemperature. According to the geothermal criterion, the balance model of oil and gas generation processes in the Bazhenov Formation [1] makes it possible to forecast center of intensive hydrocarbon formation foci from DOM: at 85 ° C – migration of source rocks into major oil generation zone. .

## 2. Characteristics of potential petroleum area

Productive complexes are the Cretaceous, Upper Jurassic, Middle Jurassic, Lower Jurassic and Paleozoic. Togur formation (tg, $J_1t$) is the source rock for pre-Jurassic, Lower Jurassic and Middle Jurassic OGC. It is confined to the Nyurolsky megadepression and tapers out in the north-westward (figure 1a).

Oil source rock for Upper Jurassic and the Cretaceous OGC is Bazhenov formation (bg, $J_3tt$) which is regional throughout the studied area. Its thickness involves: Kaymysovsk arch – from 5 to 20 m, Nyurolsky megadepression – from 20 to 41 m, North Mezhovsky megamonocline – from 8 to 30 m, Srednevasyugan megaswell – from 9 to 19 m.

Cretaceous OGC is characterized by a complex geological structure [2], where sediments in sedimentation stages have a threefold division: Neocomian, Aptian-Albian-Cenomanian, Upper Cretaceous. Petroleum potential is associated with Neocomian sediments overlying Bazhenov formation oil source rock and overlain by mature clay deposits. Potential and oil-and-gas bearing are clinoform slope sediments (Achimovsky formations), shelf zone sediments and its edges. The following oil fields, average in reserves, are being developed: South Cheremshansk (index 1), South Myldzhinsk (index 2) and Myldzhinsk (index 3) (figure 1a, table 1).

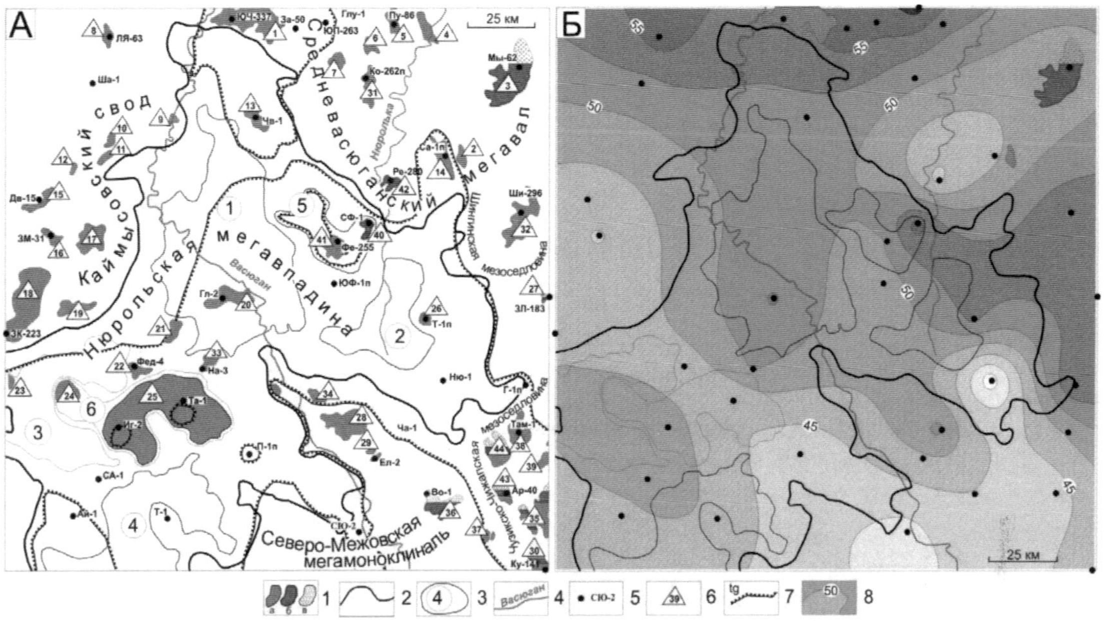


**Figure 1.** Tectonic map of Nyurolsky megadepression petroleum potential: (a) distribution of heat flow density from bottom of sedimentary mantle (b). 1. fields: a – oil, b – condensate, c – gas; 2. of Nyurolsky megadepression border 3. – structure of III order and its number: 1 – Kulan-Igayskaya depression, 2 – Tamradskaya depression, 3 – Osevoy syncline, 4 – Tamyansky syncline, 5 – Festival bar, 6 – Igolsko-Talovoye dome–shaped uplift; 4 – river system; 5 – well log where paleotemperature modeling was performed; 6 – field arbitrary number; 7 – Togur formation distribution zone border; 8 – isoline values of calculated heat flow density, mWt/m$^2$. Map (b) shows the field of Cretaceous deposits in OGC.

**Table 1.** Characteristic features of Nyurolsky megadepression field and framing structures with Cretaceous oil and gas complex deposits

| Field | Number (Figure 1a) | OGC | Phase | Horizon, layers |
|---|---|---|---|---|
| South Cheremshansk | 1 | Cretaceous | Oil<br>Gas | $A_{2\text{-}7}$, $A_{9\text{-}10}$, $B_{4\text{-}10}$, $B_{12\text{-}13}$<br>$A_{2\text{-}7}$, $B_{0,\ 4\text{-}13}$ |
| South Myldzhinsk | 2 | Cretaceous | Oil<br>Gas | $B_{9\ \text{-}13}$<br>$B_{9,\ 10}$, $B_{11\text{-}12}$, $B_{13}$ |
| Myldzhinsk | 3 | Cretaceous<br>Upper Jurassic | Gas condensate<br>Gas condensate | $A_{2\text{-}7}$, $B_{0,\ 4\text{-}10,\ 12,\ 13}$<br>$J_1$ |

## 3. Research methodology

Thickness of shelf deposits and Achimovsky reservoirs is the summing of homogeneous sediment thicknesses of 7 cyclites in the Neocomian sedimentation basin.

Thermal history matching of Bazhenov formation sediments was performed by computer paleotemperature modeling [3]. Paleotemperature calculating scheme consists of two stages. The first stage of "observed" temperature $T_i$ distribution in the well includes the calculation of the heat flow $q$ through the surface of the underlying formation, i.e. the solution geothermic inversion . In the second

IOP Conf. Series: Earth and Environmental Science **21** (2014) 012011 doi:10.1088/1755-1315/21/1/012011

stage the value $q$, is used in forward solution, i.e. – immediate calculation of temperature $U$ in given points of sedimentary sequence $Z$ at predetermined geological time $t$.

To estimate the relative density distribution magnitude of generated Bazhenov oil resources,the conditional integral index $R$ was calculated by the formula [4]:

$$R=\sum_{i=1}^{20}\left(U_i t_i \cdot 10^{-2}\right), \qquad (1)$$

where $U_i$ – calculated temperature of oil generation, °C; $t_i$ – tduration of oil generation, millions of years; number of time intervals $i$ = 1, ..., 20, defined by the number of suites during their formation stages.

## 4. Research results

The development of Neocomian sediments in Nyurolsky megadepression were identified and defined, as well as its positive framing structures (figure 2). The maximum thickness of shelf sediments (over 80 m) are mapped in the north-eastern part of the megadepression, the Achimovsky (80 m) – in the north-eastern slope of Kaymysovsk arch. Instead of Achimovsky formations one can observe a large "bay" in the south-western part of the studied area and “horseshoe” zone to the east.

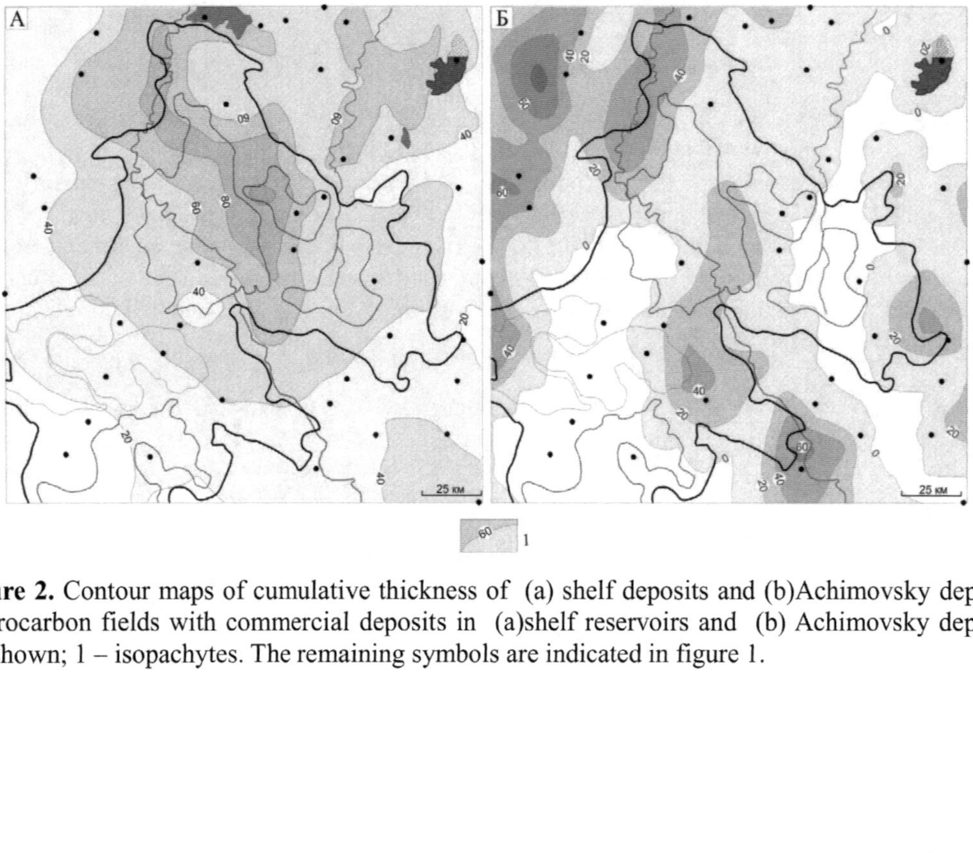


**Figure 2.** Contour maps of cumulative thickness of (a) shelf deposits and (b)Achimovsky deposits Hydrocarbon fields with commercial deposits in (a)shelf reservoirs and (b) Achimovsky deposits are shown; 1 – isopachytes. The remaining symbols are indicated in figure 1.

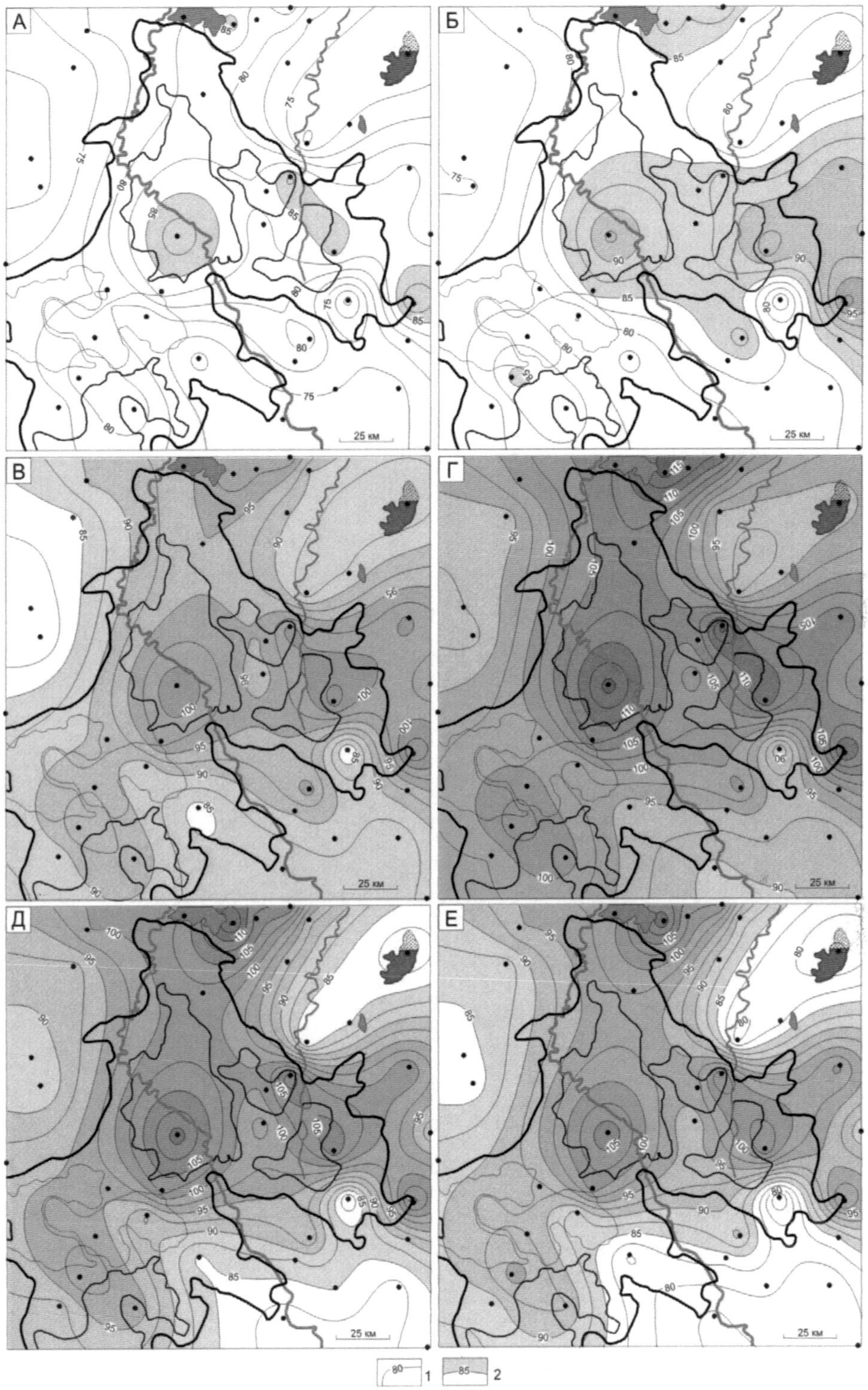


**Figure 3.** Contour maps of geotemperature distribution and source location position intensive generation of Bazhenov oils (a) 91.6 million years ago ; (b) 86.5 million years ago; (c) 61.7 million years ago; (d) 37.6 million years ago; (e) 24.0 million years ago and (f) present day. 1 – isotherm, °C; 2 – source contour. The remaining symbols are indicated in figure 1.

IOP Conf. Series: Earth and Environmental Science **21** (2014) 012011 doi:10.1088/1755-1315/21/1/012011

Paleotemperature simulation was carried out for 39 well logs (figure 1a). The following references were used formation temperatures measured during well testing and paleotemperatures recalculated from vitrinite reflectivity values .

According to heat flow values of geothermic inversion from 39 well logs, ***the contour map of heat flow density distribution values*** was plotted (figure 1b). As a result of forward solution the paleotemperatures were determined in Bazhenov formation sediments at starting/completion time of each suite formation. Figure 3 shows the contour maps reflecting the dynamics of the thermal history of Bazhenov sediments.

The first oil generation source emerged 91.6 million years ago (at the end of the Pokursky suite formation $K_{1-2}pk$) in the southern part of Kulan-Igayskoy depression, in the eastern parts of the Festival bar and the Tamratskoy depression, in the joint depression zones with positive framing structures (figure 3a). In 5 million years (at the end of the Ipatovskaya suite formation $K_2ip$) the sources combined, embracing the central and eastern part of the studied area (figure 3b). Over the next 25 million years (at the end of the Gankinskaya suite formation $Pg_1$-$K_2gn$) the sources continued generating and their size increased (figure 3c). Maximum heating of the sedimentary thickness was during the Chegan formation ($Pg_{3-2}hg$) 37.6 million years ago (figure 3d). Figure 3e (end of the Nekrasov suite formation $Pg_3nk$) illustrates the starting period of sedimentary thickness cooling caused by climatic condition changes in the Oligocene, which continues till now (figure 3f). Relative Bazhenov oil density distribution of shelf deposits and Achimovsky reservoirs (figure 4) was plotted.

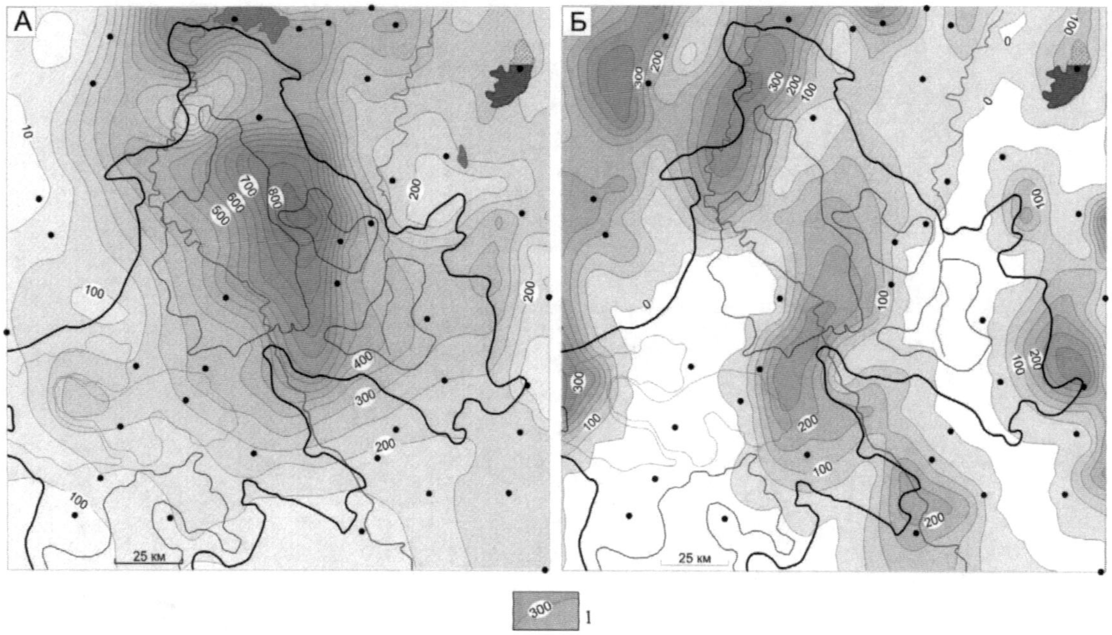


**Figure 4.** Contour maps of relative density distribution of oil resources in (a) shelf sediments; (b) in Achimovsky deposits of Nyurolsky megadepression. 1 – isopachytes of relative density values, c.u. The remaining symbols are indicated in figure 2.

## 5. Conclusion

The integrated investigation approach of the territory ( analysis of the geological structure, allocation of intensive generation zones and oil migration) allowed evaluating the prospects of Achimovsky (clinoform) and shelf reservoirs of Neocomian Nyurolsky megadepression. The zoning of the reservoirs by the relative resource density defines the priority prospecting targets in Lower Cretaceous sediments. This zone includes the western part of the Festival bar and the south-eastern part of Kulan-Igayskoy depression.

## References

[1] Burstein L M, Zhidkova L V, Kontorovich A E and Melenevskii V N 1997 Model katageneza organicheskogo veshchestva (na primere bazhenovskoy svity) *Geologija i Geofizika* **6** 1070–78 [in Russian]

[2] Danenberg E E, Belozerov V B and Brylina N A 2006 *Geologicheskoye stroyeniye i neftegazonost verkhneyursko-nizhnemelovykh otlozheniy yugo-vostoka Zapadno-Sibirskoy plity (Tomskaya oblast)* (Tomsk: TPU) p 291 [in Russian]

[3] Isaev V I and Fomin A N 2006 Centers of oil generation of Bazhenov- and Togur-type oils in the southern Nyurolsky megadepression *Russian Geology and Geophysics* **6** 734–745

[4] Lobova G A, Popov S A and Fomin A N 2013 Lokalizatsiya prognoznaykh resursov nefti yursko-melovykh neftegazovykh kompleksov Ust-Tymskoy megavpadiny *Neftianoe Khoziaystvo* **2** 36–40 [in Russian]

IOP Conf. Series: Earth and Environmental Science **21** (2014) 012012 doi:10.1088/1755-1315/21/1/012012

# Definition of facies conditions by granulometric analysis on the example of horizons $JK_{2\text{-}5}$ in Em-Egovskoe field (Western Siberia)

**L Kudryashova[1] and V Belozerov[2]**
Tomsk Polytechnic University, Tomsk, Russia

E-mail: [1]KudryashovaLK@tpu.ru, [2]BelozerovVB@hw.tpu.ru

**Abstract.** The significance increase of recoverable oil reserves in reservoirs of deep horizons requires more detailed study of their formation conditions. Granulometric and mineralogical investigations of terrigenous sediments allow obtaining the basic ideas on their sedimentation conditions. In this case the combination of the existing methods enables to address some issues of facies analysis. The integrated investigations of the horizons $JK_{2\text{-}5}$ of Tumensk formation in Em-Egovskoe area, Krasnoleninskoe field (208 samples) have been performed. Based on granulometric and mineralogical studies, the authors made the conclusion that the sediments studied had been formed within the estuary. The results of the study show that the sedimentation conditions may be different under the same dynamic accumulation conditions and similar particle granulometric distributions. The efficiency of facies prediction can be enhanced only through the integrated analysis.

## 1. Introduction

In recent years, the hydrocarbon deposits of deep horizons with complicated geological structure and ambiguous reservoir filtration-capacitive properties are the main sources of oil and gas reserves. Improving the efficiency of prospecting, exploration and development of the discussed hydrocarbon deposits requires qualitatively new geological modeling methods.

Economic feasibility of the development of such deposits is based on the reliability of reservoir extension forecast and analysis of formation conditions, affecting the choice of the technological development scheme. One of the methods in predicting structural features and spatial development of oil-bearing formation is facies analysis of its deposits, which enables to study a number of corresponding properties within each facies.

The granulometric studies are of great importance in facies analysis of sedimentary formations. As reported by many researchers, this is due to the fact that the peculiarities of clastic particle distribution to fractions indicate sedimentation environment in the paleogeographic sense (L Rukhin, R Folk, G Friedman and others) or the features of sedimentation dynamic structure (P Passega and others) [1].

## 2. Methodology of lithofacies analysis

Many researchers, particularly L Rukhin, B Sakhu, F. Shepard, B Greenwood, J Klovan, R Young and others, came to the conclusion that certain statistical characteristics of sediment fractional composition, as well as their different pairs cannot identify sedimentation environment [1]. It is necessary to take into account that similar dynamic accumulation conditions and, of course, similar granulometric particle distribution may indicate to different environments. Therefore, the combination of granulometric data genetic interpretation techniques is required; it allows predicting more accurately the sedimentation environment.

For preliminary genetic classification of precipitation and granulometric rock characteristics it is recommended to apply the generalizations of Fyuhtbauer and Mueller where the environments are allocated to the level of facies groups – continental, marine, transition [2, 3].

In lithology, there are various ways to divide precipitation based on fractional composition and, also, to distinguish different facies – for example, genetic and dynamogenetic diagrams of L Rukhin, R Passegi, H Friedman, G Rozhkov and others.

One of the relatively simple diagram has been developed by K Berlikke, which is based on the different sorting ratios and the asymmetry character of sand and silt rock fraction distribution. The study results have shown that coastal marine or lacustrine sand deposits have negative asymmetry, while river and aeolian have positive asymmetry and the mode of distribution is located in fine-grained fraction [4].

The most satisfactory results in determining the genesis of recent sediments were obtained by dynamogenetic diagrams of R Passegi and G Rozhkov. Despite the different approaches to determining sedimentation conditions, the results obtained in these two diagrams for aquatic sedimentation environment were basically similar [5].

The dynamic diagram C–M, developed by R Passegi, specifies transportation of clastic materials by particle rolling, saltation or as homogeneous or graded suspension. Large- and fine-grained fractions are transported independently. The analysis results of various precipitations in the diagram C–M form a complex configuration, which has been divided by points N–S into segments that correspond to certain conditions of sedimentation.

The dynamogenetic diagram of G Rozhkov is based on the principle of mechanical differentiation (different in intensity) of sand and silt particles in different facies. Naturally, this phenomenon is also associated with a variety of ways to transport different sized clastic particles and it is mainly determined by the energy levels of the transportation dynamic forces and sedimentation environment [5].

The authigenic minerals, which are indicators of certain sedimentation environments, are of great importance in genetic interpretation. For example, D Goldsmith and his colleagues defined the phase relations in the system Ca–Mg–Fe–Mn–$CO_2$, which helps to define the sedimentation conditions. Thus, pH environment influences calcite and dolomite sedimentation; whereas the formation of Fe and Mn carbonates is further dependent on the redox balance of the sedimentation environment. Therefore, they associate the cause of siderite formation with the environment typical to stagnant pools or areas subjected to tides and ebbs, as well as estuaries with high productivity of organic matter [6].

Thus, having analyzed the modern approaches to genesis determination based on particle size and mineralogical analyses, we can conclude that different sedimentation environments may have similar characteristics. The combination of the existing granulometric studies methods resolves some issues of facies analysis. In this paper, the authors propose a complex use of methods according to the following plan.

1. Generalized description of the environment by Furchtbauer and Muller [2, 3];
2. Definition of deposit genesis by sorting ratio and features of asymmetry (K Berlikke diagram );
3. Determination of sedimentation conditions by asymmetry and excess ratio ( Rozhkov dynamogenetic diagram );
4. Identification of the way to transfer clastic particles in aqueous environment (R Passegigenetic diagram ) ;
5. Authigenic mineral composition analysis in rocks.

In the case of such complex studies, each subsequent method clarifies the sedimentation conditions and it enables to determining the precipitation formation conditions more accurately.

The proposed method was tested on the upper-middle Jurassic sediments of Em-Egovsk area in Krasnoleninsk field (Western Siberia).

Basic productivity is associated with Tumensk formation horizons where reservoir rocks (horizons $JK_{2-5}$) are represented by sand and silt varieties (variations) with thickness-dependent clay bands. There are fine-grained sandstones, silty with predominant sand (0.12-0.30 mm) and silt (0.05-0.10 mm) dimension.

Low reservoir properties of productive sandstones are defined by several reasons. The main one is probably a high degree of compactness and transformation by epigenetic processes. Wide development of authigenic and epigenetic carbonate minerals, primarily siderite, and their presence in pore space destroy significantly the reservoir properties of sandstones and, especially, of siltstones. High content of clay and fine silt fractions impacts negatively the reservoir properties in the reservoir rocks.

## 3. Granulometric and mineralogical analyses

The authors analysed the results of granulometric analysis of 208 samples from horizons $JK_{2-5}$ Tumensk formation which were selected from 7 wells of Em-Egovsk field. The paper introduces the 1819 well results as a definite example.

The granulometric composition of the reservoir rock is determined by Ltd. "Sibgeotcentre" through two methods: transparent petrographic thin sections and mechanical sieve (up to 0.05 mm fraction), and decantation (finer fractions). It should be noted that if granulometric analysis of the samples taken from the Jurassic sediments is based on thin section data, it is more reliable than mechanical and decantation methods, as Jurassic sandy-silty rocks are well-sealed and firmly cemented that can be hardly separated by mechanical method.

As an initial step of graphical processing of the results, the authors built histograms of granulometric composition in order to study and to illustrate the changes of rock properties along the cross-section. Then, the cumulative curves were built. They are important both for graphic display of analytical data and determination of a set of parameters characterizing rock structure. The authors calculated the main granulometric characteristics in the following samples: average particle size ($X_{average}$), standard deviation (sorting coefficient $S_0$), asymmetry parameter (A), excess measure (E), median (Md). Let us consider as an example of the results obtained for samples from well № 1819 of Em-Egovsk field (table 1).

**Table 1.** Main granulometric characteristics (well № 1819 of Em-Egovsk field).

| № sample | Depth, m | Lithology | $X_{average}$ | $S_0$ | Md, mm | A | E | Effective porosity in cores, Kp, % | Bulk density, g/cm$^3$ |
|---|---|---|---|---|---|---|---|---|---|
| 27 | 2259.65 | Silty clay | 0.08 | 1.85 | 0.088 | 0.024 | 0.733 | 5.88 | 2.585 |
| 38 | 2266.25 | Sandy siltstone | 0.11 | 1.40 | 0.110 | 0.165 | 1.004 | 8.30 | 2.503 |
| 39 | 2266.95 | Argillaceous siltstone | 0.10 | 1.46 | 0.093 | -0.102 | 0.997 | 8.69 | 2.468 |
| 40 | 2267.20 | Argillaceous siltstone | 0.09 | 1.53 | 0.095 | 0.143 | 0.856 | 9.81 | 2.470 |
| 41 | 2267.95 | Argillaceous siltstone | 0.09 | 1.53 | 0.098 | 0.193 | 0.820 | 7.99 | 2.521 |
| 42 | 2268.07 | Argillaceous siltstone | 0.08 | 1.58 | 0.090 | -0.001 | 0.961 | 7.55 | 2.553 |
| 43 | 2268.86 | Argillaceous siltstone | 0.09 | 1.55 | 0.090 | -0.007 | 0.985 | 6.85 | 2.550 |
| 45 | 2270.90 | Siderite-silty clay | - | - | - | - | - | 5.38 | 2.651 |

Sorting coefficient ($S_0$) is determined by P Trask formula – the ratio of the third quartile to the first [4]. According to the Trask classification, precipitations in the studied samples are mostly well-sorted.

Asymmetry and excess were determined by Falk formula, using grain size with a certain percentage taken from the cumulative curves [5].

The next step is the data analysis obtained by the method described above.

According to Furchtbauer and Muller generalizations, the studied rocks were mainly formed in river sedimentation environment (channel and riverine shallows) because sorting was mostly > 1.2 and asymmetry < 1.

The studied rocks refer to turbidities according to sorting and asymmetry ratio (by K Berlikke).

At the same time, when speaking about turbidities, the method of their accumulation is understood in different ways: either as a result of shoreline moving and sedimentation depth changing occurred at relatively shelf shallow (less than 200 m) or as a result of sedimentation from turbidity currents in the bottom part and at the foot of continental underwater slope at a relative depth (1200 m or more) of the sedimentation basin [7]. Therefore, it is necessary to clarify precipitation genesis.

The next step is to build and analyze the dynamogenetic Rozhkov and Passegi diagrams (Figures 1, 2).

It follows from the analysis of point location on the diagram of G.F. Rozhkov that some samples are classified as marginal-marine facies (field VII, lower part), and other samples are classified as coastal facies of vast open areas (field VIII). In Passegi diagram the samples are classified within the area of river sediments (secondary channel).

Association of authigenic (kaolinite + siderite + pyrite) says for sedimentation environment transition. Thus, the predominance of clay component as kaolinite indicates the formation of the continental environment, but significant content of chlorite and mixed layered formations indicate the marine environment and this should be noted, as well. Significant content of siderite is an indicator of continental or coastal marine environment.

It is also necessary to draw attention to the presence of not only debris timber, leaf prints, root prints, plant prints, identifying continental sedimentation regime, but also oolitic limestone typical for shallow coastal areas with high flow turbulence [3], and rostrum of belemnite, the presence of which may be associated with their resedimentation under sea condition. This proves that transitional environment of sedimentation in which the complex of tidal plains and deltaic systems are singled out. Based on the structural features, the deltas are divided into tidal (estuarine), wave (blade) and river (retractable).

## 4. Conclusions

According to the proposed method, the sedimentation environment can be fully referred to the conditions typical for estuaries. It is the estuary that can explain such ambiguous results typical for different facies.

In general, estuary is a portion of the joint influence of the river and the sea, where the clastic material, on the one hand, is transported by the river and, on the other hand, it is transported from the sea by tidal currents and partly by rough sea currents. This is proved bycoastal marine environment presented in Rozhkov diagram. Owing to this fact both plant roots and transported debris of belemnites are observed in sediments. It should be also noted that pelitic deposits (primarily clay) and sand, depending on the conditions, [8] mainly precipitate in estuaries. As shown in table 1, the interval studied is folded by siltstone.

Tidal and deltaic channels within tidal plains and estuaries form sedimentation environment similar to those in river channel conditions [8], which may explain the result obtained by Furchtbauer and Mueller generalizations as well as the availability of samples in the field of secondary channels in the genetic Passegi diagram.

IOP Conf. Series: Earth and Environmental Science **21** (2014) 012012 doi:10.1088/1755-1315/21/1/012012

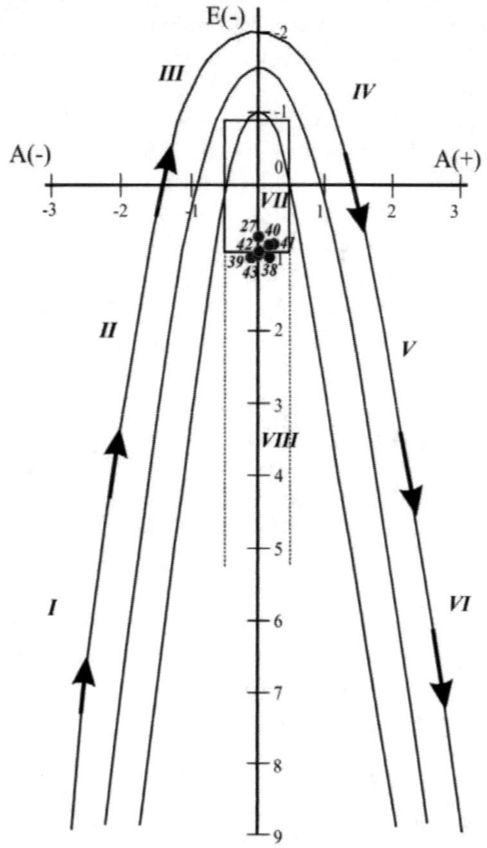


**Figure 1.** Rozhkov dynamogenetic diagram (well № 1819 of Em-Egovsk field).

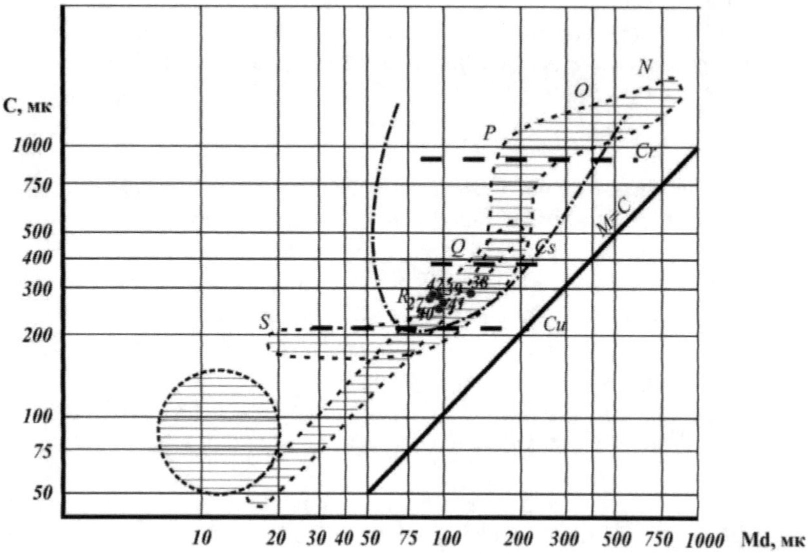


**Figure 2.** Passega genetic diagram to determine transportation ways of clastic particles in aqueous environment (well № 1819 of Em-Egovsk field).

Surface water often contains dissolved substances in the form of colloids. When penetrating marine basins, such colloids are destroyed with sea water which is electrolyte. Colloids are coagulated, and then they fall to the bottom, giving rise to marine sediments. Fe, Al, Mn hydroxides may be accumulated in the form of such colloidal chemogenic precipitation. Thus, iron hydroxide colloids are the most unstable. They are coagulated just at the river confluence into the sea [9]. This explains the presence of siderite in the samples.

Another data analysis result is the classification of rocks as turbidities according to Berlikke diagram. Considering the fact that material sedimentation in the estuary is strongly influenced by tides, the currents associated with the tides can be characterized by different velocities. In general, tidal current velocity decreases from output to the top of estuary. Typically, the power of flows associated with ebb is higher than the power of tidal currents. The tides form periodically varying sedimentation conditions [8]. In the case of two different water masses contact, the water mass of higher density tends to flow and to spread below the less dense one. Such density flows are well known in the ocean, where they are caused by the difference in temperature and salinity. In a similar way, turbid and suspended-enriched layers of high density can descend and move below pure water with lower density [3]. Therefore, lateral and vertical mixing of water occurs in estuaries [10], i. e. the collision of two divergent dynamic flows and precipitations, which are close to turbidity characteristics.

Thus, based on the analysis results obtained by determining the origin according to granulometric and mineralogical studies, it is possible to conclude that the deposits studied were formed within the estuary.

It should be noted that all the described methods in determining precipitation origin do not provide a clear description of facies sedimentation environment. They allow dividing precipitation into those formed in the saltwater pool, in the continent or in the transition zone. For more accurate definition of facies, it is necessary to carry out a complex study of the area by selecting appropriate sedimentation model based on not only on granulometric analysis, but also on the additional data obtained by core sampling (faunal, petrographic, texture and etc.), well logging (facies electrometric analysis) and well testing (filtration-capacitive).

## 5. References

[1] Romanovsky S I 1977 *Sedimentological basis of lithology* (Leningrad: Nedra) p 408

[2] Krasheninnikov G F, Volkova A N and Ivanova N V 1988 *Facies lithology with the basics: guide to laboratory studies* (Moscow: Moscow State University Press) p 214

[3] Reineck H-E and Singh I B 1981 *Depositional sedimentary environments (with reference to terrigenous clastics)* (Moscow: Nedra) p 439

[4] Kuznetsov V G 2007 *Lithology. Sedimentary rocks and their study: textbook for universities* (Moscow: Nedra-Business Centers) p 511

[5] Grossgeim V A, Beskrovnaya O V, Gerashchenko I L and other 1984 *Methods of paleogeographic reconstructions (in the search for oil and gas)* (Leningrad: Nedra) p 271

[6] Pettidzhon F, Potter M and Seaver R 1976 *Sands and sandstones* (Moscow: Mir) p 536

[7] Markiewicz P V 2004 "Turbidities" and "flysch" without explanation – the dangerous terms *Bulletin of the Far Eastern Branch of the Russian Academy of Sciences* **4** 95–105

[8] Gradzinsky R, Kostetskaya A, Radomski A and Unrug P 1980 *Sedimentology* (Moscow: Nedra) p 640

[9] Bakumenko I T 2001 *Mineralizing processes: textbook* (Novosibirsk: Novosibirsk State University Press) p 80

[10] Leeder M R 1986 *Sedimentology. Processes and Products* (Moscow: Mir) p 439

# Properties of the organic matter of the Bazhenov suite in the Yugansk Megadepression based on Rock-Eval Pyrolysis

**I Goncharov[1], R Kashapov[2] and V Samoilenko [2]**
[1]Tomsk Polytechnic University,Tomsk, Russia
[2]JSC "TomskNIPIneft" Tomsk, Russia

E-mail: [1]goncharoviv@nipineft.tomsk.ru

**Abstract**. The article provides the Rock-Eval analysis data of rock samples and a broad description of the organic matter of the Bazhenovsk **suite** of the Yugansk megadepression. The authors have **designed** and compared kinetic models for immature and mature organic matters.

## 1. Introduction

Nowadays, the Bazhenov suite is widely recognised as the main oil-and-gas source formation in West Siberia. In spite of the fact that there had been a great interest in this formation during the 70s and 80s of the last century, even today this interest is predominate. A 50-year study resulted in the thorough analysis of its petroleum potential. The progress in the global shale oil production in recent years demonstrated a growing interest in further exploration and production of the Bazhenov suite.

The presence of organic matter is the governing factor and main principle in evaluating oil source rock. In this case, the Rock-Eval pyrolysis is the best standard method to identify the organic carbon content in the rock, its potential residual generation, and the catagenetic organic matter maturity. The range of standard parameters determined by this method describes broadly the organic matter properties in rocks. Pyrolysis in combination with kinetic, geological and other data is the starting point for basin modelling.

This article describes the organic matter in the Bazhenov suite of the Yugansk megadepression based on pyrolysis data.

## 2. Location and geological setting

The Yugansk megadepression (Figure1) is in the central part of West-Siberian oil-and-gas bearing basin and is confined to the Middle Ob oil-and-gas bearing region. The largest West Siberian oil deposits (Samotlorsk, Fyodorovsk, etc.) are located here. This main oil and gas potential region consists of Jurassic and Lower Cretaceous sediments.

The Yugansk megadepression forms a complicated internal structure (Figure 2), which is divided into two depressions (Fainsk and Kulunsk), two troughs and Yuzhno-Kinyaminsk swell. The southern megadepression edge involves two ridges separated by ravines, while on the south-east facing edge- a monocline (Yuzhno-Vartovsk) and terrace (Negusynsk). The Yarsomovsk trough is situated in the north of the megadepression.

The Bazhenov suite (Yugansk megadepression) contains carbonate-siliceous clay sediments enriched by organic matter which range from the Volgian of Upper Jurassic to the Lower Berriasian substage (Lower Cretaceous). For a more detailed examination 198 core samples were retrieved from 13 wells within the Bazhenov formation section.

IOP Conf. Series: Earth and Environmental Science **21** (2014) 012013 doi:10.1088/1755-1315/21/1/012013

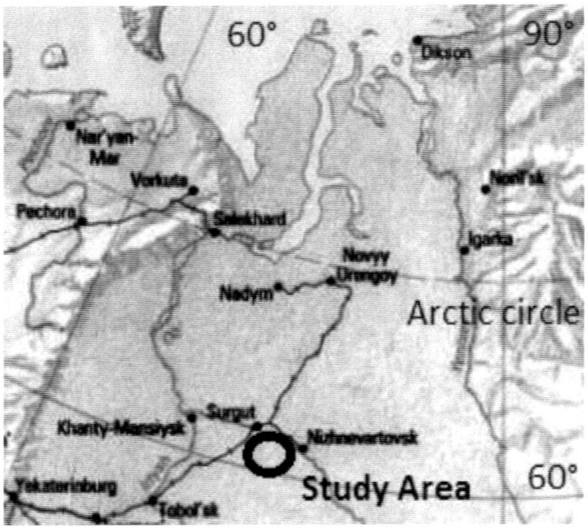


**Figure 1.** Geological setting

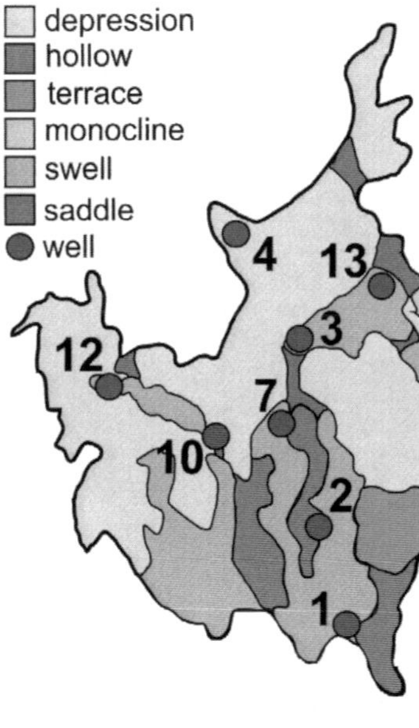


**Figure 2.** Tectonic map of the Yugansk megadepression and well location.

## 3. Methodology

The Rock-Eval 6 Turbo tool is the latest version of the Rock-Eval product line, being the best method for its high analysis accuracy [1]. Rock-Eval analysis is a quick and common method for initial evaluation of geochemical characteristics of source rocks.

The samples were pulverised, and about 30 mg of each sample were analysed by using the Rock-Eval 6 Turbo (Bulk Rock method). Thus $S_1$, $S_2$, $S_3$, $T_{max}$, TOC and MinC values were obtained. $S_1$ is the amount of free hydrocarbons (mg HC/g rock) in rock pore space at 300°C. Peak $S_2$ is the amount of hydrocarbon released during thermal cracking of kerogen (mg HC/g rock) and heavy hydrocarbons at 300–650°C and heating rate 25°C/min. (temperature-programmed pyrolysis). This parameter represents the residual oil-generating potential. The temperature complying with the maximum amount generated hydrocarbons $S_2$ is termed $T_{max}$. $S_3$ is the $CO_2$ amount formed as a result of the decomposition of carboxyl groups and other oxygen-containing kerogen compounds at 300 to 400°C. Total organic

IOP Conf. Series: Earth and Environmental Science **21** (2014) 012013 doi:10.1088/1755-1315/21/1/012013

carbon (TOC) and mineral carbon (MinC) content is determined by the oxidation of organic matter after pyrolysis (850°C in air). Hydrogen index (HI) is the $S_2$ value ($S_2$/TOC) expressed in mg HC/g TOC. Oxygen index (OI) is related to the oxygen amount in kerogen and is the normalised value ($S_3$/TOC). Production index (PI) shows the level of thermal maturation $PI=S_1/(S_1+S_2)$.

Chloroform was used in sample extraction for kinetic investigation in the Soxhlet extractor during 72 hours. After extraction, the samples were dried at 70°C for 8 hours and furthered analysed in the Rock-Eval 6 Turbo (Optkin method) at heat rate of 5, 15 and 25°C/min. Kinetic parameters were calculated by the Optkin 3.0.3 program.Optkin is based on the kinetic model of Tissot and Espitalie which predicts the amount of hydrocarbons generated by the primary cracking of kerogen when temperature increases through time [2].

## 4. Results and discussion

The average Rock-Eval parameter values of rock samples from each of the 13 wells are presented in table 1.

| Well №№ | Quantity of samples | Depth (m) | $S_1$ (mgHC/g rock) | $S_2$ (mgHC/g rock) | PI | $T_{max}$ (°C) | HI (mgHC/gTOC) | OI ($mgCO_2$/g TOC) | TOC (wt%) | MinC (wt%) |
|---|---|---|---|---|---|---|---|---|---|---|
| 1 | 10 | 2480-2490 | 3.16 | 67.65 | 0.04 | 423 | 652 | 1 | 10.42 | 1.37 |
| 2 | 7 | 2713-2718 | 5.26 | 87.31 | 0.06 | 423 | 636 | 0 | 13.63 | 0.86 |
| 3 | 23 | 2924-2941 | 5.26 | 70.87 | 0.07 | 430 | 668 | 0 | 10.61 | 0.65 |
| 4 | 4 | 2846-2857 | 8.32 | 87.30 | 0.09 | 432 | 638 | 0 | 13.73 | 0.82 |
| 5 | 3 | 2794-2795 | 3.35 | 44.12 | 0.08 | 433 | 6.33 | 0 | 6.64 | 0.31 |
| 6 | 4 | 2780-2782 | 4.16 | 42.29 | 0.09 | 434 | 631 | 0 | 6.62 | 0.17 |
| 7 | 47 | 2846-2874 | 7.25 | 69.48 | 0.10 | 434 | 619 | 0 | 11.12 | 0.99 |
| 8 | 17 | 2867-2878 | 6.43 | 51.87 | 0.11 | 435 | 621 | 0 | 8.24 | 0.82 |
| 9 | 19 | 2830-2844 | 6.29 | 61.29 | 0.10 | 436 | 624 | 0 | 9.79 | 0.26 |
| 10 | 20 | 2962-2981 | 8.58 | 73.11 | 0.10 | 436 | 602 | 0 | 12.18 | 0.75 |
| 11 | 14 | 2739-2749 | 5.97 | 48.75 | 0.11 | 437 | 608 | 0 | 7.91 | 1.61 |
| 12 | 20 | 2964-2974 | 9.11 | 63.92 | 0.13 | 438 | 537 | 0 | 11.81 | 2.76 |
| 13 | 10 | 2812-2825 | 7.44 | 60.08 | 0.12 | 439 | 568 | 0 | 10.46 | 0.54 |

**Table 1**. Average values of Rock-Eval parameters for rocks in the wells under study

Bazhenov suite rocks in all the studied wells have an excellent generation potential ($S_2$). Low OI is typical for organic matter accumulating in reduction conditions. HI–$T_{max}$ Diagram (figure 3) shows that the samples are influenced by the evolution of I-II type kerogen which generates oil [3].

According to catagenesis level, rocks from Wells 1-3 are immature. This fact is confirmed by low PI (PI< 0.10) and $T_{max}$ (423–430°C), which responds to catagenesis level $MC_1^1$ [4]. Rocks from wells № 4-13 are mature and have already entered the main oil generation phase ("oil window"). Their $T_{max}$

varies from 432 to 439°C, which responds to catagenesis level $MC_I^2$ [4]. Increased PI values (PI≈0.10) prove this fact.

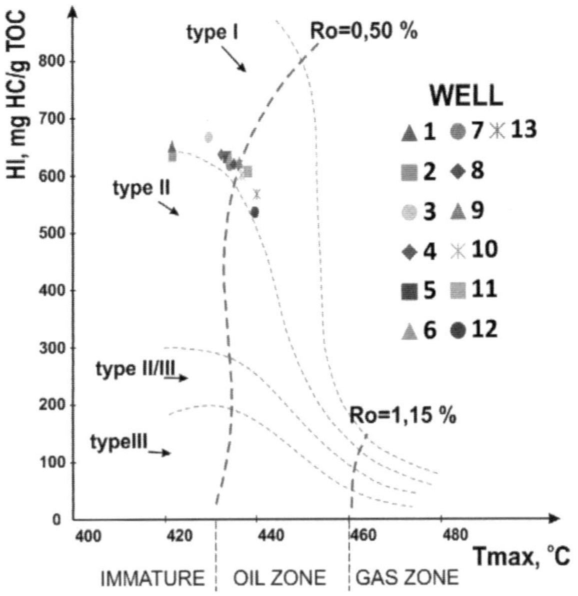


**Figure 3.** HI–$T_{max}$ diagram.

Figure 4 shows the distribution of energy activation of an immature sample from well № 2, and a mature sample from well № 13. Comparing the histograms in figure 4, it can be observed that the mature sample has already reached its generation potential, which, in its turn, corresponds to the chemical bonds with activation energy 48 kcal/mol and equal to 130 mgHC / gTOC or about 20% of initial 660 mgHC / gTOC.

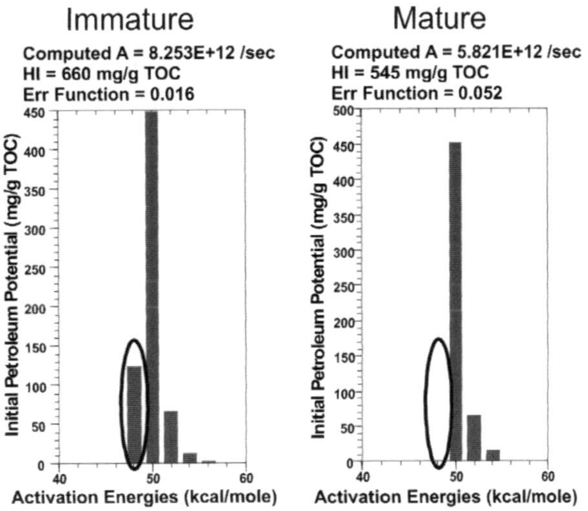


**Figure 4.** Activation energy distribution.

The dependence of oil source potential in immature rock from well № 2 to temperature increase (figure 5) proves the fact that within the temperature range of 430-440°C, kerogen embraces 14-23% of its oil source potential.

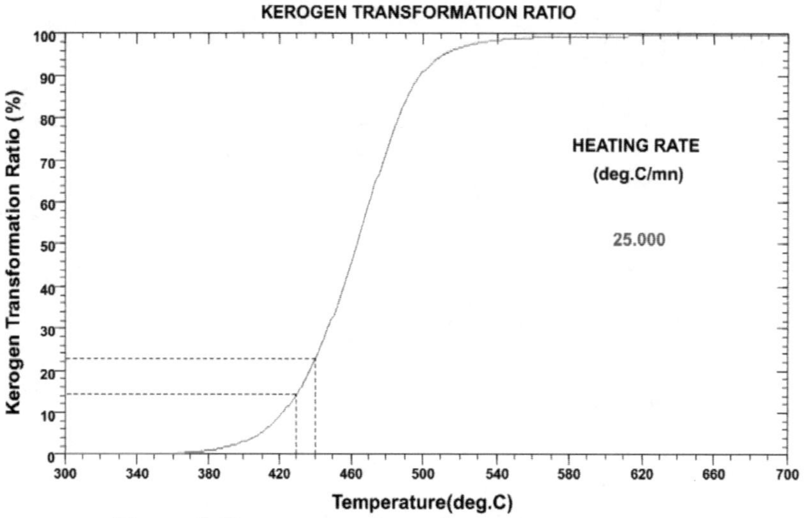


**Figure 5.** Kerogen transformation – temperature ratio.

## 5. Conclusions

Based on the results of Rock-Eval pyrolysis of rock samples retrieved from Yugansk megadepression, it was concluded that

- all examined source rocks have not only an excellent generating potential ($S_2$) but also high total organic carbon (TOC);
- rock organic matter is related to kerogen type I-II, which is the major source in the generation of liquid hydrocarbons;
- catagensis level alters from $MC_1^1$ to $MC_1^2$;
- the most mature source rocks in the investigated area have only reproduced 20% of its oil-generating potential.

## References

[1] Behar F, Beaumont V and Penteado H de B 2001 Rock-Eval Technology: Performance and Developments Revue IFP **56** (2) 111-134,111-134

[2] Peters K, Walters C, and Moldowan J 2005 *The biomarker guide* (Cambridge, U.K.: Cambridge University Press)

[3] Espitalie J, Marquis F and Drouet S 1993 Critical study of kinetic modelling parameters, in: Dore A G et al. Basin modelling: Advances and Applications (Elsevier, Amsterdam, Norwegian Petroleum Society) 233-242

[4] Goncharov I, Oblasov N, Samoylenko V and Fadeeva S (2013) Catagenesis of organic matter of the Bazhenov Formation rocks in the south-east of West Siberia (Tomsk Region) *J. Oil and Gas Industry* **10** 32 - 37

IOP Conf. Series: Earth and Environmental Science **21** (2014) 012014 doi:10.1088/1755-1315/21/1/012014

# Effective Planning of the Future of the Arctic

**A Sentsov[1], Yu Bolsunovskaya[2] and L Bolsunovskaya[3]**
Tomsk Polytechnic University, Tomsk, Russia

E-mail: [1]sentsov@tpu.ru, [2]ju_al@inbox.ru, [3]bolsunovskl@tpu.ru

**Abstract.** The problems of the Arctic region have become the most important ones in the world. Political risks hinder the industrial development of the region. This paper addresses the problem of planning and modeling the future of this region. It presents the problems of developing a model of the future due to the ideologies and strategies of two main actors in the Arctic, the United States and the Russian Federation. The effects of a bipolar perception of the future of the region and of the whole world are shown. A model of the effective planning of the future of the Arctic region is proposed.

## 1. Introduction

The Arctic region with its natural resources and high economic potential attracts many companies and has become an important area of transnational development. According to recent reports "all across the Arctic, changes in climate will create new vulnerabilities for infrastructure and present new design challenges" [2]. Global warming can greatly change the Arctic region. So, "despite uncertainties in the magnitude of expected global warming over the next century, one consistent feature of extant and projected changes is that Arctic environments are and will be exposed to the greatest warming" [12. P. 449]. The climate and nature changes in the region are under study. For example, "as a part of the national Norwegian monitoring program, long-term surveys of contaminants have been carried out" [6. P. 281], another survey "emphasizes a need to investigate nonstationary climate response of Alaskan coastal forests to warming in other tree species" [11. P. 211].

The international status of the Arctic is fixed in international agreements. The region itself is divided into five sectors of responsibility between the U.S., Russia, Norway, Canada, and Denmark. But the most important problem of development in the Arctic is that "given the sensitivity of Arctic development, there is greater risk from changes in regulation or investment frameworks, following either a change of political leadership or a specific risk event, even one in which the company is not itself implicated" [8].

Consequently the problems of the future of the Arctic and the planning of that future lead us to the problems of the future of all the Arctic countries and even those of the whole international community.

## 2. Data sets

The Arctic contains enormous unexploited energy resources, primarily oil and gas. According to the United States (U.S.) Geological Survey, the oil reserves of the Arctic (both offshore and onshore) are 90 billion barrels [4]. The United States Geological Survey (USGS) of 2008 presents "the first-ever wide-ranging assessment of Arctic oil and gas resources, estimating the region's undiscovered and technically recoverable conventional oil and natural gas resources. Of the 33 Arctic sedimentary "provinces" that the USGS evaluated, 25 were found to have a greater than 10% probability of having oil or gas deposits larger than 50 million barrels of oil equivalent" [1. P. 2].

However, recent surveys show that "disagreement over the legal status of the Northwest Passage and potentially over the status of Russian Arctic waters could lead to claims that double standards are being applied" [8]. For the Russian Federation the Arctic continental shelf is a region of high geostrategic and long-term economic interests of the state. This fact makes of this territory an independent object of national policy [3].

In areas neighboring the Arctic are located the components of the Russian and American missile warning systems. Also there are American air-defense interceptors in Alaska, and Russian analogs located on the coast of the Arctic Ocean. There is a Russian nuclear testing area on Novaya Zemlya (an archipelago in the Arctic Ocean in the North of Russia and the extreme Northeast of Europe).

Some of the Arctic countries conduct military exercises in the region. In 2013 Russia conducted military exercises using the cruiser "Peter the Great" and nuclear submarines "Orel" and "Voronezh"; launches of cruise missiles were also conducted. In turn, in 2013 the U.S. carried out the "Arctic Challenge" training exercises in the Gulf of Bothnia and in the Barents Sea with the participation of the air forces of Finland, Sweden, and the United Kingdom.

## 3. Problems of the future in political and social reality

The sense of the future is organically inherent in a man as a thinking, living being. This sense is located on the basic axis of coordinates on which a person is oriented in the world. It is included in individual, social, and therefore in political time. The temporal horizon of the modern society is determined by the future [9]. That is why the planning and the programming of the future are so important for our modern social structures.

The constructivist approach to social reality supposes that in the consciousness of a human being this reality is completed by projects. The future can be defined as a projective completion of a present reality. Therefore, the individual's temporal dimension is determined not only by sense of social time and also by culture. Thus, this temporal dimension is in turn a part of that culture. The question is about the political culture as well as about culture in general. A model of the future has a key position in contemporary communication and in formation of a modern dynamic society.

Therefore, the concept of the future is an important basic notion for an individual as well as for the society in general. This concept forms one of the elements of the orientation scheme "time – space". It brings meaning and dynamics for the actions of individuals and society.

According to the typology of British scientist Peter Laslett [6], there are four types of changes in society categorized according to their speed:

- fast, e.g. political changes, fashion or judgment changes, changes in international relationships;
- medium, e.g. economic, communication, technical, demographic, and institutional changes;
- slow, e.g. constitutional, religious changes; and
- very slow, e.g. normative changes, changes of productive relations and social structure.

The problem is that the future belongs to the category of slow and very slow changes while the political transformations and events (elections, regional conflicts etc.) belong to the fast ones. Thus, this situation provokes discordance of political cycles and problems in understanding of the future.

## 4. Representation of the future in the U.S. and Russia ruling political parties' programs

The planning and modeling of the future of the Arctic are inextricably connected with the models of the future proposed by the official ideology of the Arctic region countries. Although there are five countries present in the Arctic, due to the general international situation and global processes, the most important actors in the Arctic are the U.S. and Russia. Planning of the future in the Arctic region depends on the strategies and the programs of these two countries.

Additionally, the dominant party determines to a large extent the official ideology and national doctrine of a country. So when speaking of American and Russian strategies for the future, we examine the ideologies of the dominant parties that are reflected in their official program documents. For the U.S. we will examine the Democratic Party platform "Renewing America's Promise" [5] and for Russia, the Electoral program of United Russia party [7].

The structure of a political party program represents the programming image of the future. Analysis of goals, values, and objectives declared in the political party platforms allows us to conclude that the governing party of the U.S. proposes the future on the basis of the principles and values of the past, while United Russia creates an image of the future mostly by reference to the prospects viewed in the present situation. Nevertheless, the United Russia political program contains some measures of the restoration kind, in regard to Soviet ideology. In this way the party is trying to spark the interest of a wide range of voters.

Over the years the ruling political parties pretended to possess total control of the future. But nowadays, in our postmodern society the political parties have to renounce their claim to this control. Citizens are free to choose the definitive model of the future represented in one of the parties' programs. Thus today the most important goal of a party is to create a model of the future that will be the most popular with the masses.

Ideology and doctrine remain in the political parties' programs. But each party tends to create and to propose to the citizens goals and values that will combine as high as possible a vision of a personal future for the majority of the population with the vision of their country's future.

Both the Democratic Party and the United Russia formally declare their ideology to be centrist. However, detailed examination of their platforms leads us to the conclusion that these parties are both of the conservative type. Such a political orientation is rather natural for the dominant parties that rule their countries during a period of economic crisis when the demand for a liberal model of the future is almost zero.

In spite of the fact that changes in ideologies and doctrines are slow and even very slow, modern political reality shows that during a period of serious economic and social shocks, the goals and the values of the parties are subject to accelerated changes. And the voters appreciate these changes.

## 5. Interpretation of the results

The research demonstrates that the Democratic Party of the U.S. creates its image of the future through sources and values of the past, while the United Russia presents its model of the future through sources and values of the present day. Also, the Democratic Party in its program associates the future of the country with the condition of the whole world. Its model of the future includes almost all the countries and regions of the world. The United Russia program presents only ideas of cooperation with limited number of neighbors.

The models of the future proposed by the U.S. and the Russian Federation do not facilitate safe and effective planning of the future of the Arctic region.

As a result of the political and military actions of the countries, "Arctic development is often politically contentious, with sometimes opposing interests and perspectives between local, national, and international levels. Political support for development will continue to represent an uncertainty for businesses seeking to invest in Arctic projects" [2].

Also, "beyond the Exclusive Economic Zone (EEZ), countries must demonstrate that the seabed is a "natural prolongation" of the continental shelf in order to claim seabed rights. These practices have led to several overlapping claims between countries. For example, Russia has made a claim that its shelf extends all the way to the North Pole along the Lomonosov Ridge, a claim that the Canadians (as well as others) vigorously reject" [1. P. 5].

Today the political and geopolitical risks for the Arctic remain significant. The global balance of power in the Arctic region is in some way unstable.

## 6. Conclusion

U.S. and Russian planning and strategies in the Arctic region can not guarantee a safe and stable future there. The Arctic countries must create an effective planning system that will assure the sustainable development of the Arctic region. The ideological opposition of the two leading countries of the region could destroy the future of the Arctic.

The analysis of program documents of the political parties and the response of the voters during the last elections (2008 and 2012 in the U.S., 2007 and 2011 in Russia) suggest that there is a strong need for a political model of the future. At this stage the political parties cannot fully satisfy this need. Modern society stands in need of long-term prospects of the future.

In order to ensure peace in the region a new model of the future has to be created. In the modern global world the most successful political concepts are those offering not only a close relationship between the future of each individual and the future of his or her region and country, but also a further link with the future of the world and humanity in general.

Therefore the key to the effective planning of the future of the Arctic region is in a model of the future that takes into account the mutual interests of all the actors of the region, and presents a multi-polar image of the future of the Arctic.

## References

[1] *Arctic oil and gas* 2013 (Ernst and Young Limited) p 16

[2] Arctic Opening: Opportunity and Risk in the High North // Chatham House – Lloyd's Risk Insight Report. 01 April 2012. – Electronic date. – Chatham House, 2014. – URL: http://www.chathamhouse.org/publications/papers/view/182839 (access date: 15.05.14).

[3] Bolsunovskaya L M, Bolsunovskaya Yu A 2012 The impact of risks on investment potential of Arctic continental shelf of the Russian Federation *Bulletin of Tomsk Polytechnic University. Humanities and Social Sciences Technologies* **6** (321) 44 – 47

[4] Circum-Arctic Resource Appraisal: Estimates of Undiscovered Oil and Gas North of the Arctic Circle // USGS Science for a changing world. USGS Fact sheet 2008 – 3049. – Electronic date. – Menlo Park, CA, 2008. – URL: http://pubs.usgs.gov/fs/2008/3049/fs2008-3049.pdf (access date: 15.05.14)

[5] Democratic Party Platform 2008 "Renewing America's Promise" // Democratic Party. – Electronic date. – Washington, 1995-2012.
URL: http://www.democrats.org/about/party_platform (access date: 25.02.14)

[6] J B Ørbæk, I Tombre and R Kallenborn 2004 Arctic-Alpine Ecosystems and People in a Changing Environment *Arctic, Antarctic, and Alpine Research* **36** (3) 281 – 283

[7] Laslett Peter 1983 *The world we have lost* (Methuen: Routledge) p 353

[8] Party program proclamation // United Russia: party official site. – Electronic date. – Moscow, 2005-2014. – URL: http://er.ru/party/program (access date: 12.04.14)

[9] Political and Reputational Risk Factors in the Arctic // Lloyd's Report 2012: Arctic Opening – Opportunity and Risk in the High North. – Electronic date. – Lloyd's, 2012. – http://www.arctis-search.com/Political+and+Reputational+Risk+Factors+in+the+Arctic (access date: 15.05.14)

[10] Sentsov A E 2012 Concept of the future in ruling political parties programs: comparative aspect *Historical, Philosophical, Political and Law Sciences, Culturology and Study of Art. Issues of Theory and Practice* **6(20)** Part I 167 – 169

[11] S K Jarvis, G C Wiles, S N Appleton, R D D'Arrigo and D E Lawson 2013 A Warming-Induced Biome Shift Detected in Tree Growth of Mountain Hemlock [*Tsuga mertensiana* (Bong.) Carrière] along the Gulf of Alaska *Arctic, Antarctic, and Alpine Research* **45** (2) 211 – 218

[12] T T Høye, E Post, H Meltofte, N M Schmidt and M C Forchhammer 2007Rapid advancement of spring in the High Arctic *Current Biology* **17** (12) R449 – R451

XVIII International Scientific Symposium in Honour of Academician M. A. Usov: PGON2014 IOP Publishing
IOP Conf. Series: Earth and Environmental Science **21** (2014) 012015 doi:10.1088/1755-1315/21/1/012015

# Toward the problem of oil and gas bearing capacity of the East Tom-Kolyvan structural zone (Western Siberia)

**N F Stolbova and P O Maerkov**
Tomsk Polytechnic University, Tomsk, Russia

**Abstract**

The vast depression in the east Tom-Kolyvan folded zone (West Siberia) has been identified by the geophysical data. The well which uncovered 4000 m deep profile of the Jurassic and Paleozoic deposits has been drilled. The relevance of the research is the oil\gas-bearing capacity evaluation of the discovered depression in this West Siberia area.

**Introduction**

The target of the research is to obtain geologic characteristics of the profile deposits. Lithological, petrographical, bituminological, X-ray fluorescence- microscopy, geochemical, nuclear-physics methods have been used during the analyses.

By the comprehensive research data the lithologic column and lithologo-geochemical Paleozoic profiles of the deep well have been made. According to the obtained results the following conclusions have been drawn:

- the most ancient deposits of the depression are introduced by the volcanogenic-clastic rock sequence of the andesite-basalt composition with the moderately alkaline reduction;
- the younger rocks relate to the sedimentary sequence formed during various facial conditions: marine, coast-marine and lacustrine-swampy;
- in the marine environments the clayey deposits of the domanik facies are formed. These are characterized by the kerogen of type II, halo shows of the singenetic bitumoids of the mixed composition (from asphaltenous to oily-resinous), and by the liquid hydrocarbon seeps;
- in the coast-marine and lacustrine-swampy environments the aleurite mudstones with the coal streaks are formed;
- the crust of weathering deposits are in the closure of the profile;

The conclusion for further research of the deposits has been made. This is due to possible oil\gas-generation domanik deposits extent in the vast area and future oil\gas field exploration.

## Methodology

To address the problem of oil and gas-bearing capacity of the east Tom Kolyvan structural zone, a parametric well Vostochnaya in the south-west of the West Siberia plate was drilled. It is located in the megadepression of about 300 km$^2$, to the depths of 4007 m.

The relevance of the work and the goal are due to the need to obtain a detailed lithologo-petrographical and geochemical information on the deposits of the Paleozoic basement in the center of the depression. Currently it is the only deep drilling object in the new oil\gas – bearing area.

To achieve the goal, 138 core samples were selected from the basement rocks in the range of 3400 - 4007 m. All of them were examined visually and with the use of a binocular microscope. Petrographic thin sections were made out of 42 samples. For the detailed studies the following methods were used: petrographic, nuclear geochemistry, bituminologic, X-ray fluorescence. The selected set of methods allows understanding the processes of sedimentogenesis, postsedimentogenesis changes of the studied deposits, as well as tracing the fluids migration, including hydrocarbon ones. For all specimens were given the lithologic and petrographic descriptions, besides, the names of rocks by the modern classifications and classification of [4-5]. Petrographic thin sections were studied by the polarized light microscope POLAM 213M, by the ultraviolet light microscope MICMED-2 and LÛMAM-2M. Elementary and chemical composition of basaltoids in clastic material was identified by the analyzer XGT-7200. All samples were studied on the basis of the research nuclear reactor of Tomsk Polytechnic University by the delayed neutrons method. The concentrations of uranium and aluminum were identified, and lithogeochemical profiles were made.

The results of petrographic and nuclear-geochemical studies have shown that the basement in the major part, mainly, in the low part of the profile is characterized by a volcanogenic-clastic sequence of the basic rocks. From the depth of 3394m it is accompanied by the interlayers of high-carbon content, mainly argillaceous rocks.

Volcanogenic-clastic sequence is represented by the tuffs and consists mainly of basaltoid fragments. The textures of the clastic rocks are psammitic psefitic, and aleurite. The upper profile section is made of coarse rock fragments varieties, the low profile section - is of less coarse fragments. These are dense, uniform, well - lithified. Basaltoid rocks are often almond-shaped. Volcanic rocks are predominantly ophitic: poikilophitic, doleritic which occur in the upper part of the profile; anamesite, hyalo-ophitic textures are common in the low part of the profile. The changed plagioclase, pyroxenes and relict minerals occur in the rock composition. Besides, there are minerals that replace them and take part in the amygdules composition: chlorides, siliceous minerals and titanium minerals.

It is an important fact that potassium feldspars occur in a number of basaltoid rock composition. They are found in the plagioclase interstitials and partly replace them (figure 1-a). The presence of potassium feldspar is confirmed by the X-ray fluorescent analysis (figure 1-b). The determined amount of potassium can pertain the basic rocks to moderately alkaline, mainly potassium ones.

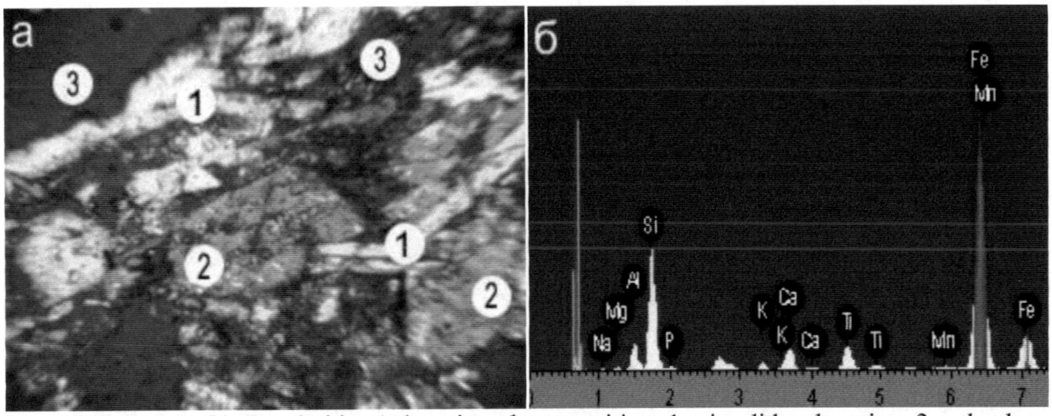


**Figure 1 (a, b).** Basaltoids: a) the mineral composition: 1-microliths alteration. 2-orthoclases. 3-chlorites (HC. 10 x, N +); b) element composition of basaltoid fragments: high level of potassium and calcium, sodium.

IOP Conf. Series: Earth and Environmental Science **21** (2014) 012015 doi:10.1088/1755-1315/21/1/012015

Judging by the amount of potassium $K_2O$ (average content 2, 64%), some basalt belongs to moderately alkaline, i.e. trahybasalt [5].

At the top of the studied Paleozoic basement profile there revealed intervals (3541.5 ...3544.7, 3546.0 ...3548.6 m.) of the basic sedimentary rocks. They are characterized by the high-content carbon aleurite mudstones with thin coal streaks and remains of plant detritus. In the sample from the depth of 3396.9 m the spores are visible (figure 2-a), which indicate the humus substance taking part in their formation.

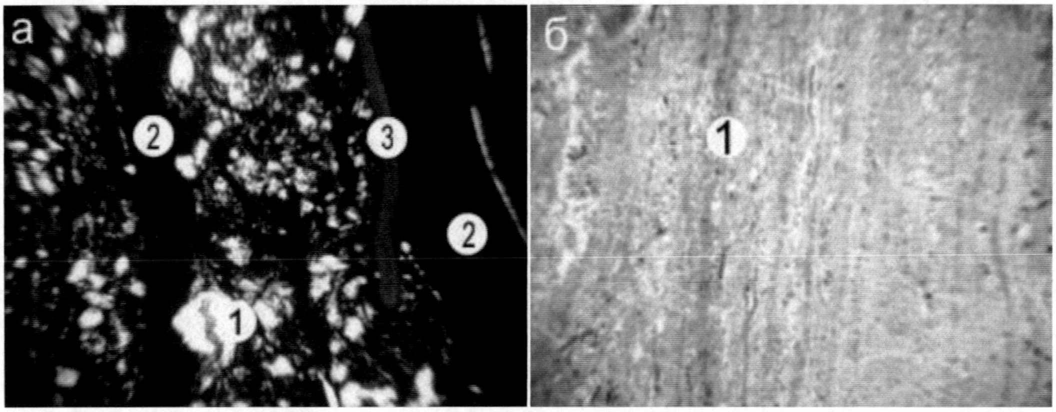


**Figure 2 (a, b)**: a) Mudstones: aleurite mudstone (1) with the coal streaks (2) and (3) the spore relics (HC. 10 x N |); b) mudstone with thin interlayers of the phosphate substance (1) (HC. 3 N |).

The thin interlayers of high-carbon content mudstone in the interval of 3541.5.3548.6 m are characterized by saturation. These should be attributed to oil-generated domanik deposits by a number of features [2]. Very thin stratification is one of the properties resulting from the climate aridization. (figure 2-b). The second feature is the clayey composition with microclasts. The main feature is the presence of oil-generated kerogen of type II [6-7]. It is red, amorphous geopolimer with a halo of a bituminous substance. Singenetic shows of type II kerogen and bitumoids are in figure 3. Bitumoids are characterized by a complex mixture of various components: from oily-resinous to asphaltenous ones. (figure 3-b).

It is important to emphasize that in the interval of 3541.5.3550.0 m. oil in a liquid form is found. According to Gribov et.al. research [6], the oil is heavy ($\rho$ = 0.921 g/cm$^3$ at t = 22° C), high viscosity oil ($\eta$ = 0.3 MPa, 38 × c), paraffin-base oil (T freezing = 40° C), with the dominance of methane. When testing the horizon with domanik deposits and increased output of the liquid hydrocarbons is represented by the letter "M". The horizon refers to the type of domanik deposits with the thickness of 8.5 m. For the low-thickness domanik deposits it is a traditional one.

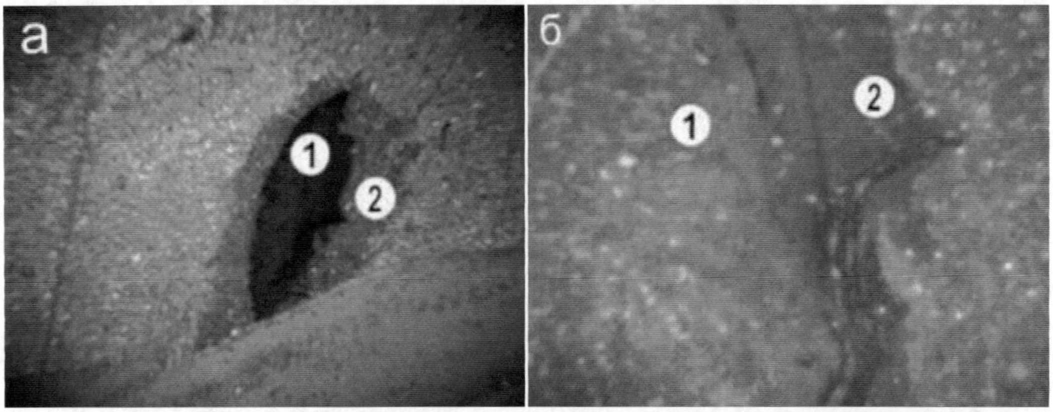


**Figure 3 (a, b).** Mudstone:a) mudstones with the inclusion of type-II kerogen (1) accompanied by a halo of a bituminous (2) (HC. 10 x N +); b) bitumoids in a halo of type-II kerogen: 1-oily- -resinous, 2- asphaltenous (HC. 100 x).

The presence of type-II kerogen is easily identified by nuclear-geochemical core sample investigations. Rocks with the increased uranium concentrations up to 6 g/t are in the range of 3541.5.3549.5 m. and belong to domanik deposits.

By the XRF analysis in the horizon M two different environmental facies of mudstone formation are identified: mudstones with organic humus in the range of 3396.9.3430.0 m. and mudstones with organic sapropel, type II kerogen in the range of 3541.5.3544.7m. By XRF analysis in the organic humus mudstones formed in the oxygen–rich environments (that contain $O_2$ in different quantities) there is less amount of P, S, Cl, Ti, Cr, Mn, Fe, Cu, Ir, Bi. Sapropel mudstones formed in the sharp reduction environment are characterized by the increased amount of chemical substances. This accounts for the conception of heavy metals accumulation in the absence of hydrogen during the process of diagenesis. [7].

The research of bitumoids distribution peculiarities in rocks of the profile was carried out by two methods: fluorescence microscopy [8] and by the chloroform extract analysis [9]. The results of the second method are in the lithogeochemical profile. Their shows provided the identification of the thickness in the upper section of the Paleozoic profile. Several anomalous concentrations of bitumoids in the interval of 3395-3555m have been identified: at the depth of 3396.6 m, and in the intervals of 3401.0 ...3403.4 m., 3422.4 ...3423.6 m., 3430.1 ...3432.0 m., 3484.3 ...3488.8 m., 3541.5 ...3549.5 m. Bitumoids shows are also confirmed by the fluorescence microscopy method. This method identifies higher quantities of hydrocarbon fluids for the rock marked intervals. The bitumoids largely correspond to medium and heavy, from oily-resinous to resinous-asphaltenous that are most characteristic for syngenetic bitumoids. In the low section of the profile the bitumoid shows are insignificant. They are characterized by the light hydrocarbons distributed through the thin cracks in the rocks, and are referred to the epigenetic bitumoids. (figure 4-a).

The pores and fractures were studied by the microscopic method. The presence of the interstitials was identified through all rocks of the profile. Pores are of irregular shape, they look like leaching pores (figure 4-b) which are not interrelated. At the same time, the schistosity and jointing zones are determined in the profile. Friction planes with bitumoid characteristics occur in some samples of the intervals.

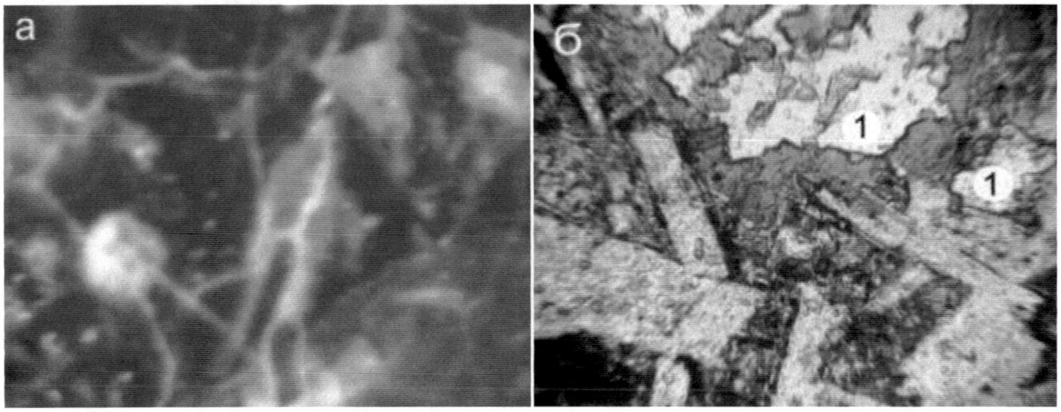


**Figure 4 (a, b)** peculiarities of the interstitials: a) cracks with traces of light bitumoids; b) leaching voids in basaltoids (1) (HC. 10 x, N |).

**Conclusion**

The given research concludes: among the basaltoid sequence deposits the interlayers of high-carbon content sedimentary rocks (mostly mudstones in the horizon M) occur. Among them there are

IOP Conf. Series: Earth and Environmental Science **21** (2014) 012015 doi:10.1088/1755-1315/21/1/012015

carbonaceous mudstone (mudstones with coal streaks) and domanik deposits with type II kerogen. The identified thickness of the domanik deposits is 8.9 m, with the large promising areas waiting for the core analysis to be performed. Thus, Paleozoic basement sedimentary deposits are considered to be the center of oil\gas generation in the area, and the eastern territory of Tomj Kolyvan structural zone – the promising one.

## References

[1] Gribova I S, Gorbachev V I and Kuzmin D 2013 New data about the prospects of oil on the right bank of the Tomsk Oblast by the results of drilling and exploration of Eastern Pajduginskaya well *proceedings of the VII all-Russian lithological meeting* 1 94

[2] Neruchev S G, Rogozina E S, Beketov V M etc 1986*In the deposits of the domanik type* (Lenningrad: Nedra) p 247 [in Russian]

[3] Ollie I A 1975 Organic substance and bituminosity of sedimentary deposits in Siberia: the luminescence-microscopic method (Moskva: «Nauka») p 130 [in Russian]

[4] Švanov V N, Frolov V, Sergeeva E I et al 1998 *Systematics and classification of sedimentary rocks and their analogues* (St. Petersburg: «Nedra») p 352

[5] Bogatikov O A, Morozov A F and Petrov O V 2009 Petrographic code of Russia. Igneous, metamorphic, metasomatic, impaktnye of education (Saint-Petersburg: izdatel'stvo vsegei) p 160 [in Russian]

[6] Stolbova N F pillars Ym 1986 Petrographic and geochemical features of buried organic matter. Bazhenov formation *organic matter in modern and fossil sediments* 278-280

[7] Stolbova N F pillars Ym 1990Using radiography in Petroleum Geology 17 *proceedings of international symposium on radiography* 95

[8] Teodorovich G I, Polonsky B Y, Andriyanova A G and others 1960 Mineralogical and geochemical facies and formation conditions of the Devonian sediments of the oil-producing West. Bashkiria and East Tartary (Moscow: ACADEMY of SCIENCES of the USSR) p 148 [in Russian]

[9] Florovskaya V N and Ovchinnikova L I 1970 Microscopy bituminous substances (Moskva: MGU) p 85 [in Russian]

# Modeling of inclined fracture network and calculation of fracture effect on seismic signal spectrum

**A A Shatskaya, M M Nemirovich-Danchenko and D A Terre**
Tomsk Polytechnic University, Tomsk, Russia

E-mail: alek.sashka@mail.ru

**Abstract.** The paper considers modeling technique for the medium with a network of inclined fractures and calculations for completing seismic tasks for such a medium. The network of plane-parallel fractures has controlled dip angle and fluid saturation. The time section for a model with water-saturated fractures is produced. The comparison of incident and transmitted signal spectra is made.

## Introduction

Oil and gas reservoirs are known to constitute structurally heterogeneous geological formations which consist of a solid matrix and voids filled with fluid. The void space structure determines reservoir properties which are considered while choosing the optimal technique for reservoir development. The algorithms of land seismic surveys and borehole seismology are being currently developed to obtain more accurate and reliable information concerning the formation microstructure and spatial distribution of formation parameters. The resolving power of seismic methods provides generalized information about the effect of inhomogeneities on a wave field. Despite the fact that physical mechanisms of elastic wave interaction with inhomogeneities of different types and dimensions are investigated to a great extent, the complexity of internal structure of existing rocks makes it necessary to apply simplifying assumptions. In particular, when wave propagation through a rock is described the medium is assumed to be homogenous with "'effective" acoustic properties [6]. The influence of inhomogeneities on a total wave field is revealed through the difference between characteristics of an effective model and those of integral solid matrix without void space.

The development of underexplored oil-gas bearing formations of Paleozoic period is a prospective area to increase mineral raw material base of West Siberia. Paleozoic deposits are mainly composed of carbonated rocks characterized by a complex cavernous-porous- fractured reservoir type [5]. Moreover, it was proved that the characteristics of a wave field are significantly dependent on tiniest fractures «opening» of which is measured by first micrometers, while the dimensions are sufficiently smaller than those of first Frenel zone section [1]. Microfracture size and shape are major factors which influence field formation of scattered waves when they propagate or reflect (are scattered) immediately off the specific surface of fractured formations.

## Method and model description

Physical modeling is one of the most widespread approaches which are applied to deal with the issue. Installations for investigation of samples with various fracture networks are made. A wave field produced by such networks is detected and analysed [4]. For instance, the effect of reservoir fracturing on a seismic wave has been studied since 1987 in the Laboratory for Applied Experimental Geophysics at Purdue University, USA, and over 30 articles were published in different periodicals.

However, it should be noted that physical modeling has a number of disadvantages. They are difficulty of parameter measurements, low flexibility of technique and technology, as well as distortions and limitations caused by equipment. It is possible to avoid these effects by using modern mathematical modeling. Nevertheless, one should consider that the accuracy of synthetic traces depends on appropriateness of a chosen medium model and on completeness of the equation system used to describe the propagation of all known wave types. The given paper considers the model of hypoelastic wave described in [2, 8] where its similarity to an actual geological medium is shown. The calculations are performed using the finite-difference method which is also applicable to cope with forward seismic tasks for various medium models. This method is proved to be useful as a result of comparison of theoretical ray surfaces and data of numerical modeling.

The necessary stage of any modeling is comparison of calculated data with the real one. In the given research the input parameters are sample characteristics which are listed in [4], which gives a detailed description of installation and the procedures of the physical experiment, the obtained results are further compared. The sample used in the experiment is an aluminum bar, in which 20 fractures are sawed using 3 mm spacing. This arrangement is approximately equal to a quarter wavelength produced using the frequency of 0,5 megacycles. figure 1 shows the acoustic wavefront recordings obtained from the intact sample F0 (a), and the fractured samples F3 one of which is dry (b) and the other is water-saturated (c).

Based on the sample description, the mathematical model the parameters of which coincide with data of the physical experiment was produced. The seismic fields which propagate through the fracture network were calculated. The model of a medium with dry fractures is given in figure 2a. Figure 2b is obtained when the parameters of water fluid are taken into account. Considering the unsaturated model in both experiments it is possible to observe the absence of distinct phase alteration, time-dependent wave diffraction. Provided that water is present, the images of seismic fields are also similar. There are quite distinct trace amplitudes which are more clearly expressed in the central zone than at the margins.

Considering the technique for carrying out the physical experiment one may notice that the installation construction makes it possible to arrange fractures parallel to one of its axes (vertical or horizontal cuts). In a real medium, tectonic stress causing rock discontinuity changes parameters of layer occurrence, consequently, fractures are located at an angle to the surface. Mathematical modeling can help to consider this peculiarity and produce models with inclined fractures.

We developed the generator of models with controlled dip angles and fracture saturation parameters. For the range of dip angles from the first semi-quadrant$0 \leq \alpha \leq \frac{\pi}{4}$ it is convenient to use the following formula for determination of current coordinates of computational cell in the source section.

$$x_c = x\cos(\propto) + w,$$

$$y_c = x\sin(\propto) + w + (N-1)h \qquad (1)$$

Where $\propto$ is a dip angle of a directional vector of a fracture network to the axis OX, N- number of fractures, h – fracture spacing, w – thickness of an individual fracture. The latter two parameters are determined by cell units. For the angles from the range $\pi/4 \leq \alpha \leq \pi/2$ fractures are drawn not from the axis OX, as it was done in the formula (1), but from the axis OY. According to vertical seismic profile data the angle of 20° is significant angle for the analysis of weakness tangent lines [7]. Therefore, it is used for the development of model for inclined fractures (figure 3).

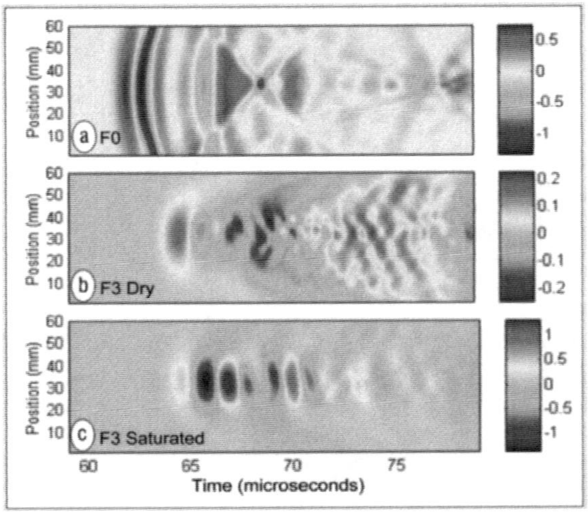


**Figure 1.** Time section obtained as a result of physical experiment: a) an intact sample, b) dry fractured sample, c) a sample with water-saturated fractures

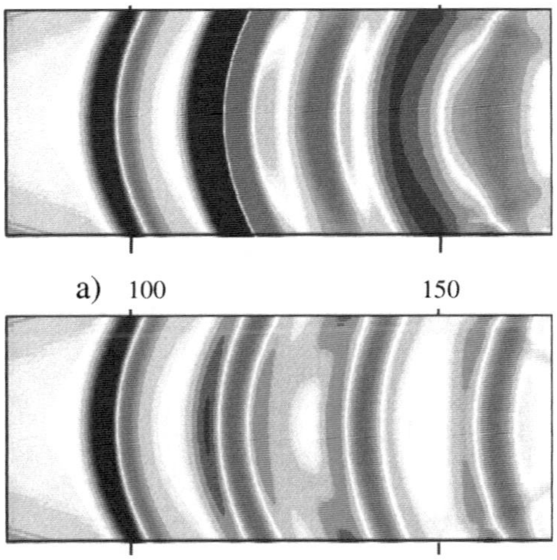


**Figure 2.** Synthetic time section with the same parameters: 2a) dry fractured sample model, 2b) water–saturated fractured sample model

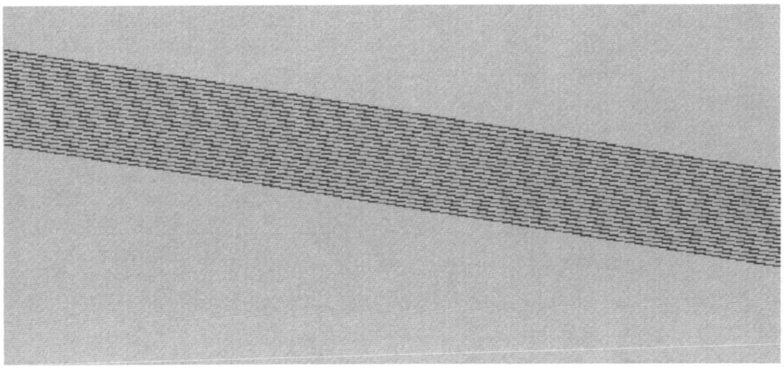


**Figure 3.** The model of a medium with network of inclined fractures:
1 – host medium, 2 – fluid in the fractures.

The figure shows that the represented model contains 22 fractures the thickness of which is 3, the spacing is 5 (see the formula 1). The host medium parameters correspond to average values for a terrigenous section; in particular, Vp=3000 m/s, Vs =2000 m/s, ρ=2000 kg/m$^{3.}$ There are the following fluid parameters: Vp=1500 m/s, Vs =0 m/s, ρ=1000 kg/m$^3$, They correspond to characteristics of water (shear velocity component does not occur in a liquid medium). For this model the time section covering plane-wave incidence, excitation and its recording on OX line was calculated. The original seismic wavelet was calculated using Riker formula [3]:

$$\dot{U}_y|_{CD} = F(t) = -2\pi f\sqrt{e}\ (t-t_0)\cdot e^{-2\pi f(t-t_0)^2},$$

where $\dot{U}_y$ is -drift velocity Y component. This impulse is characterized by smoothness of function, its derivatives and spectrum.

**Results**

For modeling the entire medium is supposed to be perfectly elastic, whereas in the fracture surroundings, i.e. in the zones of stress concentration, it is accepted to be elastoplastic [9]. The fragment of the obtained section is given in figure 4. Figure 5 shows the comparison of Fourier amplitude spectra of incident signal and wave transmitted through the fracture network.

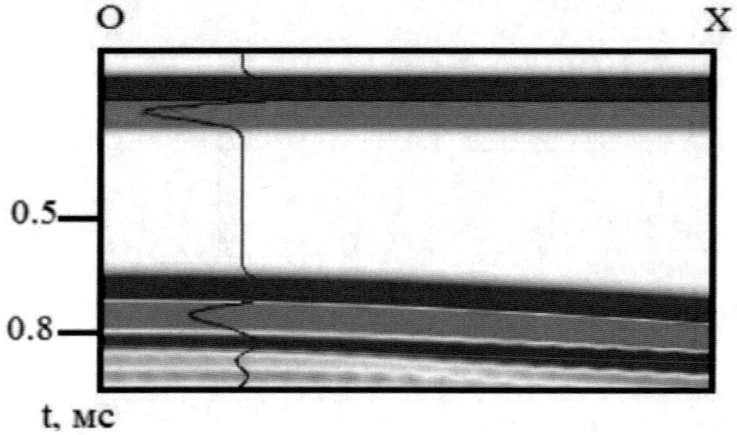


**Figure 4**. Synthetic time section. An individual trace is distinguished.

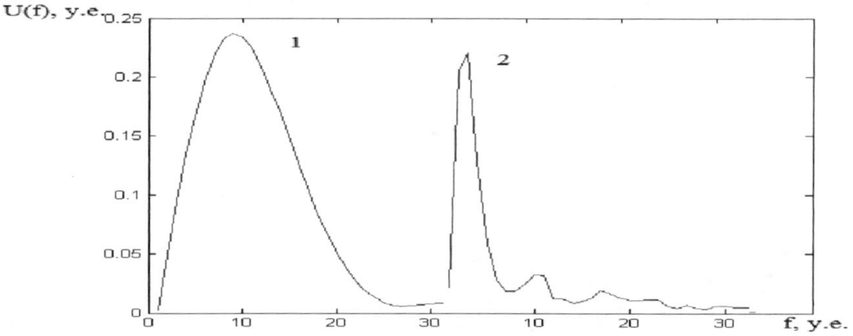


**Figure 5.** Spectrum of incident signal and transmitted signal: 1 – original signal spectrum, 2 – transmitted signal spectrum.

The analysis of spectra reveals visual displacement of carrier frequencies to a low-frequency zone of spectrum and bandwidth reduction for both saturated and unsaturated fractures [8]. Furthermore, the similar regularities in spectrum transformations are observed through physical modeling [10]. In the right section of the curve it is indicated that the drastic decrease of total share of high frequencies causes increase of effect of individual high frequency bands affecting both diffraction characteristics for the entire fracture network and individual fracture vibrations.

The performed calculations indicate that general attenuation of high frequencies of spectrum which is revealed for a wave transmitted through a fractured unit results from the impact of acoustic impedance (if the effect of friction mechanisms is excluded). In its turn, this effect is connected with a set of heterogeneous surface waves which occur at the interfaces. The detected regularities can be further applied to reveal zones of absorption in seismic sections taking into account fracture network orientation through the analysis of seismic attributes.

**References**

[1] Karaev N A, Kozlov E A, Karaev G N, LukashinY P, Prokator O M and Semenov V M 2010 Physical modeling supports testing of seismic techniques for porous-fractured reservoir characterization: EAGE, Conference, Exponded, Abstracts B 47

[2] Nemirovich-Danchenko M M 1998 A model for the brittle hypoelastic medium: application to computation of deformations and failure in rock *Phys. Mesomech* **1**(2) 101-108

[3] Nielsen P, If F, Per Berg and Ove Skongaard 1995 Using the pseudospectral method on curved grids for 2D elastic forward modelling *Geophysical Prospecting* **43** 369-395

[4] Pyrak-Nolte L J 2007 Fracture anisotropy: The role of fracture-stiffness gradients The Leading Edge **26** (9) 1124-1127.

[5] Bagrintseva K I and Dmitrievsky A N 1989 Theoretical basics for forecasting zones of high–capacity carbonate reservoirs in deposits of various facies *Sedimentary basins and oil-gas capacity* (Moscow: Nauka) 136-146.

[6] Glubovskikh S M and et al. 2011 Theoretical and experimental studies of oriented vertical fracturing effect on effective acoustic properties of rocks *Seismic technologies* **4** 60-67

[7] Dugarov G A 2013 *Evaluation of effective parameters of media with oriented fractures in linear slide modeling based on data about anisotropy of velocity and seismic wave absorption: Author's abstract/ Synopsis of a thesis* (Novosobirsk: INGG SO RAN) p 18

[8] Nemirovich-Danchenko M M and Stefanov Y P 1995 Application of finite-difference method in Lagrange variables for numerical calculation of wave fields in structurally complex media *Geology and geophysics* **11** 96-105

[9] Nemirovich-Danchenko M M 2013 Detection of multiple fracturing of solid medium based on evaluation of spectral density of reflection energy *Phys. Mesomech* **16** (1) 105-110

[10] Seismic stratigraphy 1982 *Uses in oil and gas prospecting* ed. C. Payton (Moscow: Mir) **1** p 375

# Formation evaluation and well-test analysis for complex interpretation of reservoir permeability distribution

**M O Korovin[1] and V P Merkulov**
Tomsk Polytechnic University, Tomsk, Russia

E-mail: [1]korovinmo@hw.tpu.ru

**Abstract.** Data on heterogeneity of porous space and reservoir filtration properties determines the choice of most effective development strategy. The comparative analysis of well-logging and well-test results was carried out to determine the value of filtration heterogeneity. The similarity of the results obtained using different methods makes it possible to predict orientation of enhanced permeability.

The main aim of the maximizing oil production is to choose the most appropriate and efficient development strategy. This is not an easy task but a 3-D simulation model which shows the real hydrocarbon production rate can be applied to cope with the task. However, it is necessary to take into account that the simulation model is based on a geological model, within which reservoir properties are distributed in a particular way, with the most important property being permeability, especially, its distribution through the effective porous space.

There are two basic methods applied to determine the permeability anisotropy: a special core analysis and well-test analysis [1]. But these methods are time-consuming and expensive. Another method to obtain data is formation evaluation analysis. Well-logging data is recorded at every drilled well and it is possible to evaluate permeability values through the analysis of a cross-section. To predict the spatial anisotropy of reservoir heterogeneity it is essential to provide a special analysis of the well-logging data but with obligatory comparison with core and well-test data. The integration of various methods is aimed at reaching the maximum accuracy and efficient forecasting of permeability distribution.

This condition determines the specific requirements to the area of study. It should be a "standard deposit" which shows the explicit evidence of filtration heterogeneity confirmed by development operations, well-tests, well-logging and core analysis data. From this point of view XYZ deposits located in West Siberia meet the requirements to the full extent and have a sufficient set of logging and well-test data. Standard logging, gamma ray logging, self-potential, induction logging, neutron logging, lateral logging, core analysis and tracer investigations were conducted within the entire area of the field. Using this suite of methods it is possible to detect reservoirs, evaluate porosity, permeability, and prove shaliness model validity [2]. The log-inject-log tests which are aimed at determination of permeability value and tracing fluid volume during the flow from an injection to a production well are carried out as well.

The geological structure of the field is characterized by siliciclastic formations with changing lithological composition of Mesozoic and Cenozoic platform mantle as well as pre-Jurassic basement. The total thickness of discordant deposits which occur on the basement comprises about 3 km. The north-eastern stretch of the geological bodies predetermines spatial anisotropy of reservoir properties.

IOP Conf. Series: Earth and Environmental Science **21** (2014) 012017 doi:10.1088/1755-1315/21/1/012017

From the stratigraphic point of view the net-pay reservoir is composed of Upper Jurassic deposits and can be located through detection of reference horizon – the Bazhenov suite (which is characterized by the extremely high gamma-ray values). The reservoir is subdivided into four sublayers ($J_1^1$, $J_1^2$, $J_1^{my}$, $J_3$), which occurred under different sedimentation conditions (figure 1). The $J_3$ horizon is also divided into three sub-horizons ($J_1^{3A}$, $J_1^{3B}$, $J_1^{3C}$) which are characterized by deterioration of the reservoir properties from top to bottom.

In the period from 2003 to 2009, tracer fluid tests were conducted on the area of the field in order to detect the directions of fluid flow from injection wells to production wells. The main parameter acquired from these tests is time required for a tracer fluid volume to flow from an injection well to a production well. These data make it possible to calculate the phase permeability of the reservoir.

The integrated analysis revealed that $J_1^3$ is the most producing formation. The formation sub-division corresponds to reservoir property changes observed through the vertical section of the reservoir. The shaliness model was chosen using the self-potential method and the least shaly layer was taken as reference one. The porosity was calculated using the neutron method [3, 4]. The next step of interpretation procedure is concerned with the choice of the equation that will help to predict the permeability variations within the well section.

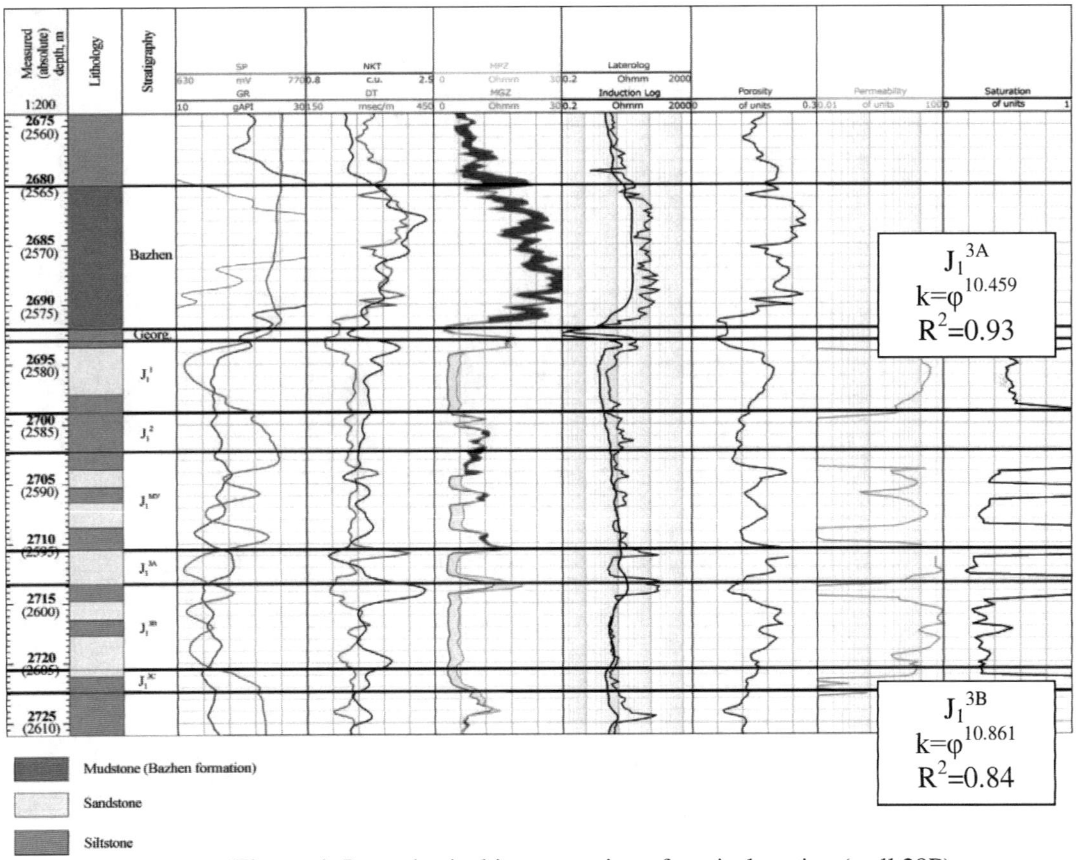


**Figure 1.** Petrophysical interpretation of vertical section (well 29P).

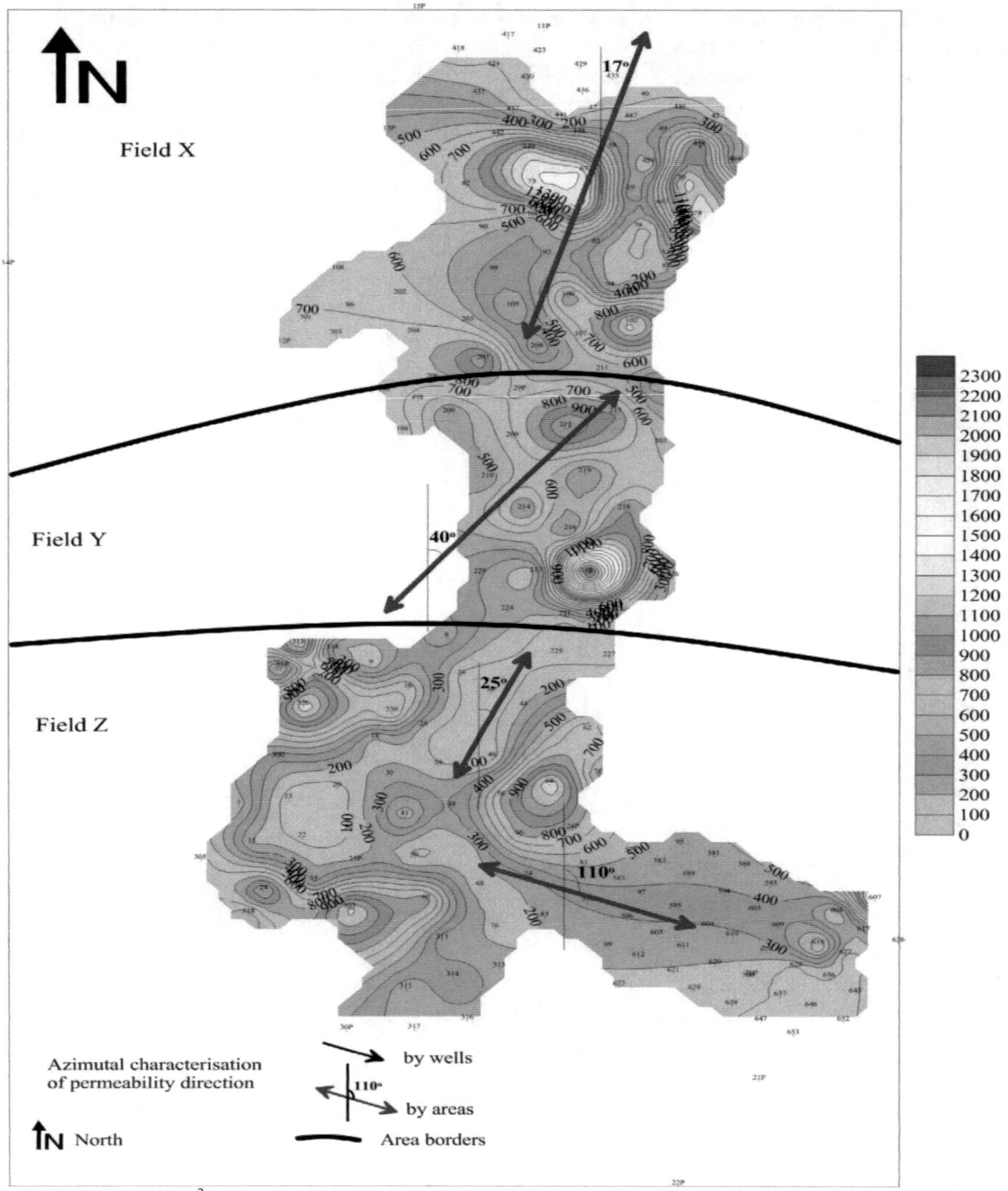


**Figure 2.** $J_1^3$ permeability map and vector characteristics of isoline directions (permeability in mD).

To calculate permeability one must consider the correlation between permeability and porosity parameters obtained from the core analysis. Layers $J_1^{3A}$ and $J_1^{3B}$ show the high correlation which, in its turn, give an opportunity to apply the appropriate equations for permeability evaluation (figure 1). The correlation coefficient convergence suggests the identity in permeability distribution depending on porosity. Moreover, it is necessary to note that a closer connection between the parameters is observed for sub-layer $J_1^{3A}$. The obtained values characterize the section of the zone where well intersects the layer.

Production wells reflect the characteristic dependence between the volumes of extracted fluid and numerical data of permeability calculated with the help of petrophysical investigations. The map (figure 2) generated on the basis of data about the permeability distribution indicates spatial petrophysical heterogeneity of the layer (a special system of isolines). If the map is produced using the vector distribution of the azimuthal permeability orientations, the similar effect is observed as well.

It is possible to use vector analysis for permeability isoline orientation determination. The produced bar graph (figure 3) indicates not only heterogeneity but also allows calculation of orientation- angle characteristics of enhanced filtration [5, 6]. The segments with maximum values on the graph correspond to general directions detected by tracer tests. It is necessary to note that permeability heterogeneity is equally distributed in opposite directions [7, 8, 9].

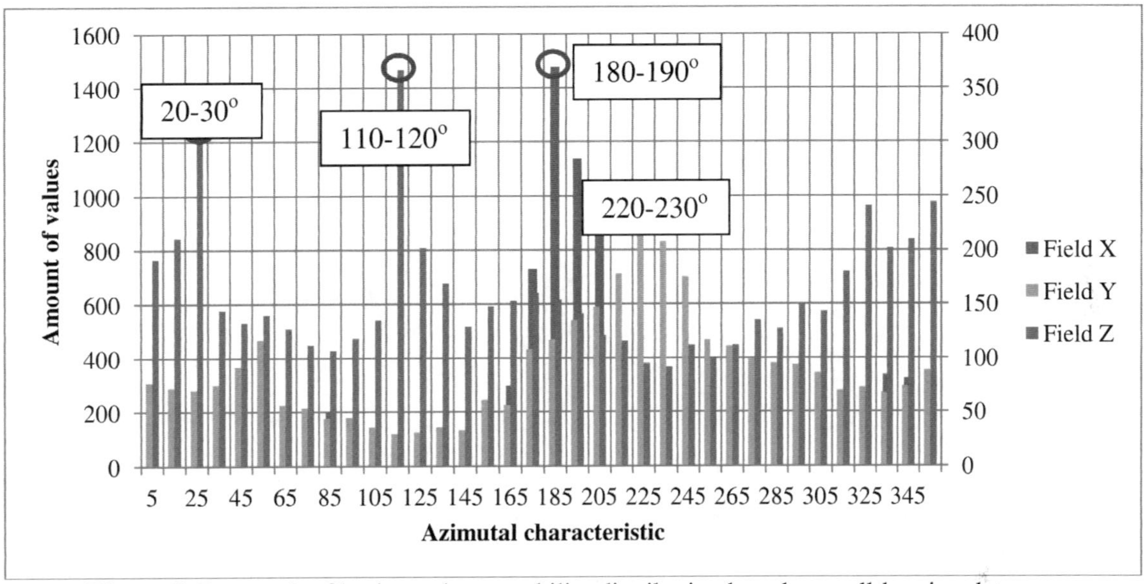


**Figure 3.** Bar graph of horizontal permeability distribution based on well-logging data.

The distribution of permeability in layer $J_1^3$ in proximity to the well 66 is presented in figure 4. Production wells are radially placed relative to the injection well. This type of displacement allows evaluation of the numerical characteristics of tracer research. Furthermore, the figure shows that the predominant direction of fluid extraction is north-eastern (south-western) one. The similar data are obtained as a result of phase permeability analysis.

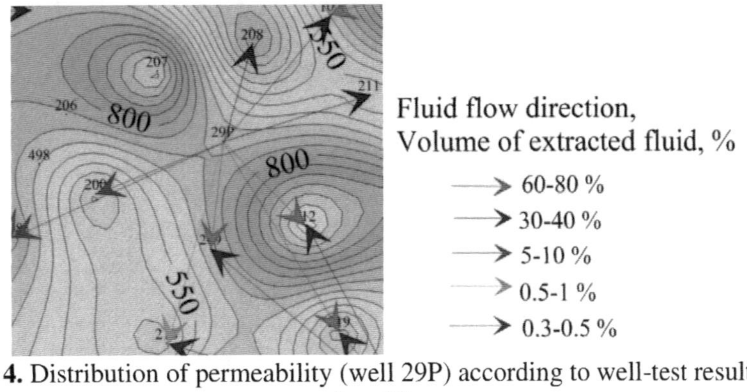


**Figure 4.** Distribution of permeability (well 29P) according to well-test results (permeability in mD).

Elliptical approximation was applied to evaluate predominant filtration. As it is seen in figure 5, models produced with different methods correlate with each other by azimuthal characteristics. The absolute values of correlation sections (ratio of maximum to minimum ellipse axes) are nearly equal (1.92 according to formation evaluation data and 1.95 well-test data). The correlation section correspondence indicates that well-tests confirm formation evaluation data.

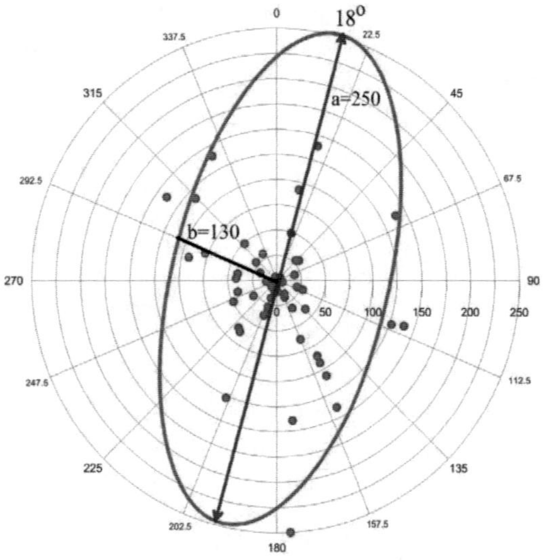


**Figure 5.** Elliptical approximation of permeability distribution obtained through well-logging (left) and well-test analysis (right).

The gradients of permeability distribution throughout the study area (figure 6) show the directions of approximating ellipses with respect to minimum and maximum values of gradients. Compression ratio in this case constitutes 2.14 and 1.9 and well-test data discrepancy makes 10% and 3%. Azimutal location of the major ellipse axis (minimum gradients) corresponds to the distribution obtained in well-tests (figure 5).

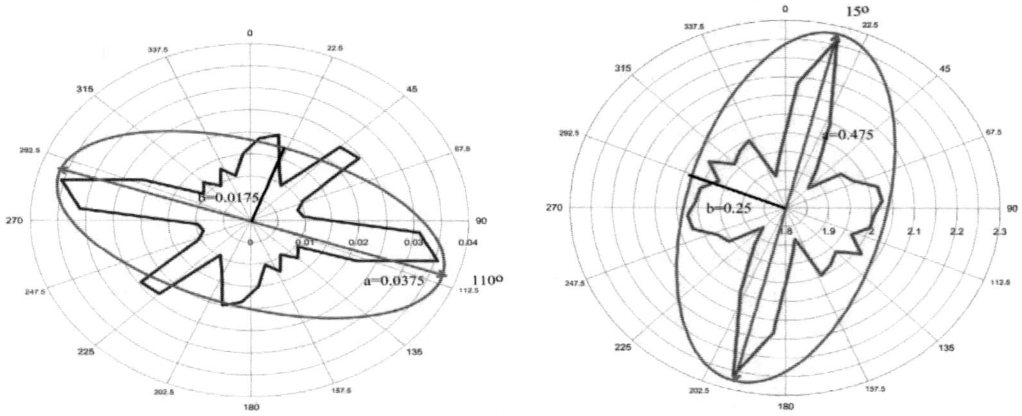


**Figure 6.** Distribution of permeability gradients based on elliptical approximation results: minimal values (left), maximal values (right).

The comparison of data obtained through well-tests (log-injection-log) which are plotted on the regional map (figure 2) reveals coincidence in direction trends of high filtration. Moreover, reservoir flow characteristics studied through log-injection-log tests also confirm the results of vector analysis based on formation evaluation data.

Thus, formation evaluation data can be applied for interpretation and determination of potential azimuthal characteristics to detect zones of enhanced permeability. Additional vector analysis of permeability gradients and elliptical models give further confirmation to well-logging results. The forecast reliability is determined by the chosen technology of formation evaluation interpretation and accuracy of produced permeability maps.

## Conclusions

1. The results of well-tests are confirmed by areal permeability distribution obtained through petrophysical justification of formation evaluation data interpretation.
2. The initial conclusions about permeability heterogeneity on the area are advisable to make at the stage of well-logging data acquisition and processing. The necessary condition is capability to produce detailed maps of stratigraphic layers.
3. Vector analysis of permeability maps and gradient zones allows for reliable data on azimuthal and numeric characteristics of permeability heterogeneity for further geological and reservoir modeling.

## References

[1] Donald H, Irving F and Bergamini G 1963 The effect of stress on permeability of sandstone cores. California Research Corp., La Habra, Calif. *The Society of petroleum engineers*. Feb 4, SPE 531

[2] *Interpretation of the results of geophysical studies of oil and gas wells*. Directory under 1988 ed D Sc Dobrynin (Moscow: Nedra) p 476

[3] Vendelshteyn B and Rezvanov R 1978 *Geophysical methods for determining the parameters of oil and gas reservoirs* (Moscow: Nedra) p 318

[4] Akkuratov O, Aronov V and Atyasheva E 2003 *Guidelines on the calculation of geological reserves of oil and gas volumetric method* ed Petersilye V, Poroskun V and Yatsenko G (Moscow-Tver: VNIGNI SPC "Tvergeofizika") p 478

[5] Glavnova E, Merkulov V and Glavnov N 2010 Comparative analysis of methods for determining the horizontal permeability anisotropy *Proceedings of the TPU* **317** (1) 127-132

[6] Merkulov V and Krasnoshchekova L 2014 Assessing the impact of oil and gas reservoirs filtration anisotropy fields for modeling *Gas industry* **703** 22-27

[7] Korovin M Specialized core analysis for the study of the anisotropy of oil and gas reservoirs *Bulletin of Tomsk Polytechnic University* **324** (1) 87-92

[8] Korovin M 2013 Investigation of the Acoustic and Petromagnetic Anisotropy of Oil Fields Rocks *75$^{th}$ EAGE Conference & Exhibition incorporating* SPE EUROPEC. London Extended abstract. Publication date: 10.06.2013. URL: http://earthdoc.eage.org/publication/publicationdetails/?publication=68815 (дата обращения: 30.09.2013)

[9] Korovin M 2011 Investigation of the Spatial Heterogeneity Petromagnetic Properties of Oil Fields Rocks *73$^{rd}$ EAGE Conference & Exhibition*, Extended abstract. Vienna, Publication date: 23.05.11. URL: http://earthdoc.org/publication/publicationdetails/?publication=50945 (date of access: 30.09.2013)

IOP Conf. Series: Earth and Environmental Science **21** (2014) 012018 doi:10.1088/1755-1315/21/1/012018

# Degree of Areal Drainage Assessment Using Digital Elevation Models

**V Pokrovsky [1,a], D Pokrovsky[2,b], E Dutova[1,c], A Nikitenkov[1,d] and A Nazarov[1,e]**
[1]Tomsk Polytechnic University, Tomsk, Russia
[2] Tomsk state university of architecture and building, Tomsk, Russia

E-mail: [a]vdp@sibmail.com, [b]dsp@sibmail.com, [c]dutova@sibmail.com, [d]corestone@mail.ru, [e]enazarov@tpu.ru

**Abstract.** The methodology of mapping and areal drainage evaluation was discussed. Generalized operations have been described, such as GIS technologies for topographic data. They resulted in maps for erosion network, density and depth of erosion network, as well as land surface sloping and further total evaluation as a drainage parameter.

## 1. Introduction

The widely-used term "drainage," as applied by the majority of researchers, is a characteristic feature reflecting a set of conditions that determine the speed and volume of surface and ground water outflow from an area. This term is applied to describe and identify those factors concerning engineering protection design of flooding processes, determination of soil structure for the spacing and management of prevailing vegetation types, development of zoning schemes in determining melioration measures and so on. However, there are no clear guidelines or recommendations to determine and assess the drainage degree within this or that territory. It seems more advisable and appropriate to consider the quantitative characteristics so as to identify those set of factors stipulating the degree of drainage and defining their size and importance within the expert systems.

Accordingly, the main factors defining the degree of drainage could be classified into two groups, each of which includes natural and technogenic components.

The first group reflects properties of the massif in respect to ground water infiltration and is defined by the characteristic features of geological section and filtration properties of rocks, as well as the adverse impact due to technogenic factors (barrage, drainage systems etc.).

The second group depends on the pattern of natural or technogenic inverted relief determining the horizontal and vertical position of drainage baselines, and, respectively, the possible formation, structure and dynamic component of the unloaded filtration flow.

Significant and expensive full-scale studies are required to identify the specific features of the geological section and to determine the filtration parameters of geological components. Commonly-used approaches in analyzing the terrain structure are rather time-consuming, however, all necessary information about terrain parameters may be obtained on the basis of digital models (DEM).

The up-dated methodical basis and software in designing and applying digital terrain models (DEM) via modern geoinformation systems (GIS), allowing a totally new level in analyzing and solving these issues. The introduction of highly effective spatial resolution DEM and computerization promoted the development of sophisticated software and hardware systems in analyzing the hydrological and morphometric characteristics of different territories within the framework of geoinformation systems.

To design regional DEM SRTM-3 data released by the United States National Aeronautics and Space Administration (NASA) [4-7] could be used as a dataset.

Previously, processing of such high-altitude data was completed by GIS i.e. Ilwis 3.5 to design the hydrographic network and its morphometric characteristics. To plot final cartographic data ArcGIS 9.3 was used. Designed regional DEM has been used to study the underground runoff characteristics of Kuznetsk Alatau.

The modeling of regional DEM is based on interpolating digital contours of large-scale topographic maps. Such a database was used to assess the drainage area of Tomsk.

## 2. Application technique of GIS in analyzing digital elevation models

The procedure in building terrain digital maps could be divided into three stages (table 1). The first two of which are supporting, while the third is the major one

The first stage involves the mapping of digitized contour map into point features, setting integrated height grids and assessing territorial coverage quality based on obtained data. The resultant sequence is transporting the obtained DTM data to GIS.

**Table 1.** Main assessment stages of territorial drainage based on DTM.

| Stages | | |
|---|---|---|
| | 1 | transport of DEM data to GIS |
| | | ↓ |
| | 2 | designing DEM without local topographic lows<br>mapping surface slope areas<br>mapping drainage accumulation values |
| | | ↓ |
| | 3 | mapping erosion network<br>mapping erosion network density<br>mapping daytime slope surface<br>mapping river network incision depth<br>mapping drainage coefficients |

The second stage includes data processing created by DTM. To calculate the runoff areas throughout the territory the undrained relief depressions are eliminated by their alignment with the surrounding area (using the tool “fill sinks”). In this case, undrained relief depressions could be considered as accumulation object of surface- and transit of underground runoffs.

Based on the aligned DEM the flow direction within each unit cell is defined. Slope values are calculated for the central cell in block size 3 x 3. Flow direction is determined by determining the steepest slope (tool “flow direction”, method for determining the flow direction-"steepest slope"). In addition to the flow direction, it is possible to characterize DTM cell exposure. Further, cumulative drainage values are calculated for each elementary DEM cell. These values represent the number of cells introducing its drain into the estimating cell under the influence of gravitational forces, i.e. characterizing the catchment area. This procedure is implemented by the tool «flow accumulation».

The third and final stage involves the mapping of erosion network density and depth, as well as territorial day slope surface. The resultant map includes the integrated information.

Mapping erosion network is the subtraction of all runoff values from the matrix of cumulative values, which are less than the threshold value that is the minimum catchment area where surface-stream flow waters are formed with steady runoff throughout the year. In our case, the threshold is 300 elementary grids, corresponding to the catchment area of about 2.5 $km^2$. Then the identification of individual streams is performed on the database of merge points and their vectorization. To perform these operations “drainage network extraction tool " was used. Accuracy of mapping (peak of junction, river network configuration) via GIS based on SRTM data is assessed in comparing analogous by elements in traditional topographic map of corresponding scale.

Subsequent mapping and calculations were conducted in relation to one grid area.

The day surface slope map is build using the tool "Slope", which is similar to the tool "flow direction". Calculation assessment of slope is the slope between the estimated grid center and the adjacent centers. The highest value is regarded as the estimated slope grid value.

To estimate the erosion network density the investigated territory was divided into square grids where the side of the elementary grid is 100m. Within each grid the total length of drains was calculated and is divided by its area. The resulting value is assigned to a grid.

Cutting erosion depth network map is based on a statistical analysis of the terrain within the grid and plotted during the assessment phase of erosion density. Cutting erosion depth network is estimated as the difference between the maximum and minimum relief point, which value is assigned to the estimated grid.

These three terrain characteristic features- surface slope, cutting depth and density of erosion network could be used as an integral assessment of the relief role and be termed as drainage coefficient.

Density and depth of erosion network characterize the rate of erosion area irregularity. In this case, it indicates the relationship of surface and underground waters and determines the potential discharge of surface water and underground water through the erosion network. The surface slope value indirectly reflects the intensity of this discharge through hydraulic gradients.

The resultant map of drainage area, expressed as a system of scores, is compiled by using the "field calculator". The map elements correspond to the selected drainage groups: not drained, poorly drained, moderately drained and well drained.

## 3. Drainage network in Tomsk

We used the above-mentioned methods to examine the territory of Tomsk. This research has been conducted for a long time [7-8]. The drainage coefficient scheme of Tomsk in accordance with presented methodology techniques is presented in figure 1.

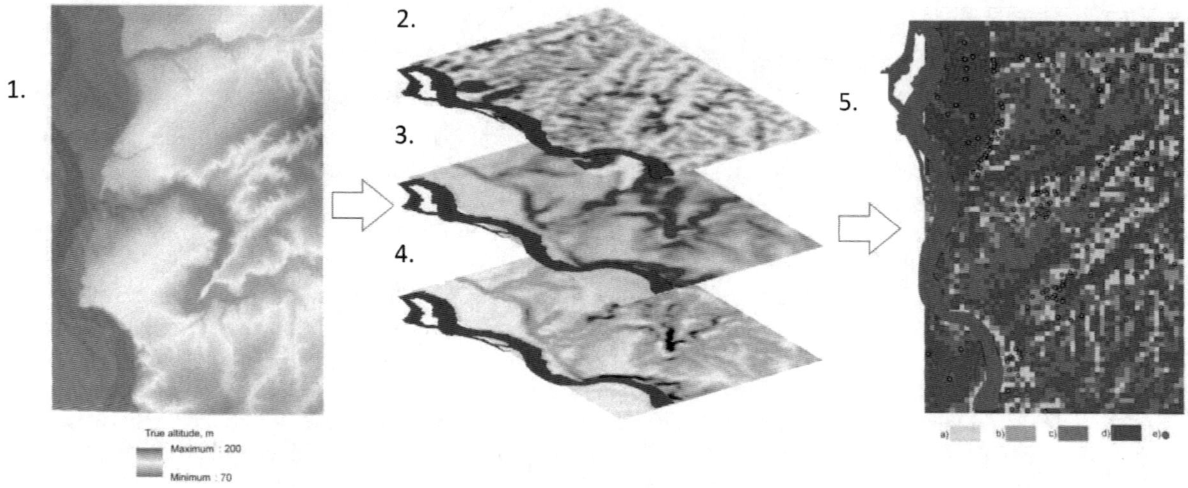


**Figure 1.** General drainage estimation scheme of Tomsk territory.
1. -digital elevation model; 2. -map of erosion network density; 3. -map of surface area slope; 4. -map of erosion network depth; 5. -map of drainage rate.

Tomsk is not only an old cultural and industrial city, but also a student town. It is located in overmoistened taiga zone of West Siberia on Tom River. The city has a complicated infrastructure. Modern multi-storyed buildings are scattered throughout the city, however, the main concentration is in the center of Tomsk and its outskirts. The historical area includes historical and cultural monuments both wooden and stone. Such stone and wooden architecture monuments should be mentioned: temples, buildings of famous architects as K.K. Lygin, A.D. Kryachkov, P.F. Fedorovsky, A.K. Bruni and others; constructions with wood carvings of extraordinary beauty and complexity. Preservation of this cultural heritage is one of the most important targets of the town-planning policy.

Like many cities and towns in the world, Tomsk is confronted by one and the same geological problem – anthropogenic underflooding. Anthropogenic underflooding involves the rise of shallow groundwater levels stipulated by the town's activities and under the influence of natural and anthropogenic factors, including territorial geological structure and drainage system, underground water intensity and its poor water flow and evaporation.

Numerous municipal service facilities provoke possible technogenic changes of hydrogeological conditions leading to various emergency situations. For example, Lagerny Garden landslide slope and Solnechny district where multi-storyed buildings are in critical condition due to the existing underflooding process.

The territorial drainage system, filtration section type and intensity flood sources predetermine the potential underflooding level. The importance of this question for Tomsk was firstly noted by D. S. Pokrovskiy. At that time only the qualitative drainage factors were considered. In this case, the assessment criterion was considered to be the underground water depth.

Figure 2 is a resultant map of the drainage coefficients within the city territory, showing only the quantitative aspect. The elements presented on the map correspond to the marked drainage groups: not drained, poorly drained, moderately drained and well drained. The characteristics of these elements shown on the resultant map and corresponding to the marked drainage groups is presented in table 2.

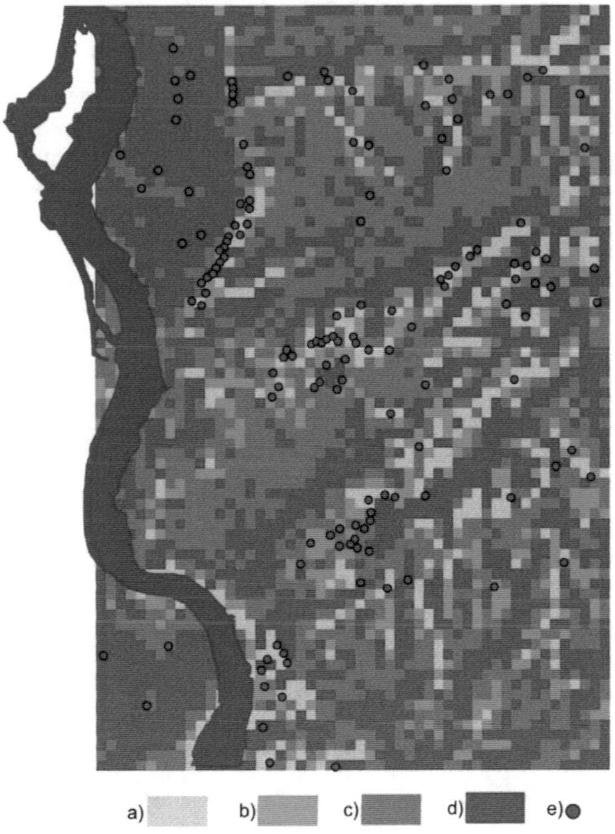


a) -well drained; b) -moderately drained; c) -poorly drained; d) -not drained; e) –spring.
**Figure 2.** Resultant map of drainage coefficient within the city territory.

To verify the suggested approaches in drainage estimation we used data based information on the mapping of springs. The configuration of drainage territorial field gradation correlates with the configuration of spring field distribution mapped by A.D. Nazarov. This fact testifies that the proposed approaches are appropriate.

**Table 1.** Indexes of drainage rate.

| | Density of erosion network, min-max | Depth of erosion network, min-max | Surface area slope, min-max | Drainage coefficient, min-max |
|---|---|---|---|---|
| | $km/km^2$ (number) | m (number) | degree (number) | (number) |
| Well drained | 0.01-0.03 (0.33-1) | 31-62 (0.5-1) | 4.8-17.3 (0.28-1) | 1.01-3 |
| Moderately drained | 0.005-0.01 (0.17-0.33) | 9-31 (0.14-0.5) | 2,4-4,8 (0.14-0.28) | 0.45-1.01 |
| Poorly drained | 0.002-0.005 (0.07-0.17) | 4-9 (0.06-0.14) | 1.2-2.4 (0.07-0.14) | 0.2-0.45 |
| Not drained | 0-0,002 (0-0.07) | 0-4 (0-0.06) | 0-1.2 (0-0.07) | 0-0.2 |

## 4. Conclusion

The research conducted on the example of Tomsk territory shows that modern geoinformation systems used in the relief analysis, allow qualitative plotting of the erosion network and determine its basic morphometric characteristics. Based on the obtained data, a territorial drainage assessment was conducted.

The research results can be applied in engineering and geological exploration and can be useful for different departments and services ensuring the vital living activity of the city:

- to monitor and efficiently control the geological environment conditions;
- to assess probability, scale and intensity of the human impact aftergrowth;
- to justify administrative and engineering decisions aimed at sustainable environmental conditions;
- to estimate the city territory cost;
- to optimize funding planning in providing restoration works of historical Tomsk.

## References

[1] Tim Ormsby, Eileen Napoleon and Robert Burke 2004 Getting to know ArcGIS (Esri Press) p 572

[2] Werner M 2001 Shuttle Radar Topography Mission (SRTM), Mission overview *J. Telecom* **55** 75-79

[3] Rodriguez E, C S Morris, J E Belz, E C Chapin, J M Martin, W Daffer and S Hensley 2005 An assessment of the SRTM topographic products *Technical Report JPL D-31639, Jet Propulsion Laboratory, Pasadena, California* p 143

[4] Coltelli M, G Fornaro, G Franceschetti, R Lanari, M Migliaccio, J R Moreira, K P Papathanassiou, G Puglisi, D Riccio and M Schwabisch 1996 SIR-C/X-SAR multifrequency multipass interferometry: A new tool for geological interpretation *Journal Of Geophysical Research* **101** 23127-23148

[5] Avseikov A S, Neifeld E A, Ryumkin A I and Tyabaev E S 1998 Geoinformation system at historical part of Tomsk city *Geoinformatics. Theory and practice* **1** 273-283

[6] Dutova E M, Nalivaiko N G, Kuzevanov K I and Kopylova J G 1998 The chemical and microbiological composition of urban groundwater, Tomsk, Russia *Groundwater in the Urban Environment. Proceedings of the XXVIIth Congress IAH* **2** 125-130

[7] Pokrovsky D S, Rogov G M and Kuzevanov K I 1998 The impact of urbanisation on the hydrogeological conditions of Tomsk, Russia *Groundwater in the Urban Environment. Proceedings of XXVIIth Congress IAH* **2** 217-223

[8] Pokrovskiy D S, Dutova E M and Kuzevanov K I 2008 Application of geoinformation technologies for estimation of hydrogeoecology conditions of developed territories *News of higher educational institutions. Construction* **3** (591) 107-112

IOP Conf. Series: Earth and Environmental Science **21** (2014) 012019 doi:10.1088/1755-1315/21/1/012019

# Financial and tax risks at implementation of "Chayanda-Lensk" section of "Sila Sibiri" gas transportation system construction project

**I V Sharf [1], N V Chukhareva[2] and L P Kuznetsova[3]**
Tomsk Polytechnic University, Tomsk, Russia

E-mail: [1]irina_sharf@mail.ru, [2]natasha@tpu.ru, [3]kuznetsova-lp@mail.ru

**Abstract.** High social and economic importance of large-scale projects on gasification of East Siberian regions of Russia and diversifying gas exports poses the problem of complex risk analysis of the project. This article discusses the various types of risks that could significantly affect the timing of the implementation and effectiveness of the project for the construction of the first line of "Sila Sibiri", the "Chayanda-Lensk" section. Special attention is paid to financial and tax aspects of the project. Graphically presented analysis of the dynamics of financial indicators reflect certain periods of effectiveness in implementing the project. Authors also discuss the possible causes and consequences of risks.

## 1. Introduction

High social and economic importance of gas projects in regional [1] and national [2] scale actualize the task of analyzing the risks that could significantly change the plans for implementation, and hence the effectiveness of the project for the construction of "Chayanda-Lensk" section of gas transport system "Sila Sibiri", which is implemented by the subsidiary of "Gazprom".

## 2. Macroeconomic risks

Macroeconomic risks emerge at lower world prices for hydrocarbons at reducing of refining margins, inflation processes and other economic phenomena. These circumstances could affect the financial stability and solvency of mining companies - service users, and therefore a temporary delay in payments is possible, as well as decrease in need for transportation of gas due to a reduction in overall demand for hydrocarbons or high price assessment for transportation service customers. Indirect impact lies is the possible negative impact on volume and structure of investment flows into this project, and therefore on its deadlines.

Thus, the analysis of capital investments aimed at the construction of major communication facilities, transport and energy facilities, auxiliary and service facilities, as well as the site preparation the level of which for the year 2013 is more than 115 billion rubles, see figure 1.

Cost analysis on the construction of basic facilities shows that the costs for the construction of gas pipeline account for 91%, water crossings 3%, the introduction of gas pipeline condition monitoring - 6%.

Further funding layout shows a high proportion (31%) of investment in the construction of access roads, electrochemical protection, power lines, which in turn are part of the transport facilities, communication and energy sector as a whole. The 32% share of capital investment foresees the construction of auxiliary and service facilities and realizing of site preparation work.

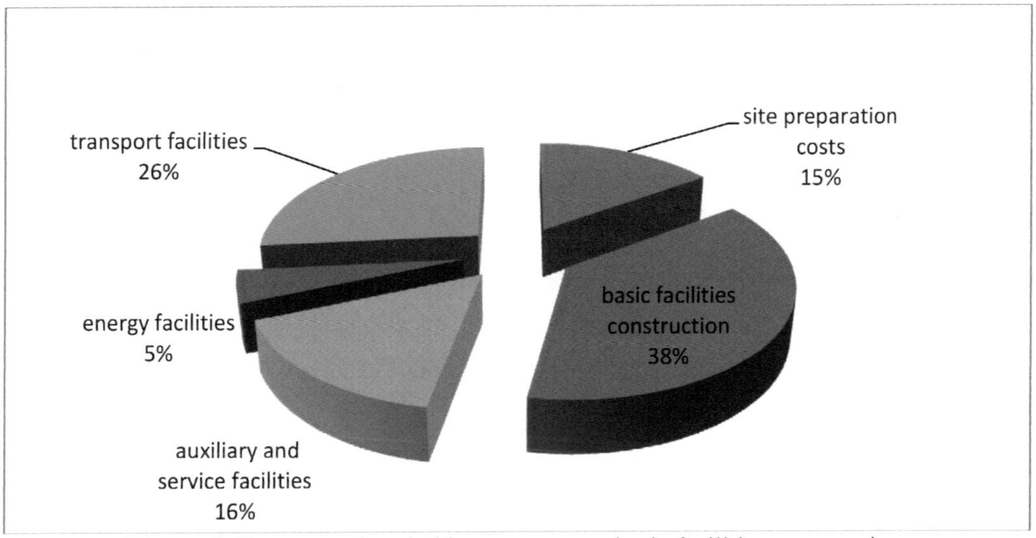


**Figure 1.** Structure of capital investments on basic facilities construction

In case of decreasing of the investment flow, from our point of view, the most vulnerable would be a completion works involving pipeline construction site improvement and the surrounding area. In addition, cheaper materials can be used, which in its turn increases the risk of accidents.

Note that hydrocarbon transportation tariff is set by the Federal Tariff Service due to assigning of pipelines to the objects of natural monopolies, which are regulated by the state. Sensitivity analysis was conducted by authors in accordance with the following efficiency criteria: the amount of investment and the cost of gas transport services. These are key indicators, as their influence is most effective on the internal rate of return, which is 12%.

The results of the sensitivity analysis are presented in table 1.

**Table 1.** Sensitivity analysis of efficiency indicator to the value of gas transport services change

| | | | The volume of investments, bln. VAT included | | | | | | | | |
|---|---|---|---|---|---|---|---|---|---|---|---|
| Internal rate of return | | | 110 | 117 | 124 | 131 | 138 | 145 | 152 | 159 | 166 |
| | | 75 | 80 | 85 | 90 | 95 | 100 | 105 | 110 | 115 | 120 |
| Fees for transportation of gas, bln. | 545 | 80 | 11,9 | 11,3 | 10,7 | 10,2 | 9,7 | 9,2 | 8,7 | 8,3 | 7,9 |
| | 579 | 85 | 12,6 | 12,0 | 11,4 | 10,8 | 10,3 | 9,8 | 9,3 | 8,9 | 8,5 |
| | 613 | 90 | 13,2 | 12,6 | 12,0 | 11,4 | 10,9 | 10,4 | 99,9 | 9,5 | 9,0 |
| | 647 | 95 | 13,9 | 13,2 | 12,6 | 12,0 | 11,4 | 10,9 | 10,5 | 10,0 | 9,6 |
| | 683 | 100 | 14,5 | 13,8 | 13,1 | 12,5 | 12,0 | 11,5 | 11,0 | 10,5 | 10,1 |
| | 715 | 105 | 15,0 | 14,3 | 13,7 | 13,1 | 12,5 | 12,0 | 11,5 | 11,1 | 10,6 |
| | 749 | 110 | 15,6 | 14,9 | 14,2 | 13,6 | 13,1 | 12,5 | 12,0 | 11,6 | 11,1 |
| | 783 | 115 | 16,1 | 15,4 | 14,7 | 14,1 | 13,6 | 13,0 | 12,5 | 12,0 | 11,6 |
| | 817 | 120 | 16,7 | 15,9 | 15,3 | 14,6 | 14,0 | 13,5 | 13,0 | 12,5 | 12,0 |

Table 1 shows that the internal rate of return (IRR) reaches 10% in two cases:

a) an increase in investment by 21%;

b) reducing the cost of gas transportation services by 17%.

Thus, the cost of gas transportation service is the most sensitive indicator of investment efficiency changes.

## 3. Financial risks

The group of financial risks can include such risks as interest rate and credit risks, liquidity risks, debt risks that emerge as a result of crisis and stagnation of national and world economy. Their importance is particularly experienced in the context of relationship between counterparties, primarily banking and real economy. Also either an increase in accounts receivable or refusal of a short-term loan from banking institutions may occur aimed to cover the insufficient amount of working capital.

Considering the effectiveness of the project [3] in the context of the two above-mentioned risks, the following deviations may be noted (table 2):

- increasing in duration of the project;
- decrease in volumes of transported gas;
- increase in operating costs;
- increase in the average cost of gas transportation;
- change in investment efficiency indicators.

**Table 2.** Basic technical and economic indicators of "Sila Sibiri" project.
Phase 1. "Chayanda-Lensk" section.

| Indicators | value |
|---|---|
| Cost-effectiveness planning horizon, years | 25 |
| The volume of transported gas, bln. $m^3$ | |
| - total | 728,3 |
| - maximum annual level | 32,6 |
| - average annual level | 29,1 |
| Length of the route, km. | 210,2 |
| Construction period, months. | 43 |
| Capital expenditures excluding VAT, total, mln.rub. | 110 000 |
| Operation costs, million. | |
| - total | 180 000 |
| - maximum annual level | 35 000 |
| - average annual level | 7 000 |
| Specific capital investments, mln. / km (excluding VAT) | 550 |
| The average annual cost of transportation of gas, rub./1000 $m^3$ | 250 |
| Investment efficiency indicator | |
| - net income, bln. | 350 |
| - net present value, bln. | 20 |
| - internal rate of return,% | 12,0 |
| - tariff for gas transportation, rub. / thousand $m^3$ | 935,3 |
| - tariff for gas transportation, rub. / thousand $m^3$ per 100 km. | 445,0 |
| - payback period, years | 10,4 |
| - payback period discounted, years | 18,5 |
| - discount rate,% | 10 |
| - maximum negative cash, bill. | 110 |
| - yield index | 1,23 |

## 4.Tax risks

Tax risks may be indirect and direct. Thus changes in oil and gas production tax, export duties on oil and oil products, excise duties may entail the above consequences as they affect the level of the financial viability of gas companies.

Note that under Paragraph 10, 11 Article 342 of the Tax Code of the Russian Federation [4] oil and gas production tax for gas-producing organization is planned to increase. Thus the oil and gas production tax for the year 2014 is 647 rubles and for the year 2015 - 679 rubles per 1 ton of gas condensate from all types of hydrocarbon reservoirs. For flammable natural gas production from all types of hydrocarbon reservoirs tax rate on mining for the year 2014 is 700 rubles, and for the year 2015 - 788 rubles per 1,000 cubic meters of gas. Discount rate, for taxpayers nonowning (during the whole tax perod) Unified Gas Supply System objects, i.e. not included in the structure of "Gazprom", is established in the amount of 0,673for the yaer 2014, and in 0,701 for the year 2015.

Analysis of gross and net margins, tax and non-tax payments clearly shows the presence of definite periods (figure 2). In the first phase rapid growth of revenues and overall costs almost with stable gross margin is observed (from 1 to 7 years from the beginning of construction in 2012). Second (from 8 to 17) and third (18 to 25 year) stages are characterizes by stable income, the cost begins to increase in the third stage, and gross profit, on the contrary, decreases. The results obtained characterize the project life cycle: growth stage (Phase 1), maturity (Phase 2) and aging (Phase 3).

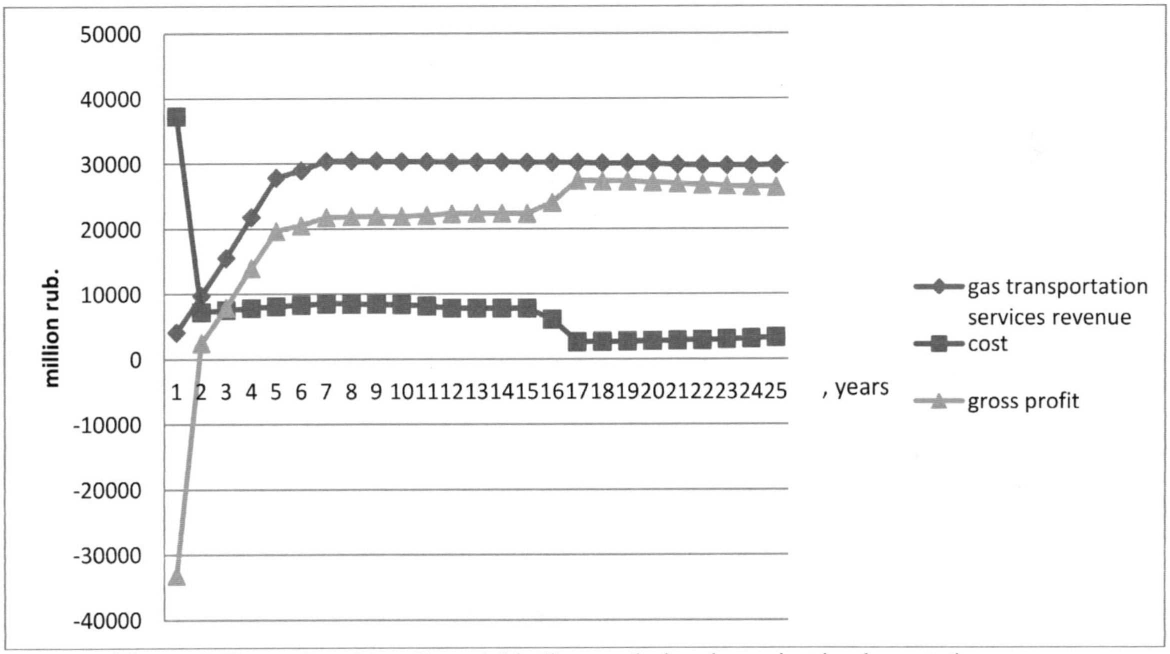


**Figure 2.** Dynamics of key financial indicators during the project implementation.

These phases as well in figure 3, illustrate the evolution of the net profit in relation to profit and income tax.

This is due to the following factors:

a) reaching of project capacity of Chayandinskoye gas condensate field and the subsequent introduction of new fields of natural gas and condensate;

b) terms of calculations, where certain parameters, such as gas transportation tariff, the income tax rate (20%), other taxes and the size of tax payments are taken as steady value;

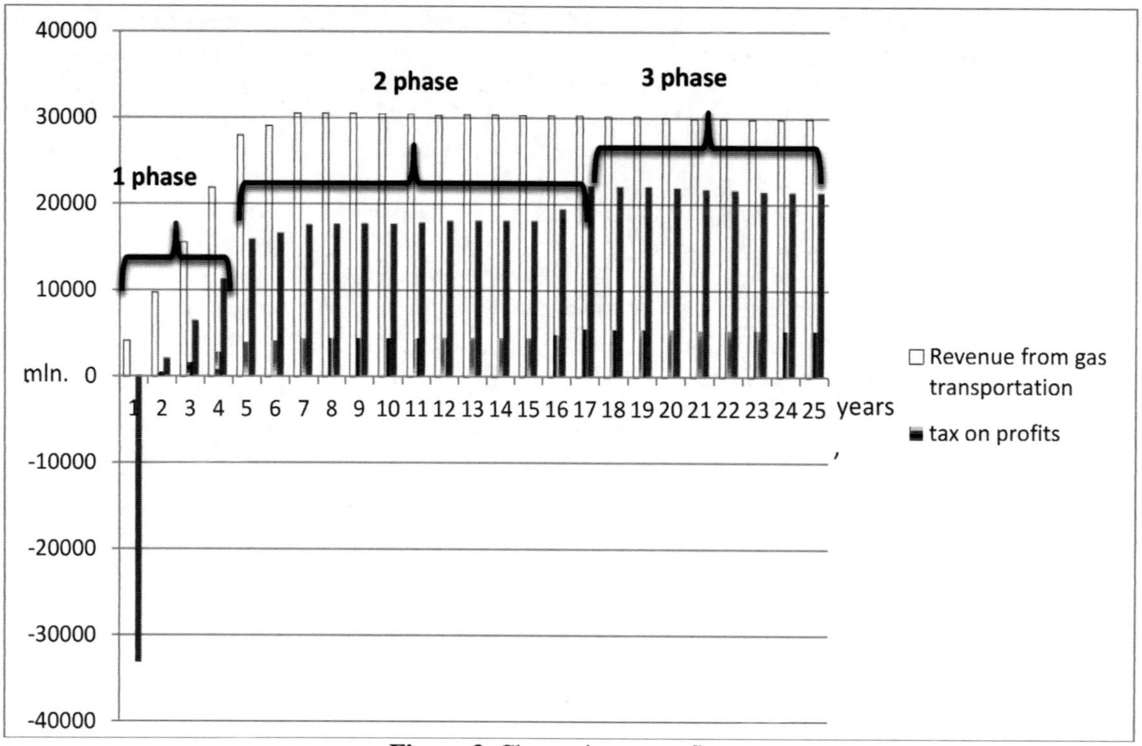


**Figure 3.** Change in net profit.

c) steadiness fo makroenergetic changes, for example, the lack of competition on the oil and gas market, in particular in the form of alternative energy sources.
Net income increased from 1% to 49% over the project duration from the first to the third period (the average value calculated as a proportion of net income from the income, according to figure 2).

As a consequence, a gradual increase in monetary income tax is observed (figure 4). Note that payments as a part of production costs (payment for environmental management, property tax) have a wavy trend, and, maximum values occur in the transitional period between the first and second stage.

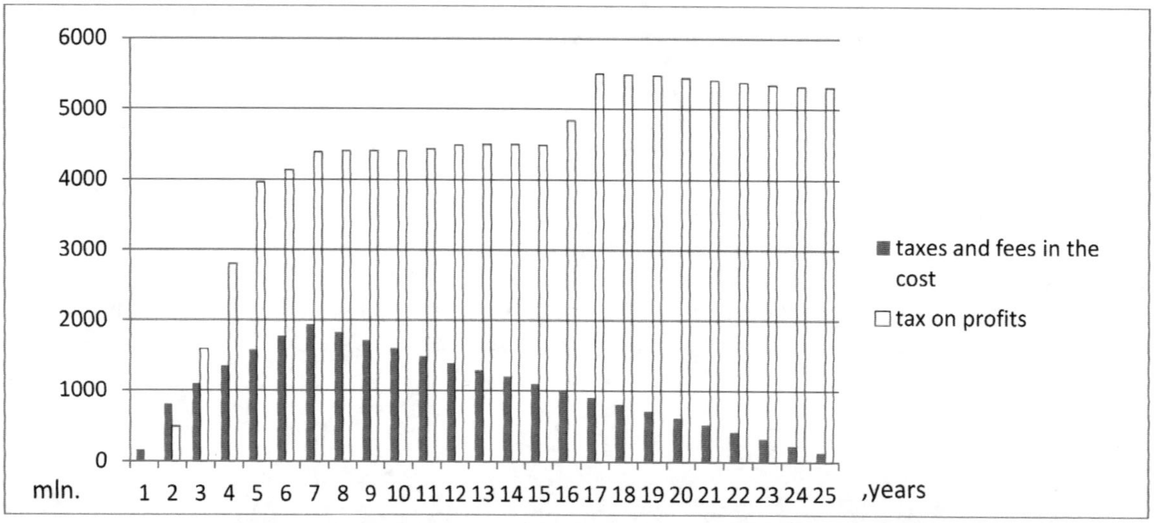


**Figure 4.** Dynamics of tax payments for the entire period.

The direct impact of the tax system is to change the tax law regarding tax exemptions. Thus, in accordance with Paragraph 3 of Article 380 of the Tax Code, the tax rate on business property tax in respect of pipelines and facilities, which are an integral part of the process, may not exceed 0.7 % for the year 2014, 1.0% - 2015, 1.3% - 2016, 1.6% - 2017, 1.9% - 2018 [2]. Note that the rate of property tax in the Russian Federation is, as a rule 2.2% - maximum permissible value.

Thus, this exemption significantly affects the amount of property taxes in production cost, and hence the dynamics of the cost itself.

## 5. Conclusion

Dual influence of the above described risks, both positive and negative, requires constant monitoring and operational control over the implementation of the project by the company's management in order to maintain and improve the effectiveness of the project.

## References

[1] Chukhareva N V, Sharf I V and Tikhonov T V 2013 Socio-economic factors of the gas transportation system of the Republic of Sakha (Yakutia) *Oil and gas : the electron. scientific. journal* **6** 416-431

[2] Main provisions of the draft Energy Strategy of Russia until 2035 *Ministry of Energy* http://minenergo.gov.ru/documents/razrabotka/

[3] Guidelines for evaluation of investment projects. ratified. RF Ministry of Economy, Ministry of Finance and the State Construction Committee of Russia June 21, 1999 **477** http://base.consultant.ru/cons/cgi/online.cgi?req=doc; base = LAW; n = 28224

[4] The Tax Code http://www.consultant.ru/popular/nalog2/3_8.html # p13551

# Strength properties of peat in Tungol Field (Tomsk Oblast)

**V V Kramarenko [1], A V Kolesnikova [1], I A Lappo [2] and I A Matveenko[1]**
[1] Tomsk Polytechnic University, Tomsk, Russia
[2] Tomsk subsidiary of Avers, Tomsk, Russia

E-mail: mia2046@yandex.ru

**Abstract.** The article is devoted to the study of non-drained peat strength in the site of Tungol oil field determined by the core sampling with the help of impeller. The authors present the research results in peat composition, physical, and strength characteristics, changes in properties with depth, revealed relationships between the indicators, the regression equation obtained on their basis for prediction of composition and characteristics. Based on the cluster analysis including all studied indicators the peat classification is performed, three main classes are distinguished, typical values of characteristics as well as recommendations for using peat of every class as construction foundation and corresponding load bearing capacity are given.

## 1. Introduction

Data on structural behavior of peat soil is applied in calculation of deposit carrying capacity and vehicle flotation at arranging deposit facilities, in this context the relevance of the research in this field is doubtless.

The study of peat shear strength characteristics and peaty soil was carried out in works by L.S. Amaryan, E.T. Bazin, Kosovo V.I., N.N. Morareskul, V.N. Bronin, P.A.Konovalov, G.V. Sorokina as well as Golebiewska A., Kanmuri, H., Kato, M., Suzuki O., Hirose, E., Landva, A.O, Ogino, T., Oikawa, H., Tsushima, M., Mitachi, T , Miyakawa, I, Radforth, J. , Sodha V.G., Helenelund K.V. [1, 4, 8-11, 14]. Great contribution to the research of peat in Western Siberia was made by M.K. Baryshnikov, M.I. Neishtadt, G.Ya. Bronzova, S.N. Tyuremnov, V.M. Yeliseyev and other scholars.

The peculiarities of Western Siberian peat have stipulated the development of regional classifications in botanical composition. Distinguishing new species in botanical classifications is explained by their heritable resistance, geographic and geomorphologic location, and peat qualitative features. Such attempts were made for both definite regions and the entire territory of Western Siberia.

Wetland zoning in Western Siberia for different purposes was performed by S.N. Tyuremnov, N.Ya. Katz, M.I. Neishtadt, M.F. Loginov, L.V. Shumilov, O.L. Liss, N.A. Berezina, G.G. Kulikova, Yu.A. L'vov, Ye.A. Romanova, V.G. Matukhin, L.S. Mikhant'eva, V.M. Altukhov and others.

In Tomsk Oblast the research in vegetation and wetland conditions was made by the workers of Tomsk State University – Yu.A. L'vov, Ye.D. Lapshina, V.A. Bazanov.

The research is based on the data obtained in the course of engineering surveys made in Tudol field in 2013 by "Avers 1" Ltd company [12]. The first geological engineering surveys on the investigated site were carried out in 1947-1948 by the Order of the USSR Government on Oil and Gas Exploration as well as Planned Geological Surveys Performance on West Siberian Plain.

In recent years the large-scale geoengineering survey has been performed in order to equip the local construction sites and oil exploration wells, as well as the sites for gathering pipeline construction, etc.

IOP Conf. Series: Earth and Environmental Science **21** (2014) 012020 doi:10.1088/1755-1315/21/1/012020

The designed site is located in Alexandrovskoye region, Tomsk Oblast, 145 km eastward from the regional center – Alexandrovskoye settlement.

The regional climate is sharply continental, with long cold winters, late spring and early autumn. The maximum precipitation (75%) occurs in April-October (436 mm), the minimum – in February. Some days nearly month precipitation can fall down. The maximum one-day rainfall takes place in warm season and can reach 62 mm. Taking into account the average monthly negative temperature in the area of Alexandrovskoye meteorological station the calculated specified depth of seasonal peat freezing is 0.75 m.

The site is located in the central part of West Siberian Plain, which is a large platform composed of Paleozoic folded basement formations. Basement formations are overlaid with sedimentary mantle of Mesozoic-Kainozoic deposition. The upper part of the section is made up of Middle Quaternary lacustrine and alluvial deposit of Suzgunskaya Suite (laQIIsz) and modern cap biogenic formations (bQH). According to the previous data, the thickness of Suzgunskaya Suite on the explored area can reach 28 m.

The geomorphologic and lithologic conditions enable interception and swamp formation on the plain interfluve area (figure 1).

**Figure 1.** Water saturation of territory Tungol field

Therefore, the explored area type is referred to watershed raised moor with specific vegetation: sphagnum moss, dwarf birch, ledum (figure 2). In terms of botanic-geographic zoning vegetation corresponds to the subzone of Middle taiga of Western Siberia. The prevailing vegetation type is mixed forest and peatland.

Among the natural processes on the investigated area, the most dangerous are high rate of water-logging process, peatland microrelief transformation as a result of peat formation, forest and peat fires.

**Figure 2.** Typical vegetation on the area of well #5, Tungol field

The area is located within the Irtysh-Yenisei hydrogeologic basin. The bog water breaks up just under frozen layer. The occurrence depth is less than 3 m. Hemic and sapric peats are water bearing. The thickness of boggy deposits within the area can reach 1.0-4.0 m, that of the access road is 2.0-6.4 m. Peat is mainly sphagnum, grass-moss with grass inclusions (mostly sedge), without and with wooden culls up to 5-10% including sunken trunks.

## 2. Methods

Work was carried out in accordance with the methods provided in existing regulatory standards [5, 6, 7]. Peats were sampled from 16 wells, the grain size was determined, the density ($\rho$), the density of the solids ($\rho_s$), the density of the dry soil ($\rho_d$), water content (w), Atterberg limits for clay soils ($w_L$ and $w_p$); the parameters of the state, such as the void ratio (e), porosity, saturated limit, plasticity limit and consistency index for clay soils, ash content ($D_{as}$) and the degree of decomposition were obtained by the gravimetric method ($D_{dp}$) for peat.

Field tests for determination of undrained shear strength for soil with natural ($c_u$) and disturbed structure ($c_{ud}$), as well as sensitivity ($S=c_u/c_{ud}$) were carried out by soil using vane shear test, according to [5, 7; analogues 2, 3]. The field soil test was performed by shear-meter impeller (figure 3).

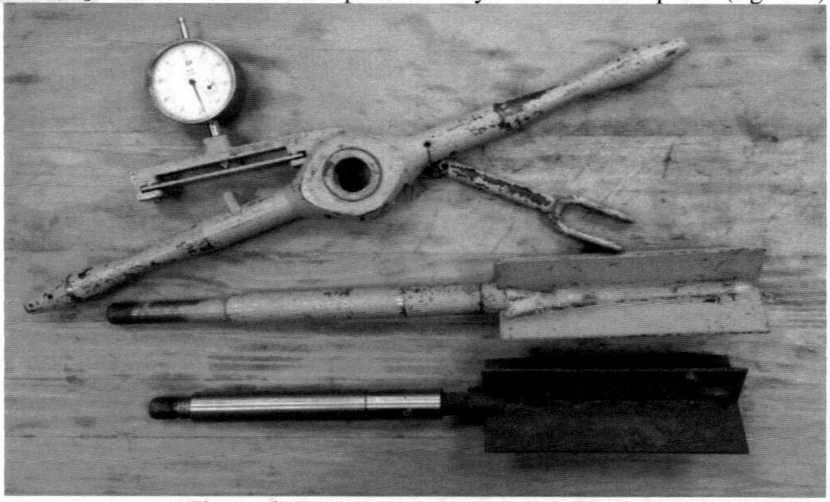

**Figure 3.** The general view of the impeller

IOP Conf. Series: Earth and Environmental Science **21** (2014) 012020 doi:10.1088/1755-1315/21/1/012020

## 3. Results and discussion

According to the field description and laboratory data, there distinguished 2 kinds of peat presenting different engineering-geologic features (table 1).

**Table 1**. Characteristics of Engineering-geologic Features.

| | |
|---|---|
| 1 kind | Hemic, medium-ash peat. Light-brown, brown color, with wooden inclusions. Thickness is up to 3.3 m. |
| 2 kind | Sapric, medium-ash peat. Brown, dark-brown color, with wooden inclusions. Thickness is up to 6.4 m. |

In the course of peat laboratory research its physical features (density, water content) as well as ash content, decomposition degree, botanical composition have been determined. The data obtained were classified and processed with Statistica software, the correlation analysis for all indicators was made (table 2), as a result of which the maximum dependence factors were stated.

**Table 2**. Correlation coefficients.

| | Depth, h | Undrened strength , $c_u$ | Undrained strength (disturbed), $c_{ud}$ | Sensiviy S | Water content, w | Ash content $D_{as}$ |
|---|---|---|---|---|---|---|
| Undrened strength , $c_u$ | 0,32 | 1,00 | | | | |
| Undrained strength (disturbed), $c_{ud}$ | 0,11 | 0,89 | 1,00 | | | |
| Sensitivity, S | -0,59 | -0,07 | 0,35 | 1,00 | | |
| Water content, w | -0,83 | -0,51 | -0,20 | 0,48 | 1,00 | |
| Ash content, $D_{as}$ | 0,04 | -0,28 | 0,24 | 0,76 | 0,08 | 1,00 |
| Degree of decomposition, $D_{dp}$ | 0,44 | 0,18 | -0,07 | -0,55 | -0,26 | -0,71 |

The degree of peat decomposition increases with depth (r=0,44), this is due to the composition of sediments, age, and its drainage during formation, which promotes decomposition. The sphagnum peat has the maximum decomposition degree. It is explained by the release of organic acids by sphagnum mosses and presence of sphagnole (phenol-based glycoside) that condition their biochemical stability having a bactericide effect and retaining the remains of bog plants. The negative correlation coefficient shows that decomposition degree decreases with water content growth, which is a typical dependence for sphagnum mosses.

The results of the correlation analysis showed that the water content decreases with depth, which is also determined by the presence of sphagnum moss in the upper section. Since natural peat density varies from 1.0 to 1.09 g/cm3, the relations turned out to be insignificant. The closest correlations were established between $c_{ud}$ and $c_u$ (table 2).

The ash content of peat does not basically exceed 5%, only several samples have increased the value of that parameter more than 20%. For sphagnum peat types the minimal values of ash content is typical that grows with the rise in water content.

Correlation analysis has shown that decomposition degree decreases with increasing density, since minimal density is typical for grass-moss peat type. Peat decomposition rate grows with depth that is connected with drainage contributing to decomposition.

Undrained shear strength grows with increase in density and occurrence depth in natural state. The value of sensitivity factor decreases with depth, therefore, the strongest peat occurs at the surface as it is less decomposed. For upland bogs a rather strong layer – peat top spit – is typical, which presents a surface vegetation cover of living and dead mosses and grass forming a springing floor of bog tree and bush roots unaffected by peat formation and comparatively easily separated from lower peat layer.

As a result the closest relationship among the properties of the following characteristics was revealed, which can be used for similar peat deposits:

$c_{ud} = 0,006\ln(c_u) + 0,037$ ($r^2 = 0,82$),

$c_u = -0,00\ln(w) + 0,069$ ($r^2 = 0,26$),

$S = -0,10\ln(h) + 0,697$ ($r^2 = 0, 28$),

$w = 1053\ h^{-0,35}$ ($r^2 = 0,64$),

$\rho = 0,5574 w^{0,0894}$ ($r^2 = 0,86$ ),

$p = 1,039\ h^{-0,03}$ ($r^2 = 0,60$),

$D_{dp} = 40,73 h^{0,286}$ ($r^2 = 0,20$),

$D_{dp} = 82,72 D_{as}^{-0,42}$ ($r^2 = 0,66$),

$D_{as} = 3,147 h^{0,13}$ ($r^2 = 0,21$).

Cluster analysis allows distinguishing 3 main peat groups based on all revealed features. The groups are characterized in detail, the average, minimal and maximal values are calculated (table 3), the building types and road capacity types are defined in terms of shear stress test and L.S. Amaryan's classification [1], and the conditions of pipelaying are stated according to "The Instruction of Road Design for Oilfield of Western Siberia" [13].

**Table 3.** Classification of peats of Tudol field.

| Characteristics | Group of the peat | | |
|---|---|---|---|
| | **A** | **B** | **C** |
| Undrened shear strength $c_u$, MPa | 0,007 | 0,010 | 0,014 |
| Undrained shear strength for soil with disturbed structure $c_{ud}$, МПа | 0,003 | 0,007 | 0,008 |
| Sensiviy S | 0,40 | 0,75 | 0,57 |
| Water content W, % | 853 | 987 | 782 |
| Density ρ, g/cm$^3$ | 1,01 | 1,05 | 1,04 |
| Ash content $D_{as}$, % | 2,35 | 5,94 | 12,25 |
| Degree of decomposition $D_{dp}$ % | 61,5 | 52,0 | 56,5 |

## Summary

According to the results obtained, group A is the least strong. It is referred to the 3-d building type – soil is squeezed off at load transmission due to insufficient strength in natural state and insufficient strengthening at thickening. In terms of pipelaying condition classification the type is characterized in the following way: bogs that are completely filled with baken peat up to mineral bed. Peaty soil is a reliable foundation for pipelines. Trafficability is designed in terms of shear strength test results (undrained strength, $c_u$, kPa). In the present instance the passibilty condition are specific, in which undrained deep peat deposits with ponds and weedy pools are typical. Organic slime is deposited on the bottom of ponds or at the depth of less than 2m. Floating mats are of different thickness. In this condition the passage of tracked drifting transport or arch-tired vehicles is possible. The road is available in winter, but not in warm winter.

Group B is referred to the 2-nd building type – soil not possessing sufficient strength in natural state, but at quick load transmission from banked earth it is squeezed off, whereas at slow load transmission it thickens and hardens to an extend that it is not squeezed, but compacted. In terms of pipelaying conditions it is a bog completely filled up with stable consistency peat up to the mineral bed. Peat can be used as a carrying foundation for pipelines. The passibilty conditions are hard: low-drained and undrained peatland filled up with peat, afforested. Organic slime underlays on the bottom of ponds or at the depth of less than 2 m. Floating mats are of different thickness. The passage of special drifting transport with specific pressure of less than 15 kPa is possible. It is necessary to remove trees before it.

IOP Conf. Series: Earth and Environmental Science **21** (2014) 012020 doi:10.1088/1755-1315/21/1/012020

Among the three groups described here, group C is the most stable. It is referred to the 1-st building type: soil possesses sufficient strength in natural state and at load transmission from banked earth it is compressed regardless of load transmission rate. In terms of pipelaying conditions it is a bog completely filled up with stable peat up to mineral bed. Peat can be used as a carrying foundation for pipelines. In terms of passage conditions the group is referred to the average one: previously drained peatland with removed tree stratum, low-drained small deposits of depth less than 1.5 m. The passage of tracked drifting tractors and off-road vehicles as well as arch-tired transport is possible.

Thus, we have characterized the peat strength features of Tungol oil field, revealed the correlations between properties (density, water content, ash content, and decomposition rate) on the basis of which we obtained the regression equations. As a result of cluster analysis their classification was performed. 3 groups were distinguished. The research results can be applied for the peat deposits of similar composition.

## References

[1] Amaryan L S 1969 *Strength and deformation of peat soils.* (Moscow: Nedra) p 192
[2] ASTM STP 1014 on Vane Shear Strength Testing in Soils
[3] Eurocode ENV 1997-3, part 3
[4] Golebiewska A. 1983 *Vane testing in peat* Proc. 7th Danube ECSMFE Kishinev **1**113–117
[5] GOST 20276-99 Field methods for determining the strength and strain characteristics
[6] GOST 30416 2012 Laboratory testing. General requirements
[7] GOST 30672-2012 Soils. Field testings. General requirements
[8] Hanrahan E. T. 1954 An investigation of some physical properties of Peat. *Geotechnique* **1**(3) 108-123
[9] Huat B B K 2004 *Organic and peat soils engineering* 119 -120
[10] Kramarenko V V and Savichev O G 2014 *Engineering-geologic survey: research methods of peat soil* (Tomsk: Tomsk Polytechnic University Press) p 281
[11] Landva A O 1980 Vane testing in peat *Canadian Geotechnical Journal* **17**(1) 1-19
[12] Report completed geotechnical investigations on arrangement Tungolskogo field, LLC" Avers 1 "Tomsk Branch 2013 p 113
[13] VCN 26-80 The Instruction of Road Design for Oilfield of Western Siberia p 68
[14] Helenelund K V 1967 Vane tests and tension tests on fibrous peats, *Proc. Getech. Conf.*, Oslo **1** 199–203

IOP Conf. Series: Earth and Environmental Science **21** (2014) 012021 doi:10.1088/1755-1315/21/1/012021

# Parameter Calculation Technique for the Waste Treatment Facilities Using Naturally-Aerated Blocks in the Bog Ecosystems

**K V Akhmed-Ogly, O G Savichev[1], O G Tokarenko, E Yu Pasechnik, M V Reshetko, N G Nalivajko and M V Vlasova**
Tomsk Polytechnic University, Tomsk, Russia

E-mail: [1]osavichev@mail.ru

**Abstract**. Technique for the domestic wastewater treatment in the small residential areas and oil and gas facilities of the natural and man-made systems including a settling tank for mechanical treatment and a biological pond with peat substrate and bog vegetation for biological treatment has been substantiated. Technique for parameters calculation of the similar natural and man-made systems has been developed. It was proven that effective treatment of wastewater can be performed in Siberia all year round.

## 1. Introduction

Effective wastewater treatment is one of the key objectives in solving the problem of preservation and restoration of the disturbed environment components. A lot of attention is paid to the solution of this problem, thus, new methods of treatment are being developed, layout of the waste treatment facilities and their integration in the natural and man-made complex [1, 2] are being optimized as well, though, nowadays the last aspect is insufficiently elaborated both from theoretical and practical point of view. As a result, quite expensive waste treatment facilities are being built in Siberia, they don't always provide a necessary level of wastewater treatment from organic and biogenic substances (probability of emergencies in the work of biological artificially-aerated treatment facilities in severe weather conditions is rather high), while regional natural features which enable natural aeration of the wastewater and reduction of the pollutant concentration using aquatic and bog vegetation are not properly used. Special work is surely carried out in this direction but mostly in the European Union, the USA, also in the European part of the Russian Federation [2]. Similar investigations with regard to Siberian conditions are carried out in Tomsk State University, but, unfortunately, the results obtained are rather empirical and give no way to methodology for design of the waste treatment facilities with naturally-aerated blocks and adsorption of pollutants with hydrophilic vegetation.

Taking this into consideration the investigations concerning the development of methodology for creation of the natural and man-made system intended for nature protection purposes, including treatment facilities integrated in the natural biogeochemical cycle, have been carried out over a period of years in Tomsk Polytechnic University. In particular, a technique (adapted to the conditions in Siberia) for domestic wastewater treatment in the small residential areas and oil and gas facilities in the shallow biological ponds with peat substrate and bog vegetation by means of sorption of pollutants on the peat particles, their adsorption by vegetation and organic-matter degradation was offered earlier [3]. Results of the further stage of investigations, i.e. technique for design of such facilities, are presented below.

## 2. Domestic Wastewater Treatment Objectives and Constrains to Treatment

Any wastewater treatment technology must be adapted to the chemical composition and treatment conditions. To solve this problem data of ecological monitoring in the territory of Tomsk Oblast were analyzed, as a result, the following conclusions were made. Firstly, domestic wastewater which comes to the treatment facilities of the region contains $NO_2^-$, $NH_4^+$, $PO_4^{3-}$, Si, Fe, Cu, Zn, Pb, phenols, oil minerals, surface active substances, organic substances in the amount higher than maximum permissible based on the 5-day biochemical oxygen demand test with permanganate and dichromate as the oxidants (table 1). Increased concentration of Si and Fe can be explained by the fact that underground water is high in Si and Fe content (iron is contained both in the water being used and discharges of the facilities for iron removal; silicon comes mainly with domestic water being in use), water contamination during conditioning and transportation of domestic water causes high Pb concentration. High content of organic substances in the wastewater and products of their degradation, nitrogen and phosphorus compounds, are quite regular. Cu concentration (and also Zn and Fe) that is higher than maximum permissible for fisheries streams is commonly found in the underground, river, lake and bog waters of the region, so we can't associate it with certain anthropogenic factors.

Secondly, as a result of treatment, the content of suspended substances, dissolved $HCO_3^-$, $SO_4^{2-}$, $NH_4^+$, Fe, Al, Cu, Zn, Pb, phenols, surface active substances, oil minerals, organic substances based on the 5-day biochemical oxygen demand test ($BOD_5$) with permanganate (PO) and dichromate (DO) as the oxidants in the wastewater decreases, concentration of $PO_4^{3-}$ and Si is without considerable change, dry residue and sum of major ions of $\Sigma_{pi}$, $Na^+$, $K^+$, $Ca^{2+}$, $Mg^{2+}$, $Cl^-$, $NO_2^-$, $NO_3^-$ increase. Maximum permissible concentration (MPC) of $\Sigma_{pi}$, $Na^+$, $NO_2^-$, $NH_4^+$, $PO_4^{3-}$, Si, Fe, Cu, phenols, oil minerals, organic substances based on $BOD_5$, PV, DV is exceeded on the average. In case of $Na^+$ it can be explained by the excess decontamination of wastewater with sodium hypochlorite, in case of $\Sigma_{pi}$, $NO_2^-$, $NH_4^+$, $PO_4^{3-}$ (and other substances with lower concentration than MPC) exceeded maximum permissible concentration is caused by the inefficient removal of organic substance degradation products from wastewater, it suggests that advanced waste treatment is necessary. In practice silicon is not removed from wastewater, its concentration is almost the same before and after treatment of wastewater and underground waters from the aquifers in use. Removal of oil minerals, surface active substances, organic substances based on $BOD_5$, PV, DV is rather effective (degree of treatment is higher than 30…50%), though, it doesn't allow to rich water quality target.

Thirdly, metal concentration in the wastewater tends to decrease in the process of precipitation of suspended substances. This indicates the significance of the given stage of wastewater treatment. Fourthly, many substances in the domestic wastewater are components which are necessary for forest community development in taiga zone of Siberia, particularly in the area of domed oligotrophic bog development [4]. In particular, $NO_3^-$, $PO_4^{3-}$ and easily oxidizable organic substances (based on $BOD_5$) can be referred to such substances for the wetlands of Tomsk Oblast. Fifthly, construction of expensive artificially-aerated treatment facilities operated by qualified people in the administrative areas of the Siberian and Urals Federal Districts which are characterized by the bogginess of the territory, severe climate and vast distance from cities doesn't provide the required quality of wastewater treatment.

Thus, nowadays money is spent over balance (in relation to the results achieved) on domestic wastewater treatment from substances which are in short supply in the wetlands, the factor that affects development of the forest communities in the vast territory of Tomsk Oblast and other regions. Therefore, it is reasonable and even necessary to consider possibility of bog feeding with the treated wastewater or the use of bogs or their zones for treatment of the domestic and storm wastewater from the small residential areas and rotation villages. Considering restrictions of the environmental legislation it enables to improve forestry in Western Siberia in response to the peat accumulation increase by 1 mm per year and to reduce costs on housing and utilities sector without environmental damage. Key objectives for the wastewater treatment in such natural and man-made systems are the following: 1) precipitation of the suspended substances as an important factor for reduction of toxic trace substance concentration (Pb, Zn, etc.); 2) reduction of concentration of organic and biogenic substances as they are assimilated by bog vegetation and absorbed by peat. Giving due consideration to the regulatory requirements a typical scheme for domestic wastewater treatment involves mechanical treatment in a settling tank and biological treatment in a biological pond with peat substrate and bog vegetation.

**Table 1.** Average values of hydrochemical and physicochemical parameters of domestic wastewater and storm water run-offs in the territory of Tomsk Oblast before and after treatment.

| Parameter | Units of measure | Wastewater before and without treatment | | Wastewater after treatment | | Storm wastewater | | $MPC_{fh}$ | $MPC_{dh}$ |
|---|---|---|---|---|---|---|---|---|---|
| | | *A* | *N* | *A* | *N* | *A* | *N* | | |
| $\Sigma_{pi}$ | $mg/dm^3$ | 808.2 | 12 | 1179.7 | 17 | – | – | – | 1000.0 |
| $NO_2^-$ | $mg/dm^3$ | 0.247 | 82 | 1.178 | 139 | 0.775 | 35 | 0.080 | 3.300 |
| $NH_4^+$ | $mg/dm^3$ | 45.537 | 82 | 36.068 | 139 | 5.684 | 34 | 0.500 | 1.930 |
| $BOD_5$ | $mgO_2/dm^3$ | 175.63 | 28 | 64.55 | 90 | 31,12 | 35 | 2.00 | 2.00 |
| DO | $mgO/dm^3$ | 278.10 | 53 | 164.80 | 102 | 81.46 | 35 | – | 15.00 |
| Phenols | $mg/dm^3$ | 0.128 | 77 | 0.013 | 136 | 0.008 | 23 | 0.001 | 0.001 |
| Oil minerals | $mg/dm^3$ | 1.579 | 65 | 0.232 | 119 | 0.388 | 32 | 0.050 | 0.300 |

Note: *A* – average value, *N* – number of samples, $\Sigma_{pi}$ – sum of major ions; $BOD_5$ – 5-day biochemical oxygen demand; DO – dichromate oxibility.

### 3. Calculation of Parameters of the Mechanical Treatment Block

The main mechanical treatment is carried out in a settling tank, where suspended substances are precipitated. Types of the settling tank are chosen according to engineering analyses (mainly depending on the land topography and peat bed, water inflow and its distribution in the area, the presence of confining layer and its power): an excavated pond with bunding; an excavated pond with earth-fill dam and bunding; a pond with bunding and a dam. The bottom, bunding and the pond dam should be constructed with low permeability soils or using waterproofing [3].

Parameter calculation of the mechanical treatment unit is performed according to [5]. Total flow of wastewater which gets into the wastewater treatment plant (WTP) is determined by the customer (water consumer) depending on the population size (number of employees at the site), current standards for wastewater disposal and irregularity of wastewater inflow. Amount of water pollutants per one resident $m_{sr}$ for determination of their concentration in the domestic wastewater of the treatment facilities being designed is taken according to [5], and concentration of pollutants $S$ ($g/m^3$) is calculated by the following formula:

$$S = m_{sr} \cdot N_p / q_m,$$

where $N_p$ is a number of residents, people; $q_m$ is a calculated annual average wastewater flow rate with account for irregularity of inflow, $m^3$/day; $m_{sr}$ – amount of water pollutants per one resident.

Throughput of a single settling basin $q_{set}$ ($m^3$/hr) is estimated according to the predetermined dimensions of the facility and the required effect of wastewater clarification using a formula for horizontal-flow settling basin:

$$q_{set} = 3{,}6 \cdot K_{set} \cdot L_{set} \cdot B_{set}(u_0 - v_{tb}),$$

where $K_{set}$ is the factor of utilization of the flow part of the settling basin; $L_{set}$ is the total length of the settling basin, m; $B_{set}$ is the width of the horizontal-flow settling basin determined by correlation $B_{set} \approx (2\ldots5) \cdot H_{set}$, where $H_{set}$ is the depth of the flow part (values $L_{set}$ and $H_{set}$ are determined by way of selection corresponding to the predetermined clarification effect and bottom slope of the settling basin); $v_{tb}$ is the turbulence component (mm/s) of the flow velocity $v_w$ which is determined by the maximum daily wastewater flow and the average section area. Hydraulic coarseness of particles $u_0$ is determined by formula:

$$u_0 = 1000 \cdot H_{set} \cdot K_{set} / t_{set} \cdot \left( \frac{H_{set} \cdot K_{set}}{h_1} \right)^{n_2},$$

where $t_{set}$ is the duration of settling (sec), corresponding to the predetermined purification effect and obtained in a laboratory cylinder in a layer $h_1$; $n_2$ is an exponent dependent on agglomeration of the suspended matter in the course of its settling [5]. The quantity of sediment $Q_{mud}$ (m$^3$/hr), separated by a settling basin, is determined from the concentration of the suspended matter in incoming ($C_{en}$) and clarified water ($C_{ex}$):

$$Q_{mud} = q_w \cdot (C_{en} - C_{ex})/(100 - p_{mud}) \cdot \gamma_{mud} \cdot 10^4,$$

where $q_w$ is the flow rate of waste water, m$^3$/hr; $p_{mud}$ is the moisture content in the sediment, %; $\gamma_{mud}$ is the density of the sediment, g/cm$^3$ [5]; values $p_{mud}$ and $\gamma_{mud}$ are taken according to the comparable facilities or (as a first approximation) the published data.

## 4. Calculation of Parameters for the Biological Treatment Block

Biological treatment is performed annually in the block which is basically a bog (or a part of it) and similar in function to a shallow biologic pond with hydrophilic vegetation. The main difference of such a block (hereafter biologic pond with peat substrate BPPS) from an ordinary shallow pond is in using regional bog vegetation and additional wastewater treatment by sorption of pollutants on the particles of the peat substrate. Wastewater from the settling basin gets through a discharge sluice into a biologic pond all year round. Design parameters of the biologic pond with peat substrate are calculated considering requirements [5] for organic waste content reduction based on $BOD_{tot}$ determining minimum permissible parameters. The latter can be increased as a result of calculation of the degree of treatment from other substances which are contained in wastewater. Thereafter, there are two stages of parameter determination of BPPS, calculation algorithm is applied as follows. Firstly, condition specified in [5] is verified, according to it wastewater being treated in the naturally-aerated biological ponds mustn't have $BOD_{tot}$ higher than 200mg/l. If this condition is not satisfied additional sections of settling tanks are designed. The time of stay of wastewater $t_{lag}$ (days) in the naturally-aerated pond is determined from the formula:

$$t_{lag} = \frac{1}{k \cdot K_{lag}} \cdot \sum_{1}^{N-1} \lg \frac{L_{en}}{L_{ex}} + \frac{1}{k' \cdot K'_{lag}} \cdot \lg \frac{L'_{en} - L_{fin}}{L'_{ex} - L_{fin}},$$

where $N$ is the number of successive stages of a pond; $K_{lag}$ is the factor of utilization by volume of each stage of the pond; $K'_{lag}$ is the same, but for the last stage of the pond; $L_{en}$ is the $BOD_{tot}$ value of the wastewater coming into a given stage of the pond; $L'_{en}$ is the same, but for the last stage of the pond; $L_{ex}$ is the $BOD_{tot}$ value of the wastewater exiting from the given stage of the pond; $L'_{ex}$ is the same, but for the last stage of the pond; $L_{fin}$ is the residual $BOD_{tot}$ value determined by the processes inside the pond and taken 2 mg/dm$^3$; $k$ is the constant of the oxygen consumption, day$^{-1}$; $k'$ = -0.07 day$^{-1}$ at a water temperature 20°C [5]. Minimal total area of a biological pond with peat substrate $F_{lag}$ (m$^2$) is determined from the formula:

$$F_{lag} = q_w \cdot C_a \cdot (L_{en} - L_{ex}) / K_{lag} \cdot (C_a - C_{ex}) \cdot r_a,$$

where $q_w$ is the flow rate of waste water, m$^3$/day; $C_a$ is solubility of oxygen (natural aeration); $C_{ex}$ is the concentration of oxygen that shall be maintained in water exiting the pond, mg/L; $r_a$ is the value of atmospheric aeration in shortage of oxygen equaling 1 taken at 3…4 g/(m$^2$·day). After this the depth of a pond $H_{lag}$ (m) is determined from a formula (1), the working depth of a pond mustn't exceed: with $L_{en}$ above 100 mg/dm$^3$ – 0.5 m, with $L_{en}$ up to 100 mg/dm$^3$ – 1 m; for the deep treatment ponds with $L_{en}$ from 20 to 40 mg/dm$^3$ – 2 m, with $L_{en}$ up to 20 mg/dm$^3$ – 3 m [5].

$$H_{lag} = K_{lag} \cdot (C_a - C_{ex}) \cdot r_a \cdot t_{lag} / C_a \cdot (L_{en} - L_{ex}). \quad (1)$$

The obtained result is increased by 0.5 m (taking into account wintertime) and compared to the calculated maximum water level in a certain BPPS determined according to the maximum of these two values is used to determine the height of the bunding from inorganic soil around a selected part of the bog.

A channeling facility, which forms nonlinear flow of wastewater ($r$ in length) with limited meandering for the treatment duration increase, is constructed inside a biological pond using inorganic soil, sludge from the settling basin, tree vegetation remained in the area of biological pond during construction [3].

At the second stage permissible concentration of organic matter based on $BOD_5$ and other substances is determined from the formulas:

$$C_{w,\lim} \le C_b \cdot (\lambda \cdot f(C) - \lambda + 1) + n \cdot f(C) \cdot Z_\alpha \cdot \sigma \cdot \sqrt{\frac{2}{M}}, \quad (2)$$

$$f(C) = \exp\left( k_C \cdot r^2 / 2 \cdot \left( D - \frac{q_w}{\varphi \cdot 0{,}5 \cdot h_{ap}} \right) \right), \quad (3)$$

where $C_{w,lim}$ is a permissible substance concentration in the peat water in the control section, which is situated at a distance $r$ from wastewater discharge; $\lambda$ is a reciprocal dilution of the wastewater; $Z_\alpha$ is a critical value of Student's coefficient at significance level $\alpha$=0.05; $M$ is a series length of the hydrochemical observations at the bog under study or similar bogs where background substance concentration is calculated $C_b$; $\sigma_C$ is a mean square deviation of the substance concentration; $k_C$ is a self-purification coefficient; $D$ is a dispersion coefficient, $m^2$/s; $h_{ap}$ is an active layer of bog, м; $\varphi$ is a sector angle of distribution of impurities; $C_b$ is a background substance concentration in the peat water for the edge or homogeneous site of the main part of the bog where influents get; $q_w$ is in $m^3$/s. Equations (2, 3) are proved in [6]. If $C_{w,lim}$ is higher that permissible value (maximum possible or background depending on their relation and prevailing sources of substance influent by the method of hydrochemical balance analysis [6], value $r$ is increased in formula (3) and, consequently, area of BPPS expands till the required result is obtained. It must be mentioned that relation between values $F_{lag}$ and $r$ is linear only when $r$ is the length of BPPS.

## 5. Waste Management

Treated effluent is discharged from biological ponds in the water disposal system, where it is disinfected using a dosing tank with disinfectant solution or by means of periodic back filling of BPPS adjacent to the treated effluent discharge with lime. After disinfection the treated effluent is fed into a body of water. Sludge is removed from the settling pond in the frost-free period of the year (1...2 times) if and when it is accumulated. Sludge is moved inside the biological pond to the channeling facility. Sludge can be hauled away to the recultivated sites of the disturbed lands and waste landfills. Sludge is disinfected by mixing with quicklime. Liming to the sludge is determined in the process of treatment facility design on the basis of engineering research analysis. Sludge application rate $C_{ws}$ is determined by the formula:

$$C_{ws} = 2.4 \cdot (MPC_s - C_{bp}),$$

where $C_{bp}$ is an initial substance concentration in the peat before sludge application, mg/kg; $MPC_s$ is maximum permissible substance concentration in peat (soil), mg/kg. If $MPC_s$ is not available background concentration in the peat of the bog under study or similar bogs is used. Maximum dose is limited by the total nitrogen content which is applied with sludge and mustn't exceed 300 kg/ha a year.

## 6. Wastewater Treatment Plant Operation Conditions during Wintertime

A typical drawback of the shallow-water-type biological ponds is that wastewater treatment is grown worse or stops in winter when their surface is totally or partly frozen and aquatic vegetation dies. But as far as biological pond with peat substrate is concerned, this problem is solved as a result of sorption of substances by peat and peat microbiocenosis. To confirm this thesis samples of bog water were taken in winter (20.02.2013) at different distances from the place where domestic wastewater of the housing and utilities sector of the village Melnikovo (regional centre of Tomsk Oblast) is discharged into the Ob bog. Their chemical and microbiological content was determined (table 2).

**Table 2.** Chemical and microbiological content of the polluted peat water of the active layer of the peat deposit of the Ob bog in the area affected by wastewater discharge in the village Melnikovo as of 20.02.2013.

| Parameter | Units of measure | Distance from the dry land with wastewater discharge | |
|---|---|---|---|
| | | 50 m | 350 m |
| $NO_2^-$ | mg/dm$^3$ | 0.012 | <0.002 |
| $NH_4^+$ | mg/dm$^3$ | 35.00 | 2.34 |
| $PO_4^{3-}$ | mg/dm$^3$ | 18.200 | 0.084 |
| $BOD_5$ | mgO$_2$/dm$^3$ | 6.78 | 2.40 |
| Mineral oils | mg/dm$^3$ | 0.400 | 0.029 |
| Psychrophilic saprophyte | kl/ml | 10130 | 540 |
| Nitrogen-fixing | kl/ml | 51500 | 31200 |
| Ammonifying | kl/ml | 1000 | 100 |
| Oil-degrading | kl/ml | 1850 | 510 |

Analysis of the obtained data suggests a significant microbiological activity of the peat water under study during wintertime and makes it possible to explain this activity by the significant reduction $NH_4^+$, $PO_4^{3-}$, $BOD_5$ concentration. Therefore, the main function of the biological pond, i.e. removal of organic and biogenic substances from wastewater, is successfully performed even in winter.

## 6. Conclusion

Technique for calculation of the natural and man-made system used for domestic wastewater treatment of the small residential areas and consisting of the blocks for mechanical treatment in the settling tank and biological treatment in the biological pond has been offered. Biological pond is a part of the bog with peat (peat substrate which is formed as a result of wastewater – initial peat interaction) and bog vegetation. The use of the technique and method of BPPS design makes it possible to treat effectively domestic wastewater at considerable cost cutting and stability of operation. Direction of the further investigations concerning substantiation of the similar facility constructions involves development of technologies of transformation of the peat substrate and wastewater sludge from the settling tanks into substrate for forest melioration of the wetlands.

## 7. References

[1] Loucks D P and Van Beek E 2005 *Water Resources Systems Planning and Management. An Introduction to Methods, Models and Applications* (Turin: UNESCO Publishing) p 679
[2] *Guidebook on Modern Technologies of Natural and Wastewater Treatment and Equipment* 2001 Kruger International Consult A/S, Denmark, V F Karpuhin. Copenhagen: Ministry of Environment and Energy, Danish Environment Protection Agency p 253
[3] Savichev O G 2008 Biological Wastewater Treatment Using Bog Biogeocenosis *Bulletin of Tomsk Polytechnic University* **312** (1) 69–74
[4] Shvartsev S L, Serebrennikova O V, Zdvizhkov M A, Savichev O G and Naimushina O S 2012 Geochemistry of wetland waters from the lower Tom basin, Southern Tomsk oblast *Geochemistry International* **50** (4) 367 – 380
[5] *Kanalizatsiya. Naruzhnye seti i sooruzheniya. Construction Rules and Regulations* 2.04.03-85 2007 Moscow, Construction Design Center FGUP (Federal State Unitary Enterprise) p 87 [in Russian]
[6] Savichev O G 2010 Discharge Regulations for the Surface Water Bodies *Water: Chemistry and Ecology* **9** 35–39

# Dissolved gas composition of groundwater in the natural spa complex «Choygan mineral waters» (East Tuva)

**Y Kopylova**[1,a], **N Guseva**[1,a], **A Shestakova**[1,b], **A Khvaschevskaya**[1,a] **and K Arakchaa**[2,c]

[1] Tomsk Polytechnic University, Tomsk, Russia

[2] Research Institute Medical and Social Problems and management of Tuva, Kyzyl, Russia

E-mail: [a]unpc_voda@mail.ru, [b]stysha@sibmail.com, [c]chodura@yandex.ru

**Abstract.** The natural spa complex «Choygan mineral waters», a unique deposit of natural carbon dioxide mineral waters in Siberia, is located in the Eastern Sayan Mountains. There are 33 springs discharge in this area. Spring waters are mainly $HCO_3$-Na-Ca type. TDS varies from 300 mg/L to 2600 mg/L and temperature ranges from 7 °C (in spring 33) to 39 °C (in spring 12), pH varies from 5.9 to 8.3, and the value of the oxidation-reduction potential is from - 170 mV to 236 mV. All studied waters were divided into two groups according to their temperature and geochemical conditions: cold fresh water with oxidizing conditions and warm slightly brackish water with reductive conditions. The gas composition of the studied waters is represented by nitrogen (28-75 vol.%), carbon dioxide (6-65 vol.%), oxygen (7-19 vol.%), radon (4-948 Bq/l). The studied gases differ not only by the content but by the different sources.

## 1. Introduction

Natural carbon dioxide mineral waters are widely spread in the structures of the Alpine folded system in the western and central parts of Eurasia and constitute the natural carbon dioxide mineral waters province [1]. Natural carbon dioxide waters are rare beyond this province within more ancient geological structures. However they are found in Siberia: the Krasnoyarsk Territory (Kozhanovskoe deposit of natural carbon dioxide mineral waters), Kemerovo Oblast (Tersinskoe deposit of natural carbon dioxide waters) [1].

A.M. Ovchinnikov, V.V. Ivanov, M.I. Vroblevskiy, I.J. Panteleev, N.I. Tolstikhin, G.S. Vartanyan etc. contributed greatly to the natural carbon dioxide mineral water study in the USSR. Among them the chemical and gas composition of the Caucasian natural carbon dioxide mineral waters are well studied. The new data allow us to understand fully the mechanisms of their chemical composition formation [2, 3]. In particular, the zonal distribution of mineral waters associated with regional geological structural features is observed in the Caucasus region [3]. The area of natural carbon dioxide mineral waters is located in the axial part of the Greater Caucasus near the centers of the Pliocene-Quaternary volcano activity. The natural carbon dioxide waters originate from deep magmatic processes and carbon dioxide has magmatic and metamorphic origin. The metamorphogenic $CO_2$ significantly exceeds magmatogenic one in the Caucasus carbonated mineral waters.

Natural carbon dioxide mineral waters are replaced by nitrogen-methane thermal waters away from the axial zone of the Main Range. Nitrogen thermal waters are located in the areas having vertical geological movements with high rates and young volcanic activity. These conditions provide deep meteoric waters penetration into rocks and their heating by the regional thermal field.

In recent decades, new data on the chemical elements and gases distribution in natural carbon dioxide mineral waters of the Sikhote-Alin in the Maritime region were obtained [4, 5]. Mongolian researchers presented the information on the gas composition of mineral groundwaters in Mongolia [6]. At the same time, the natural spa area «Choygan mineral waters», a unique deposit of natural carbon dioxide mineral waters in Siberia, is located in the Eastern Sayan Mountains. Its formation is related to the latitudinal fault in the area with current and recently extinct volcanic activity in the East Sayan Mountains [7]. These mineral waters were discovered many years ago, but they are not well studied.

In this regard, the aim of the study is to examine the gas composition and conditions of carbonated cold and thermal groundwater formation in the natural spa complex «Choygan mineral waters».

Numerous researchers in Siberia, I.S. Kryzhin (1858), S.V. Obruchev (1945), V.G. Tkachuk (1955), E.V. Pinneker (1966-1967), studied Choigan springs. Geology and chemical composition description of some springs are given in their works. A comprehensive study of Choygan springs has been carried out by researchers from Tuva Institute of Complex Natural Resources and Tomsk Polytechnic University under Arakchaa K.D. supervision since the 1990s [8].

Along with this, due to the territory inaccessibility and lack of road communications Choygan carbonated waters are insufficiently investigated for therapy use. The processes of the gas composition formation and its relation with the groundwater chemical composition were not well studied. It is necessary to study the gas composition of groundwater, to identify patterns of carbon dioxide and radon distribution in springs, and their association with water chemical composition, geological and hydrogeological conditions of the study area.

Carbonated waters are different in distribution and discharge conditions, water temperature, the chemical elements concentration, TDS and gas composition. V.I. Vernadsky [9] emphasized the necessity of a quantitative study of the natural water gas composition, especially in the groundwaters. According to his opinion it is significant both for understanding of the chemical composition and identification the groundwater position in relation to the earth surface level.

## 2. Materials and methods

The natural spa complex «Choygan mineral waters» is located 1550 m above the sea level in the north-eastern part of the Tuva region on the western slope of the Eastern Sayan Mountains. There are 33 springs having various chemical composition and temperature discharge in this small area (400*700 m) (figure 1) [10, 11].

The origin of these springs is the result of a large latitudinal fault in pre-Cambrian rocks (gneisses, marbles, slates), intruded by Palaeozoic granites and diorites. Aquifer is the Upper Proterozoic Ailygskaya suite sequence represented by marbles, stratified and marbled limestones, interbedded with shales, quartzites, conglomerates, intruded by Devonian granite intrusions of Brensky sequence [7]. The hot and cold carbonated waters discharge is accompanied by the recent travertine formation. The travertine dome is brown due to iron oxides and hydroxides.

In the summer of 2013 the study of groundwater chemical composition in the natural spa area «Choygan mineral waters» was carried out by the members of «ArzhaanLab» Ltd. and Problem Research Laboratory of Hydrogeochemistry Research-Educational Center «Water» (TPU).

Electrical conductivity, temperature and pH were measured *in situ* using a portable device Water Test. Water samples were investigated by the following methods: tytrimetric ($HCO_3^-$, $CO_2$, $CO_3^{2-}$), ion chromatography ($SO_4^{2-}$, $Cl^-$, $Ca^{2+}$, $Mg^{2+}$, $SO_4^{2-}$, $Na^+$, $K^+$). Radon content in waters was studied using radiometer RGA-01, which registers the volume activity of radon 222 nuclide in liquid samples for alpha radiation.

Gas samples were taken by the vacuum method in glass bottles of 0.2 liters using a portable syringe-degasser [12].

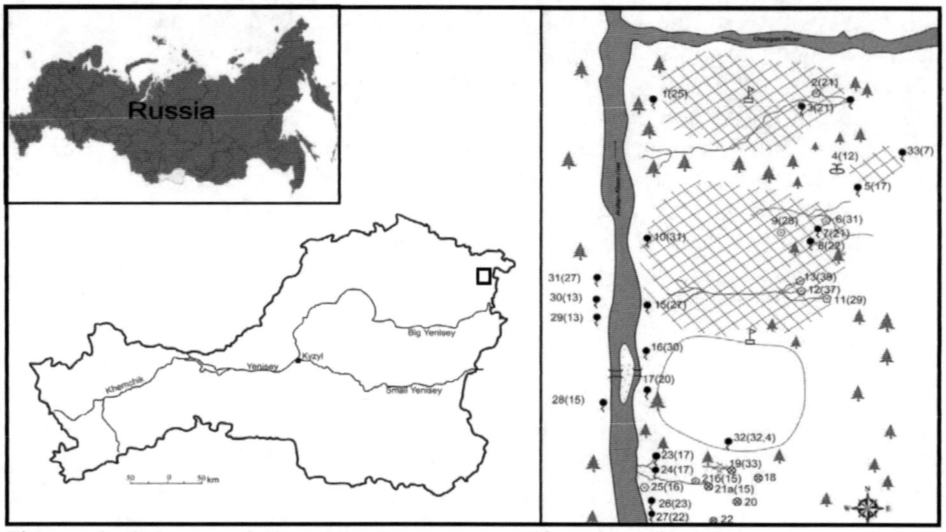


**Figure 1.** Scheme of the natural spa area «Choygan mineral waters» and groundwater sampling points locations (number in parentheses – water temperature)

The determination of the gas composition (nitrogen, oxygen, carbon dioxide) was performed on the Hardware and Software Complex (HSC) based on the chromatograph «Chromatec Crystal-2000M» with the software «Chromatec Analyst». The analysis of the real gas composition of water was performed by an engineer V.A. Shusharina in the accredited Problem Research Laboratory of Hydrogeochemistry Research-Educational Center «Water» (TPU).

## 3. Results and Discussion

Spring waters are mainly of $HCO_3$-Na-Ca type. TDS varies from 300 mg/l to 2600 mg/l and temperatures range from 7 °C (in spring No 33) to 39 °C (in spring No12). pH varies from 5,9 to 8,3, and the value of the oxidation-reduction potential - from -170 mV and to 236 mV [5].

Choygan waters were divided into two groups according to he geochemical characteristics and temperature: cold water with temperatures up to 20 ° C (springs No 2-5, 21a, 23-25, 28-29a, 30, 33) and warm water with the temperature above 20 ° C (springs No 1, 6-20, 22, 26, 27, 31, 32). The gas and chemical composition of the studied springs are shown in table 1. The results in table 1 demonstrate that the gas composition does not depend on the water temperature. Although, the oxidation-reduction properties, TDS, carbon dioxide and nitrogen ratio, depending on the oxygen concentration, change with the temperature increase.

The cold spring group is situated in the northern and south-western part of the studied area. The average temperature of these waters is 13.8 °C, and they are characterized by oxidizing (Eh 201 mV) and subacidic (pH 6.6) conditions. The cold waters are fresh (TDS up to 1036 mg/L) and refer to $HCO_3$-Ca-Na type. Most of springs discharge in the Arzhan-Khem river floodplain. There is hydraulic connection between the studied groundwater and the Arzhan-Khem River.

**Table 1.** Gas and chemical composition of Choygan mineral waters.

| Group of waters | T, °C | Eh, mV | pH | TDS, mg/l | $CO_2$, mg/l | $O_2$ | $N_2$ | $CO_2$ | Rn, Bq/l |
|---|---|---|---|---|---|---|---|---|---|
| | | | | | | vol.% | | | |
| Cold | 7.3-18.4<br>13.8 | 169-236<br>201 | 5.9-8.3<br>6.6 | 288-1545<br>1036 | 41-1488<br>669.2 | 7-19<br>13.7 | 28-75<br>55.3 | 6-65<br>31.0 | 53-520<br>194.3 |
| Warm | 20.3-38.5<br>27.3 | -170-183<br>36.5 | 6.1-6.9<br>6.4 | 1545-2646<br>2207 | 277-1251<br>728.5 | 7-17<br>13.0 | 32-70<br>54.4 | 13-61<br>32.6 | 4-948<br>180.4 |

The numerator indicates the minimum and maximum values of the index; the denominator shows the average value.

Dissolved gas composition of cold waters is represented by nitrogen (average content 55.3 vol.%), carbon dioxide (average content 31 vol.%) and oxygen (average content 13.7 vol.%). Other gases in these waters were not identified. In fact, the gas composition of cold waters with oxidizing conditions is mostly nitrogen one. Only when $O_2$ content in waters is less than 10 vol. %, $N_2$ and $CO_2$ concentrations become equal or even $CO_2$ prevails (table 2).The $CO_2$ content ranges from 6 to 65 vol.% in cold waters. The maximum $CO_2$ concentration at the minimum $N_2$ и $O_2$ contents of 28 and 7 vol. % respectively is observed in spring No 4 (table 2). $CO_2$ domination in dissolved gas composition is observed also in spring No 3 and No 21. It is worth noting the high free carbonic acid concentration of 760-1488 mg/l in these springs. On the whole in the study of Choygan mineral waters free carbonic acid concentration varies greatly from 41 mg/l in spring № 29a to 1488 mg/l in spring No 4 at the average of 670 mg/l.

The lowest $CO_2$ (6 vol. %) and the highest $N_2$ (75 vol. %), $O_2$ concentration (19 vol. %) occur in the fresh neutral waters of oxidizing conditions, discharging in spring No. 33. The gas composition of these waters is consistent with the atmosphere composition but with a small excess of $CO_2$, due to the fractured zone, providing water enrichment in free carbonic acid.

**Table 2.** Chemical and gas composition of cold waters of oxidizing conditions.

| Number | T, °C | pH | Eh, mV | $CO_2$, mg/l | TDS, mg/l | Rn, Bq/l | $O_2$, vol.% | $N_2$, vol.% | $CO_2$, vol.% |
|---|---|---|---|---|---|---|---|---|---|
| 33 | 7.3 | 8.3 | 224 | 4.4 | 351 | - | 19 | 75 | 6 |
| 30 | 10.9 | 6.7 | 228 | 188 | 1546 | 53 | 18 | 70 | 12 |
| 29a | 13.74 | 6.6 | 216 | 41 | - | 62 | 15.7 | 65 | 19.3 |
| 28 | 12.7 | 6.3 | 180 | 688 | 1204 | 354 | 17.7 | 66.1 | 16.2 |
| 23 | 16.3 | 6.5 | 230 | 253 | 1100 | 122 | 14 | 55 | 31 |
| 5 | 13.5 | 6.2 | 200 | 946 | 1018 | 126 | 14 | 55.7 | 30.3 |
| 25 | 16.4 | 6.4 | 195 | 600 | 1533 | 93 | 10 | 51 | 39 |
| 3 | 18.4 | 6.3 | 190 | 760 | 1235 | 520 | 10.6 | 43 | 46.4 |
| 4 | 11.5 | 5.9 | 188 | 1488 | 607 | 71 | 7 | 28 | 65 |
| 21 | 13 | 6.2 | 236 | 872 | 1037 | 4 | 10.2 | 39 | 50.8 |

TDS and concentrations of such major ions as $HCO_3^-$, $Cl^-$, $Ca^{2+}$, $Mg^{2+}$ grow with an increase of carbon dioxide mineral water temperature. The oxidizing conditions change to the reductive ones [11].

Warm springs with the average temperature 27.3 °C are located in the central part of the studied area. These waters are characterized by the reductive (average Eh – 36.5 mV) and slightly acidic (average pH 6.4) conditions. Warm waters of the springs belong to $HCO_3$-Ca-Na type and a high mean value of TDS is 2200 mg/l. The TDS of warm waters is higher than TDS of cold waters (table 1). The content of dissolved free carbonic acid in waters varies from 277 mg/l (spring № 8) to 1251 mg/l (spring No. 27). Dissolved gas composition of warm waters is represented by nitrogen within 32-70 vol.%, carbon dioxide 13-61 vol. % and oxygen 7-17 vol. %.

**Table 3**. Chemical and gas composition of warm waters in reductive conditions

| Number | T, °C | pH | Eh, mV | $CO_2$, mg/L | TDS, mg/l | Rn, Bq/L | $O_2$, об.% | $N_2$, об.% | $CO_2$, об.% |
|---|---|---|---|---|---|---|---|---|---|
| 1 | 22.6 | 6.2 | 62 | 1133 | 2288 | 123 | - | - | - |
| 6 | 29.5 | 6.3 | -90 | 744 | 2333 | 95 | 14.6 | 58.4 | 27 |
| 7 | 23.8 | 6.4 | -150 | 800 | 1863 | 204 | 16 | 62.2 | 21.8 |
| 8 | 25.3 | 6.3 | -170 | 277 | 2328 | 360 | 14 | 58 | 28 |
| 9 | 27 | 6.2 | -24 | 371 | 2180 | 948 | 17 | 65 | 18 |
| 12 | 38.5 | 6.3 | -54 | 691 | 2525 | 90 | 15 | 58 | 27 |
| 13 | 36.8 | 6.3 | 0 | 326 | 2586 | 122 | - | - | - |

| | | | | | | | | | |
|---|---|---|---|---|---|---|---|---|---|
| 17 | 22.4 | 6.1 | 70 | 1074 | 1545 | 74 | 16 | 62 | 24 |
| 19 | 30.9 | 6.4 | 142 | 608 | 2140 | 9 | 14 | 60 | 26 |
| 20 | 28.2 | 6.6 | 142 | - | - | - | 17 | 69 | 14 |
| 11 | 31.5 | 6.3 | 29 | 746 | 2318 | - | 12 | 52 | 36 |
| 15 | 24.9 | 6.4 | 51 | 484 | 2265 | 106 | 11 | 43 | 46 |
| 9a | 25 | 6.9 | 22 | 409 | 2136 | 84 | 9.6 | 48 | 42.4 |
| 10 | 30.2 | 6.5 | 21 | 460 | 2570 | 64 | 9 | 37 | 55 |
| 22 | 27.3 | 6.3 | 108 | - | - | - | 9 | 41 | 50 |
| 27 | 21.4 | 6.3 | 169 | 1252 | 2064 | - | 6.6 | 31.7 | 60.8 |

The gas composition of warm springs is characterized by $N_2$ predominance, but it is different in waters where the oxygen content is less than 10 vol.%. In springs No. 15, 9a, 10, 22 and 27 carbon dioxide dominates over the nitrogen and gas composition becomes oxygen-nitrogen-carbon. The gas composition formation of these springs is associated with the carbon dioxide coming along the fault zones from great depths, where the oxygen concentration is substantially lower than in the atmosphere. Nitrogen and oxygen enrichment of the studied waters occurs due to meteoric waters. According to [3], the oxygen presence in the springs in mountain areas is a characteristic feature of infiltration recharge of mineral waters. The maximum content of nitrogen and oxygen in waters of the natural spa complex «Choygan mineral waters» can reach their content in air, that is concentration of 78,09 vol.% and 20,95 vol.% respectively [13]. As for carbon dioxide, its content in the groundwaters is much higher than that in the atmosphere, containing only 0.03 vol. % of $CO_2$.

At the same time, the ratio between free carbonic acid and carbon dioxide gas is found in the studied waters. The free carbonic acid content in waters is typically greater than 300 mg/l, that is the concentration of carbon dioxide is more than 20 vol.%, so 15 mg/l of free carbonic acid corresponds to 1 vol.% of carbon dioxide. In the cold waters with oxidizing conditions this ratio between free carbonic acid and carbon dioxide is 30-40 mg/l per 1 vol.%. This ratio is reduced to 18 mg/l of free carbonic acid at 1 vol.% carbon dioxide in deeper cold groundwaters with oxygen concentration of less than 10 vol.%. The ratio between free carbonic acid and carbon dioxide is 24 mg/l per 1 vol.% in the warm waters with reduciive conditions. In deeper warm groundwaters having the content of oxygen less than 10 vol.%, the ratio is 13 mg/L per 1 vol.%.

Cold and warm waters of Choygan springs do not differ in the radon concentrations. In the cold waters the radon concentration varies from 53 Bq/l to 519 Bq/l (in spring No 3) and that is for warm waters from 4 Bq/l (in spring No 21) to 948 Bq/l (in spring № 9). The mean radon concentrations are similar both in cold and warm waters, whereas the maximum radon concentration is almost twice higher in warm waters than that in cold waters (Table 1). This is explained by the fact that radon is a radioactive chemical element. It is the product of alpha decay of radium-226, included in the radioactive uranium-238 series. The radon waters formation is related to granites emanation enriched by radioactive elements. In contrast to carbon dioxide, the radon source should be located near the springs, because $^{222}Rn$, most stable isotope, has a half life of only 3.82 days [1, 10, 14].

## 4. Summary and Conclusions

The study of 33 springs in the natural spa complex «Choygan mineral waters» has shown new data on the chemical and gas compositions. All the studied waters were divided into two groups according to their temperature and geochemical conditions: cold fresh water with oxidizing conditions and warm slightly brackish water with reduction conditions. The gas composition of the studied waters is represented by nitrogen (28-75 vol.%), carbon dioxide (6-65 vol.%), oxygen (7-19 vol.%), radon (4-948 Bq/l). The gas composition of all water groups is characterized by a clear predominance of nitrogen. However, in the springs, where the oxygen content is less than 10 vol.%, carbon dioxide prevails.

The studied gases differ not only in the content but in different sources. Carbon dioxide comes along the fault zones from great depths, where the oxygen concentration is substantially lower than in the atmosphere. Nitrogen and oxygen enrichment of the studied waters occurs due to meteoric waters. The oxygen presence in the springs in mountain areas is a characteristic feature of infiltration recharge of mineral waters. The radon waters formation is related to granites emanation.

The ratio between free carbonic acid and carbon dioxide has been identified. The free carbonic acid content in waters is typically greater than 300 mg/l at concentrations of carbon dioxide more than 20 vol.%, so 15 mg/l of free carbonic acid corresponds to 1 vol.% of carbon dioxide.

The specific features of water, chemical and gas composition, its physical properties, TDS and a unique combination of carbon dioxide and radon provide an opportunity of their use for balneotherapy. Natural carbon dioxide springs in the natural spa complex "Choygan" are natural therapy resources appropriate for organizing treatment centers and resorts.

**Acknowledgements**
The investigations were supported by State Assignment «Nauka».

**References**
[1] Kraynov S R, Ryzhenko B N and Shvets V M 2012 *Geochemistry of groundwater. Theoretical, applied and environmental aspects* (Moscow: CentrLitNefteGas) [in Russian]
[2] Ovchinnikov A M 1947 *Mineral waters* (Moscow: Gosgeolitizdat) [in Russian]
[3] Lavrushin Vv Yu 2012 *Subsurface fluids of the Greater Caucasus and its surrounding* (Moscow: GEOS) [in Russian]
[4] Chudaev O, Chudaeva V, Kharitonova N, Chelnokov G, Mao X and Wang Y 2007 Composition and origin of CO2-rich waters of the Russian Far East (Primorye) and NE China *Water-Rock Interaction - Proceedings of the 12th International Symposium on Water-Rock Interaction* **WRI-12** 489-492
[5] Kharitonova N A, Chelnokov G A, Vakh E A, Karabtsov A A and Zykin N N 2010 Geochemistry of carbonated mineral waters from the Fadeevskoe deposit of the Primorye region *Russ J of Pac Geol* **4** 78-90
[6] Ganchimeg Darmaa 2006 *Gas composition of Mongolia underground mineral waters*. Author's abstract of PhD thesis for the degree of candidate of geological and mineralogical sciences, Irkutsk, Institute of the Earth's crust [in Russian]
[7] Pinneker E V 1968 *Mineral waters of Tuva* (Kyzyl: Publishing House of Tuva) [in Russian]
[8] Arakchaa K D, Smirnova I N and Kopilova Y G 2012 Arzhan Choygan – the curative area of the Tuva region (Kyzyl: Publishing House «Anyak») [in Russian]
[9] Vernadsky V I 2003 History of Natural Waters (Moscow: Science) (in Russian)
[10] Rychkova K M, Kopylova Y G and Oyun L A 2012 Chemical and gas composition, helium isotopes in groundwater of Tuva *«Underground hydrosphere»: Materials of All-Russian conference on groundwater in eastern Russia* (Irkutsk: Publishing House of Ltd. «Georgaf», 2012) 236-240 [in Russian]
[11] Kopylova Yu G, Guseva N V, Arakchaa K D and Khvashchevskaya A A 2015 Geochemistry of carbonic-acid mineral waters of choygan natural complex (noth-east Tuva) *Russ Geol Geophys (in Press)*
[12] Zuev V A 1993 Portable field syringe degasser *Multipurpose hydrogeochemical investigations in connection with the search for mineral resources and groundwater protection* (Tomsk Polytechnic University) 185-186 [in Russian]
[13] Voitkevich G V, Miroschnikov A E, Povarennykh A S and Prokhorov V G 1977 Geochemistry *pocket book* (Moscow: "Nedra") [in Russian]
[14] Bonotto D M 2014 «$^{222}$Rn, $^{220}$Rn and other dissolved gases in mineral waters of southeast Brazil» *J. Environmental Radioactivity* **132** 21 – 30

# Characteristic features of groundwater pollution in the Poyang Lake catchment

**E Soldatova[1], N Guseva[2] and G Wang[3]**
[1,2] Tomsk Polytechnic University, Tomsk, Russia
[3] China University of Geosciences, China

E-mail: [1]2a61@mail.ru, [2]guseva24@yandex.ru

**Abstract**. Research of shallow groundwater in the Poyang Lake catchment should be considered as one of the priority targets in this region, due to the fact that shallow groundwater is formed under the influence of complex factors, including both natural and anthropogenic, and also the fact that it is related to the Poyang Lake water. Shallow groundwater of the Poyang Lake area is fresh. High concentrations of $NO_3$ ,$Cl^-$, $SO_4^{2-}$, $K^+$ and rarely, $Na^+$ are not typical of the groundwater of subtropical humid climate. Within the investigated territory the groundwater could be divided into two types - relatively pure shallow groundwater, formed primarily under the effect of natural factors and having a high content of $NO_3^-$ but rarely $K^+$, $Cl^-$, $SO_4^{2-}$, in some sampling points; and relatively polluted shallow groundwater, characterized by significant changes of chemical composition due to intensive agricultural activity. This type has higher pH-values, concentration of main components, including $NO_3^-$, $K^+$, $SO_4^{2-}$, $Cl^-$, and, accordingly, value of total dissolved solids in comparison with relatively pure shallow groundwater.

## 1. Introduction

The largest freshwater lake in China – Poyang Lake – is a unique ecosystem. The Poyang Lake basin is not only a habitat for rare animal species, but also an important economic sector of Jiangxi Province. Resources of freshwater are used for domestic needs and industry, as well as for agriculture. Vast alluvial plains, surrounding Poyang Lake, have converted the basin into one of the main agricultural regions in China, where territories within the Poyang Lake catchment are often irrigated for rice and other crops cultivation.

Due to wide spread croplands, dense population and frequent shortage of centralized water supply and sewerage within the basin, the ecosystem is subjected to significant anthropogenic effect, in this case, shallow groundwater is not an exception. Furthermore, as long as the shallow groundwater is related to the Poyang Lake water, polluted groundwater may exert a negative influence on Poyang Lake. Therefore, comprehensive investigations of shallow groundwater chemical composition within the Poyang Lake area, in our opinion, are one of the priority targets of scientific research in this region. Investigation of shallow groundwater chemical composition formation allows estimating the influence of natural and anthropogenic factors on the existing ecosystem.

## *2.* Materials and methods

### 2.1. *Study area*

The investigated area is situated in south-east China (Jiangxi Province) and is the part of the Poyang Lake catchment. It is one of the main hydrological subsystems of the Yangtze River. The area of the Poyang Lake catchment is 162225 km2 [1]. The investigated area belongs to the province of subtropical humid climate. The annual precipitation is 1400–2400 mm [2]. Irregular rainfall distribution during the year is controlled by the eastern-Asian monsoons. Wet season lasts from March – April to June, while further rainfall decreases from July to September and at, the same time, evapo-transpiration reaches its maximum, and after September the dry season is up to December. The relief of the investigated area is quite various from mountain alluvial plains in the relief depression areas to river valleys [3].

### *2.2 Sampling and analytical procedures*

Hydrogeochemical analysis in the Poyang Lake catchment was conducted in November, 2011. The investigation target was shallow groundwater. 54 groundwater samples were retrieved from household wells in the investigated area. The sampling points are concentrated primarily to the north of Poyang Lake and on the territory of the Ganjiang River basin and the Xiushui River basin (figure 1).

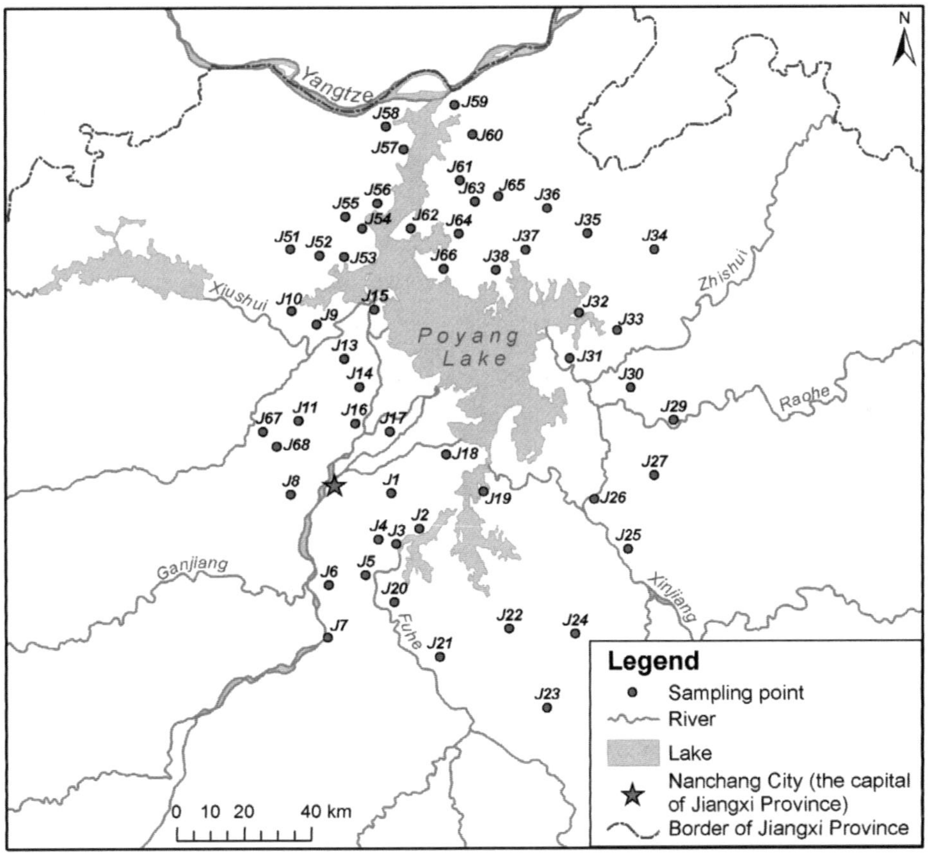


**Figure 1.** Scheme of the groundwater sampling points located in the Poyang Lake catchment.

One litre of water was collected in polyethylene bottles in each sampling point to determine the following ions- $NO_3^-$, $PO_4^{2-}$, $Br^-$, $F^-$. Electrical conductivity, temperature and pH were measured in situ. Chemical analysis of water samples was performed in the laboratory of China University of Geosciences (Beijing) using ion chromatography (Dionex-900) and titration method.

IOP Conf. Series: Earth and Environmental Science **21** (2014) 012023 doi:10.1088/1755-1315/21/1/012023

Based on the specific chemical composition of the latter and how sampling points are confined to territories of different economic and agricultural development, shallow waters were classified into various types.

## 3. Results and Discussions

Groundwater of the Poyang Lake area is basically ultra-fresh (TDS is up to 200 mg/L) and moderately fresh (TDS is 200–500 mg/L) while only in several sampling points TDS is more than 500 mg/L. The average value of TDS for investigated groundwater is 229 mg/L (table 1). Acid-alkaline properties of groundwater vary significantly, pH-value ranges from 4.75 to 7.26, i.e. geochemical conditions change from acid to neutral. The average pH-value is 6.20. pH -value increases parallel to the increase of TDS, whereas sharp pH increase could be observed in ultra-fresh waters and visa versa, decreases in waters with rather high TDS (figure 2).

| Content | pH | $HCO_3^-$ | $SO_4^{2-}$ | $Cl^-$ | $NO_3^-$ | $Ca^{2+}$ | $Mg^{2+}$ | $Na^+$ | $K^+$ | TDS |
|---|---|---|---|---|---|---|---|---|---|---|
| Min. | 4,75 | 3,25 | 1,36 | 1,23 | 0,10 | 1,77 | 0,23 | 1,04 | 0,58 | 24,0 |
| Max. | 7,26 | 353 | 148 | 102 | 206 | 98,2 | 31,6 | 52,7 | 57,4 | 768 |
| Average | 6,20 | 92,7 | 24,7 | 23,1 | 26,4 | 28,7 | 9,58 | 17,6 | 6,34 | 229 |

**Table 1.** Groundwater chemical composition of the Poyang Lake catchment, mg/L.

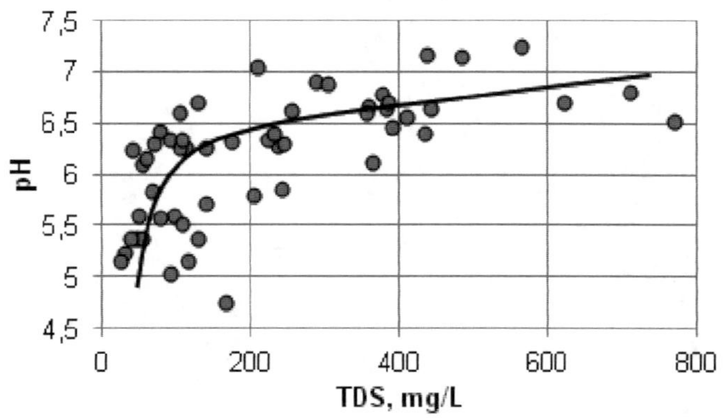


**Figure 2.** TDS versus pH-value in shallow groundwater of Poyang Lake catchment.

High concentrations of sulfate ($SO_4^{2-}$), chloride ($Cl^-$) and nitrate ($NO_3^-$) are observed, however, hydrocarbonate ion ($HCO_3^-$) is prevalent in investigated groundwaters. Increasingly relative average concentration of nitrate is registered in one third of the sampling points. Maximum concentration of $NO_3^-$ (206 mg/L) is observed in point J11. TDS increase is related to the content increase of all anions (figure 3).

The groundwater of Poyang Lake area contains approximately equal concentrations of calcium ($Ca^{2+}$) and sodium ($Na^+$), 1,77 – 98.2 mg/L and 1.04 – 52.7 mg/L, respectively. In certain sampling points (J2, J3, J6, J11, J32) abnormal high content of potassium ($K^+$) is observed, ie. 57.4, 40.5, 39.0, 38.0 and 21.5 mg/L, respectively. Content of all cations increases consistently to TDS (figure 3). In ultra- fresh groundwater it is impossible to identify the dominant cation, while in water with TDS (up to 200 mg/L) the dominant cation is calcium. Such a phenomenon could be associated with the host rock composition, which controls the cationic composition of groundwater at the initial stage of chemical composition formation [4]. Magnesium ($Mg^{2+}$) behavior is specific with slight concentration, but obviously, decreases with the increase of TDS up to 600 mg/L (figure 4).

The formation of groundwater chemical composition in investigated area occurs under the influence of factors, both natural and anthropogenic ones. Within the investigated territory there are two types of shallow groundwater – relatively pure and conditionally polluted. Comparative analysis of these two types is shown in table 2.

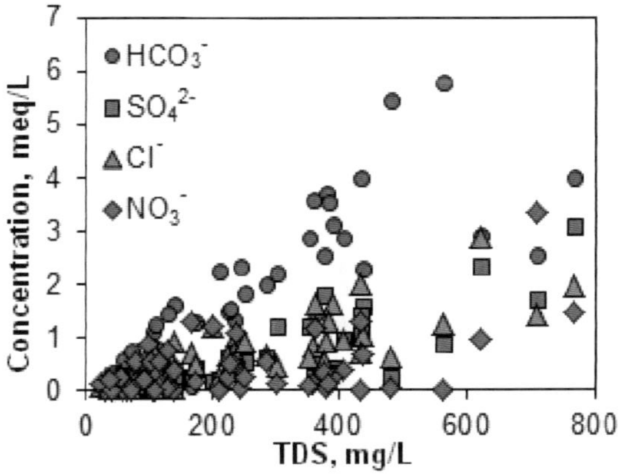


**Figure 3.** TDS versus content of main anions in the shallow groundwater of Poyang Lake catchment.

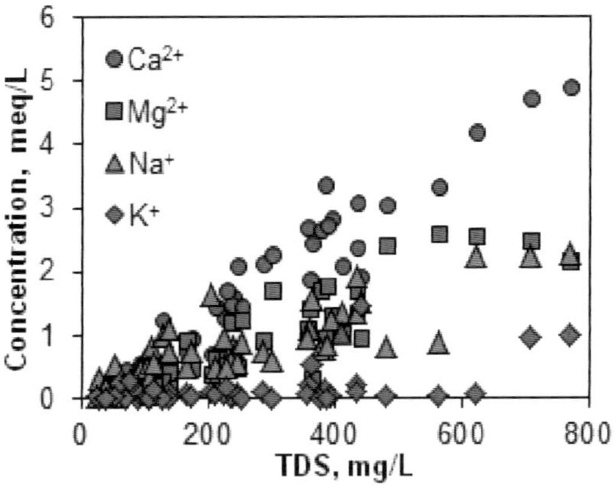


**Figure 4.** TDS versus content of main cations in the shallow groundwater of Poyang Lake catchment

| Content | pH | $HCO_3^-$ | $SO_4^{2-}$ | $Cl^-$ | $NO_3^-$ | $Ca^{2+}$ | $Mg^{2+}$ | $Na^+$ | $K^+$ | TDS |
|---|---|---|---|---|---|---|---|---|---|---|
| **Relatively pure shallow groundwater** (32 sampling points) | | | | | | | | | | |
| Min. | 4.75 | 3.25 | 1.36 | 1.23 | 0.10 | 1.77 | 0.23 | 1.04 | 0.58 | 24.0 |
| Max. | 6.70 | 143 | 35.0 | 42.6 | 80.7 | 42.1 | 14.9 | 38.0 | 12.0 | 243 |
| Average | 5.87 | 36.8 | 8.57 | 12.1 | 19.4 | 11.4 | 3.86 | 11.0 | 2.73 | 106 |
| **Conditionally polluted shallow groundwater** (22 sampling points) | | | | | | | | | | |
| Min. | 5.87 | 42.2 | 10.4 | 9.68 | 1.00 | 29.3 | 3,77 | 11.8 | 0.63 | 209 |
| Max. | 7.26 | 353 | 148 | 102 | 206 | 98.2 | 31,6 | 52.7 | 57.4 | 768 |
| Average | 6.68 | 174 | 48.0 | 39.0 | 36.7 | 53.7 | 17,9 | 27.2 | 11.6 | 408 |
| **Groundwater of tropical and subtropical humid climate province according to S.L. Shvartsev [4]** | | | | | | | | | | |
| Average | 6.4 | 109 | 7.10 | 7.35 | 1.52 | 16.6 | 8.07 | 10.9 | 2.25 | 185 |
| **Poyang Lake** (2 sampling points, according to results of field work carried out in October, 2013) | | | | | | | | | | |
| Average | 6.94 | 42.7 | 15.1 | 15.1 | 2.58 | 15.5 | 2.51 | 10.1 | 2.38 | 104 |

**Table 2.** Chemical composition of investigated groundwater types, mg/L.

Relatively shallow pure groundwater, presupposingly formed under the conditions of natural factors, is spread throughout the whole territory except in the north-eastern area. Conditional shallow polluted groundwater is largely localized in the north of the investigated area, as well as in the Ganjiang River basin and the Fuhe River basin to the south-west of Poyang Lake (figure 5).

Average concentrations of the major ions in relatively shallow pure groundwater are generally associated with the chemical groundwater composition of tropical and subtropical humid climate province (table 2). It should be noted that there is a sufficiently low content of $HCO_3$. The most significant average concentration is typical for nitrate-ion whereas the increased content of $NO_3$ can be observed to the south and to the west of the Poyang Lake (figure 5). There are also high concentrations of $SO_4^{2-}$, $Na^+$ and $K^+$. However, while increased $SO_4^{2-}$, $Cl^-$ and $Na^+$ concentrations are point spread throughout the Poyang Lake area and, in most cases, are associated with each other, high $K^+$ concentrations are confined to the south-east of the investigated territory and are not associated with the increased concentrations of other chemical elements (figure 5). Increased $NO_3$ concentrations are best correlative with high Cl concentrations and in less degree with $Na^+$ (figure 5). The described type of shallow groundwater, in general, is associated with crust weathering process and the terrain, where agricultural activity is not widespread and, therefore, pollution embraces a pin-point pattern.

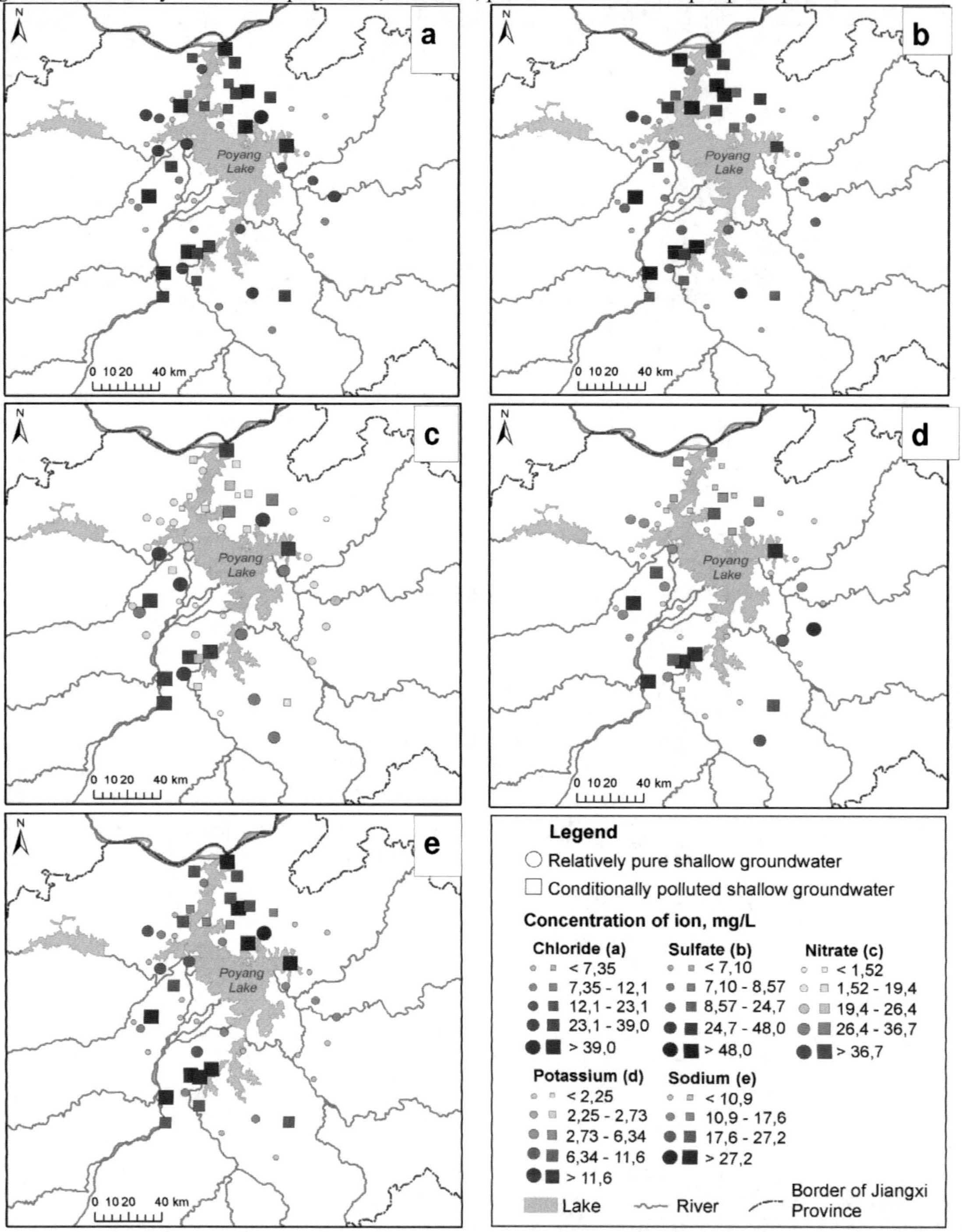


**Figure 5.** Abundance of $Cl^-$ (a), $SO_4^{2-}$ (b), $NO_3^-$ (c), $K^+$ (d), $Na^+$ (e) in shallow groundwater of Poyang Lake catchment.

Conditionally polluted shallow groundwater is characterized by significantly high pH-values which correspond mainly to neutral conditions, in comparison to relatively pure shallow groundwater. The average concentrations of the major ions and TDS are several fold higher in comparison to those in the groundwater of tropical and subtropical humid climate province (table 2). Even minimum concentrations are high enough for some components. These features highlight the fact of significant groundwater pollution. This type is found within highly cultivated areas, where flood irrigation, paddy fields and fish pounds are prevalent. Based on A Perelman's concept, the authors suggest the term “agrolandscapes” to define such territories [5]. It should be noted that the sampling points situated to the west and to the south-west of Poyang Lake in the Xiushui, Ganjiang and Fuhe basins are characterized by significant concentration content of all analyzed components, in particular, $Cl^-$, $SO_4^{2-}$, $NO_3^-$, $K^+$, $Na^+$ (figure 5). However, to the north of Poyang Lake there is a wide-spread increase of $Cl^-$, $SO_4^{2-}$ and $Na^+$ concentrations. Furthermore, concentrations of $Cl^-$ и $Na^+$ are bound between themselves in most cases (figure 5). Obviously, the formation of groundwater chemical composition of the above-mentioned water types is connected with the intensification of agricultural activity, high dense population, and potentially with industrial production. This, in its turn, results in the accumulation of not only pollution agents, but also the increase of TDS and pH-value in this type of shallow groundwater.

## 4. Summary and Conclusions

Shallow groundwater of the Poyang Lake area is basically fresh, whereas pH-value varies widely, from 4.75 to 7.26, i.e. the geochemical environment changes from acid to neutral.

According to the conditions of groundwater formation within the Poyang Lake catchment, there are two main types of shallow groundwater – relatively pure shallow groundwater and conditionally polluted shallow groundwater. The first type, relatively pure shallow groundwater, is associated mainly with the natural site, crust weathering, and characterized by subacidic conditions, relatively low concentration of the major chemical composition components and TDS, and also extensive in-situ pollution. The chemical composition of this type is formed generally under the influence of natural factors, and the sampling sites are related with less agricultural and economically developed areas. The second type, conditionally polluted shallow groundwater, is characterized by neutral conditions. The pH-value growth is in accordance with TDS increase, and especially, with $NO3^-$, $K^+$, $SO_4^{2-}$, $Cl^-$ concentration increase. It occurs in agricultural areas which are often irrigated for rice and other crops cultivation and characterized by widespread fish ponds. The chemical composition of this mentioned type indicates significant pollution, so its chemical composition is formed primarily under the effect of anthropogenic factors.

**Acknowledgements**

Authors thank Prof. Dr. S.L. Shvartsev for his helpful advice and suggestions. The investigations were supported by The Russian Foundation for Basic Research (Project No 14-05-31267) and State Assignment “Nauka” № 2014/226.

**References**

[1] Feng L, Hu C, Chen X, Li R, Tian L and Murch B 2011 MODIS observations of the bottom topography and its inter-annual variability of Poyang Lake *Remote Sensing of Environment* **115** 2729–41

[2] Wang Q, Riemann D, Vogt S and Glaser R 2013 Impacts of land cover changes on climate trends in Jiangxi province China *Int. J. Biometeorol.* 10.1007/s00484-013-0645z

[3] Ye X, Zhang Q, Liu J, Li X and Xu C Y 2013 Distinguishing the relative impacts of climate change and human activities on variation of streamflow in the Poyang Lake catchment, China *J. of Hydrology* **494** 83–95

[4] Shvartsev S 2008 Geochemistry of fresh groundwater in the main landscape zones of the Earth *Geochemistry Int.* **46** 1285–1398

[5] Perelman A and Kasimov N 1999 *Landscape geochemistry*

XVIII International Scientific Symposium in Honour of Academician M. A. Usov: PGON2014 IOP Publishing
IOP Conf. Series: Earth and Environmental Science **21** (2014) 012024 doi:10.1088/1755-1315/21/1/012024

# Equilibrium of Groundwater with Carbonate Minerals of the Water-Bearing Rocks under Anthropogenic Impact (by the example of Kishinev, Moldova)

**A N Timoshenkova[1], E Yu Pasechnik[2] and O G Tokarenko[3]**
[1] Institute of Geology & Seismology of ASM, MD-2028, Kishinev, Republic of Moldova
[2,3] Tomsk Polytechnic University, Tomsk, Russia

E-mail: [1]asia444@mail.ru, [2]paseyu@yandex.ru, [3]tog@tpu.ru

**Abstract**. The paper presents calculation results of equilibrium of groundwater in Kishenev with a variety of secondary carbonate minerals. It is shown that the groundwater-rock system is in equilibrium with some minerals, such as calcite, magnesite, dolomite, siderite, but at the same time is not in equilibrium with strontianite. It indicates that secondary mineral precipitation is possible. Specific nitrate chemical water type, which is rarely observed in nature and characterized by the presence of anthropogenic impact in this territory, in some cases is of higher saturation as compared to calcite, dolomite and magnesite due to the fact that nitrate ion content increases with the increase of calcium content.

## 1. Introduction

In recent years the interest of the world scientific community is directed to the study of geologic evolution of the water-rock interaction [1]. The water-rock-gas-organic matter system is studied by such scientists as R.M. Garrels, Ch.L. Christ, R.A. Berner, A.E. Blum, W.H. Casey, R. Hellmann, G.R. Holdren, G. Jordan, S.L.Shvartsev, V.A. Alekseev, B.N.Ryzhenko, I.K. Karpov. It can be explained by the fact that the water-rock system is universal on our planet and its geologic evolution leads to the formation of numerous geochemical types of natural water, various secondary minerals, including crust of weathering, hydrothermally altered rocks, various mineral deposits, etc. [2]. At present the theory of water-rock interaction is well-developed in geochemical terms, in terms of experimental modeling, physical and chemical modeling of the rock dissolution processes, etc. [3]. Moreover, the use of modern approaches is particularly relevant for the study of groundwater composition formation under anthropogenic impact. However, the solution to this problem is impossible without a detailed study of the thermodynamic equilibrium of groundwater with principal minerals of the enclosing rocks, carbonate in particular. The study of water-carbonate rock equilibrium is essential for understanding of evolutionary development of water and enables to determine indirectly relation between degree of water saturation to minerals and anthropogenic impact.

The main emphasis of this paper will be on groundwater as it is the least studied as compared to deep waters. In spite of the fact that nowadays groundwater aquifers are not currently used for the centralized water supply, some water bodies (pit wells and springs) are still of practical significance for individual water use in some areas of the city and in the periods when centralized water supply is impossible because of technical breakdowns [3]. Hence, the study of geochemical content of the groundwater of active water exchange zone is particularly relevant.

## 2. Geology and Hydrogeology of the Territory

Geologically the area under study is located in the south-west of the East European Precambrian platform in the center of the Moldavian platform. The crystalline basement is at a depth of 1600 ... 1800 m and is made up of the Archean and Lower Proterozoic dislocated igneous and metamorphic rocks. Sedimentary cover is in the sedimentation area of the Upper Proterozoic and Lower and Middle Paleozoic and Upper Cretaceous. When sedimentary cover comes to the end there are marine deposits of the Middle Paleogene (the Eocene), Neogene (Baden and Sarmatian stages), quaternary deposits of the Dnestr and Byk river terraces. Neogene deposits are characterized by the formation of three reef ridges, the second of which is Middle Sarmatian and go through Kishinev.

There are two types of groundwater depending on the hydrodynamic and genetic characteristics of the occurrence conditions. The first type is groundwater of the active water exchange zone which underlies mostly at a depth of 0 ... 15 m, it is rarely deeper. Water-bearing minerals are sands and sandy clays. Water horizon bottom is made up of the heavy blue-gray and greenish-gray Middle Sarmatian clays eroded in some areas. Groundwater is supplied by infiltration of atmospheric precipitation, groundwater inflow from terraces above flood-plain and by means of flood water under flooded conditions. Intermediate water is a complex of Sarmatian aquifers. Unconfined aquifers of the study area can be divided into four types: 1) aquifer in the modern alluvial deposits; 2) in the ancient alluvial deposits; 3) in the alluvial and deluvial deposits; 4) eolian and deluvial deposits [4].

## 3. Materials and Methods

### *3.1. Study area*

The study area clearly demonstrates anthropogenic impact which shows itself in the presence of nitrate pollution and its effect on the geochemical characteristics of groundwater in the Republic of Moldova as a whole. Fig. 1 shows the area under study of urban agglomeration, consisting of five residential districts, each of which is subjected to various degrees of anthropogenic impact.

### *3.2. Sampling and investigation methods*

The given investigation conducted in summer in 2013 is based on the results of groundwater sampling from water sources used by residents for drinking purposes. Totally, 85 sources were sampled, among them there are 20 ascending and descending springs with the natural discharge within the city, and 65 wells with a depth not less than 10 m. Chemical analysis of water samples was carried out in an accredited laboratory of Hydrogeology and Engineering Geology of the Institute of Geology and Seismology of the Academy of Sciences of Moldova using state-of-the-art analytical equipment and methods of titrimetry and ion chromatography.

The equilibrium calculations were made using well-known methods developed on the basis of thermodynamics of hydrogeochemical processes [3] using HydroGeo software [5]. Free energies of formation of carbonate minerals and dissolved chemical elements are taken from [3]. To determine the degree of equilibrium of therm to individual minerals, we used method of mineral stability fields developed by R.M. Carrels and Ch.L. Christ [6], and saturation index (si - saturation index) which is equal to

$$si = \lg Q/K,$$

where Q is a reaction quotient; K is a reaction constant. Saturation index increases to zero (state of equilibrium) with saturation of solution in relation to some minerals. With oversaturation its values become positive. Calculations were performed at 25°.

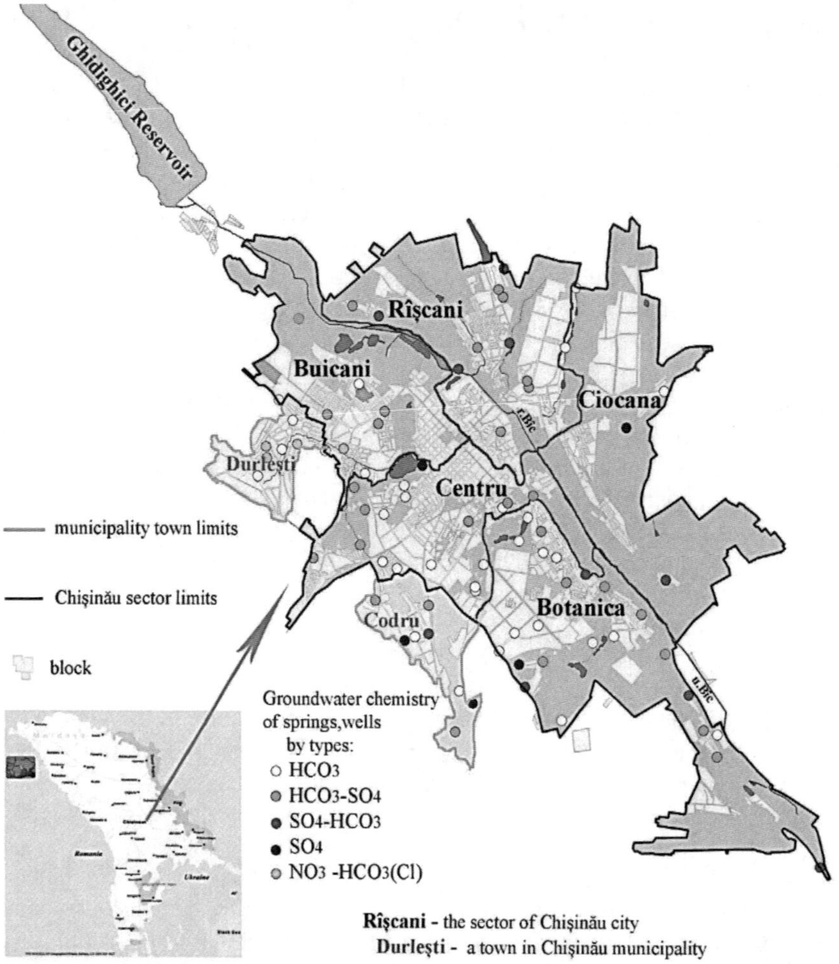


**Figure 1.**The map of the study area.

## 4. Results and Discussion

*4.1 Chemical composition*

It was found that groundwater has a specific chemical composition. The first specific feature is salinity which varies in a wide range from 0.39 to 2.63 g/l in the samples studied. Water is mainly weakly alkaline, fresh, rarely weakly salted, with an average salinity equaled to 1.1 g/dm$^3$ (table 1).

Water under study is different by chemical type. In general, the following chemical types of water with predominant content of anions with different combinations of cationic component are abundant in Kishenev (according to S.A. Shchukarev) (in brackets – % of the total number of the samples under study in descending order): *I type* – $HCO_3$–Mg–Ca, –Na–Mg–Ca, –Ca–Na, –Na–Mg, rarely–Mg–Ca–Na or–Ca–Mg–Na or – Ca, –Mg and –Na (40 %); *II type* – $HCO_3$–$SO_4$–Mg–Na, –Mg–Ca, rarely–Ca, –Na–Ca, –Na–Mg–Ca, –Na–Mg and –Na–Ca (37 %); *III type* – $SO_4$–$HCO_3$–Ca–Mg–Na, –Mg–Ca–Na, –Ca–Na–Mg, –Ca–Mg, –Mg–Ca, –Mg–Na, –Na–Mg (12 %); *IV type* – $SO_4$–Ca–Mg–Na, –Ca–Mg–Na, –Ca–Na–Mg, –Na–Mg–Ca, –Mg–Na–Ca (6 %);

The increase of nitrate ion concentration in groundwater allows us to distinguish an additional chemical type which can be called 'exotic' and characterizes specific nature of the study area. i.e. *V type* – $NO_3$–$HCO_3$–Mg–Ca, sometimes with a high content of chloride ion (5%). It was found that nitrate chemical type of groundwater is abundant mostly in the northwestern part of the city. Nitrate ion concentration tends to increase with the increase of concentrations of calcium and magnesium, and sulfate ions as well. It was found [7] that elevated nitrate concentrations are caused by agriculture.

**Table 1.** Groundwater chemical composition, town Kishinev (number of samples – 85).

| Constituent | Concentration, mg/L | | |
|---|---|---|---|
| | Min | Max | Average |
| TDS, g/L | 0.4 | 2.6 | 1.1 |
| pH | 7.0 | 8.4 | 7.6 |
| $HCO_3^-$ | 244.0 | 902.8 | 526.1 |
| $Cl^-$ | 13.6 | 408.7 | 110.8 |
| $SO_4^{2-}$ | 8.0 | 1488.0 | 270.4 |
| $NO_3^-$ | 0.4 | 994.4 | 116.5 |
| $Ca^{2+}$ | 14.2 | 383.2 | 121.6 |
| $Mg^{2+}$ | 7.8 | 222.5 | 90.4 |
| $Na^+$ | 28.0 | 305.0 | 124.9 |
| $K^+$ | 0.1 | 125.2 | 6.4 |
| $Fe^{2+}$ | 0.1 | 1.0 | 0.3 |
| $Sr^{2+}$ | 0.5 | 6.8 | 1.8 |
| $SiO_2$ | 1.9 | 35.0 | 9.7 |

It was established that areal hydrogeochemical zoning of groundwater is characterized by the abundance of hydrocarbonate water with magnesium cations and calcium or sodium and magnesium base in the elevated areas of the city. Sulfate ion appears in the water composition in the areas of much lower absolute elevation. Water with the increased concentration of sulfate ion is abundant in the territory of the Russian Federation and other countries [8]. Their occurrence is associated with the areas of continental salinization development processes [9], or areas with sulphide deposits which are abundant, for example, in Trans-Baikal [10]. The content of sulphates in the water under investigation is comparable to the concentration observed in the water areas of the above mentioned areas and reaches about 1.5 g/l.

Correlation analysis of the main chemical components enabled to find out direct relation of sodium, magnesium and calcium content with sulfates. Positive correlation of sulphates with these elements might be explained by the fact that they accumulate due to the processes of evaporating concentration [9]. Positive correlation between nitrate and chlorine (r = 0.80) indicates a considerable impact of surface technogenic processes on the formation of chloride concentration in groundwater. Relatively high correlation coefficients are detectable for pairs of sodium, magnesium, calcium, sulphate, chloride and nitrate with salinity.

**Table 2.** Pair correlation coefficient of the chemical composition of groundwater.

| r | TDS | pH | $Na^+$ | $Mg^{2+}$ | $Ca^{2+}$ | $K^+$ | $HCO_3^-$ | $SO_4^{2-}$ | $Cl^-$ | $NO_3^-$ |
|---|---|---|---|---|---|---|---|---|---|---|
| TDS | 1.00 | | | | | | | | | |
| pH | –0.34 | 1.00 | | | | | | | | |
| $Na^+$ | 0.61 | –0.26 | 1.00 | | | | | | | |
| $Mg^{2+}$ | 0.68 | –0.08 | 0.12 | 1.00 | | | | | | |
| $Ca^{2+}$ | 0.84 | –0.42 | 0.34 | 0.38 | 1.00 | | | | | |
| $K^+$ | 0.16 | –0.01 | 0.07 | –0.01 | 0.12 | 1.00 | | | | |
| $HCO_3^-$ | 0.12 | –0.17 | 0.37 | 0.21 | –0.05 | –0.02 | 1.00 | | | |
| $SO_4^{2-}$ | 0.77 | –0.28 | 0.64 | 0.41 | 0.62 | 0.14 | 0.01 | 1.00 | | |
| $Cl^-$ | 0.77 | –0.20 | 0.19 | 0.70 | 0.69 | 0.13 | –0.01 | 0.27 | 1.00 | |
| $NO_3^-$ | 0.55 | –0.10 | –0.01 | 0.52 | 0.53 | 0.03 | –0.23 | –0.03 | 0.80 | 1.00 |

Thus, the carried out geostatistical analysis of the chemical composition of groundwater shows direct anthropogenic impact on the formation of the geochemical characteristics of groundwater in Kishenev which determines groundwater distinctive features associated with the given territory.

IOP Conf. Series: Earth and Environmental Science **21** (2014) 012024 doi:10.1088/1755-1315/21/1/012024

*4.2 Equilibrium with carbonate minerals*

Thermodynamic calculations showed that most groundwater under study, despite the low values of mineralization and calcium content, reaches equilibrium with carbonate minerals such as calcite (figure 2). Sufficiently remote location of imaging points indicates oversaturation of water with calcite and in water of all chemical types.

Groundwater as well as typical water of a leach cap falls under a general idea about its equilibrium with calcite. According to this fact water with salinity of more than 0.6 g/l and pH higher than 7.4 at the same time reaches equilibrium with the calcium carbonate [2, 8] which under certain geochemical conditions can not only dissolve but also precipitate from the solution to form a secondary mineral.

Minor part of water still remains undersaturated with calcite. This group includes hydrocarbonate and hydroarbonate-sulphate types of water. The average value of salinity in this water is 0.4 g/l, content of sodium ion is 329 mg/l, calcium content is 16 mg/l at pH equaled to 7.7, whereas average values of these parameters in calcite-saturated water is much higher. In this water calcium can be accumulated in solution as a result of its leaching from water-bearing rocks under conditions of active water exchange.

Oversaturation is observed in groundwater in relation to dolomite and siderite (figures 3 *a and b*). However, despite lower free energy of dolomite formation, secondary calcite will be the first to form because of a smaller number of components in the mineral composition. Siderite characterized by fairly high energy of formation will be the last to form.

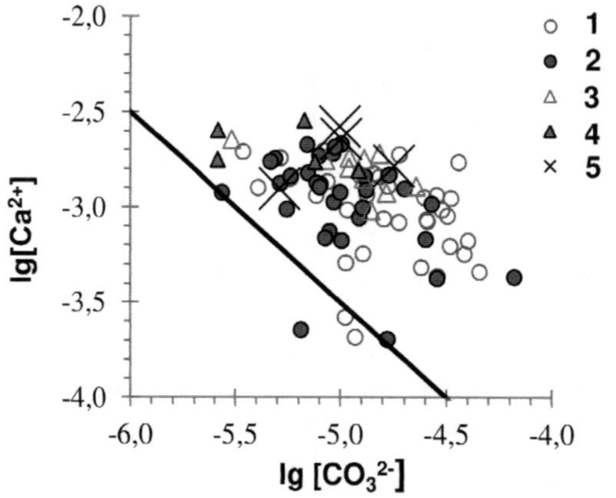


**Figure 2**. Diagram of equilibrium of groundwater in Kishinev with calcite at 25°C. Chemical types of water: 1 – $HCO_3$; 2 – $HCO_3$–$SO_4$; 3 – $SO_4$–$HCO_3$; 4 – $SO_4$; 5 – $NO_3$–$HCO_3$

## 5. Conclusion

Despite high water retention time and short time of interaction of groundwater in Kishinev with rocks there is equilibrium with many carbonate minerals. As a result of this water-rock interaction most of the elements which get into solution bind to the newly formed secondary minerals: Ca – calcite and dolomite, Mg – magnesite, Fe – siderite. Thus, groundwater under study can be attributed to carbonate forming geochemical type according to the classification of S.L. Shvartsev [1]. Observed equilibrium with calcite substantiates the early stage in the groundwater evolution which is characterized by relatively low salinity (less than 1 g/dm$^3$) as a result of hydraulic connectivity with surface water and atmospheric precipitation. Equilibrium to water carbonates shifts towards undersaturation at these very points. Since water continues to interact with aluminosilicates which are abundant in this area, the carbonate formation processes characterize groundwater composition in Kishinev at this stage.

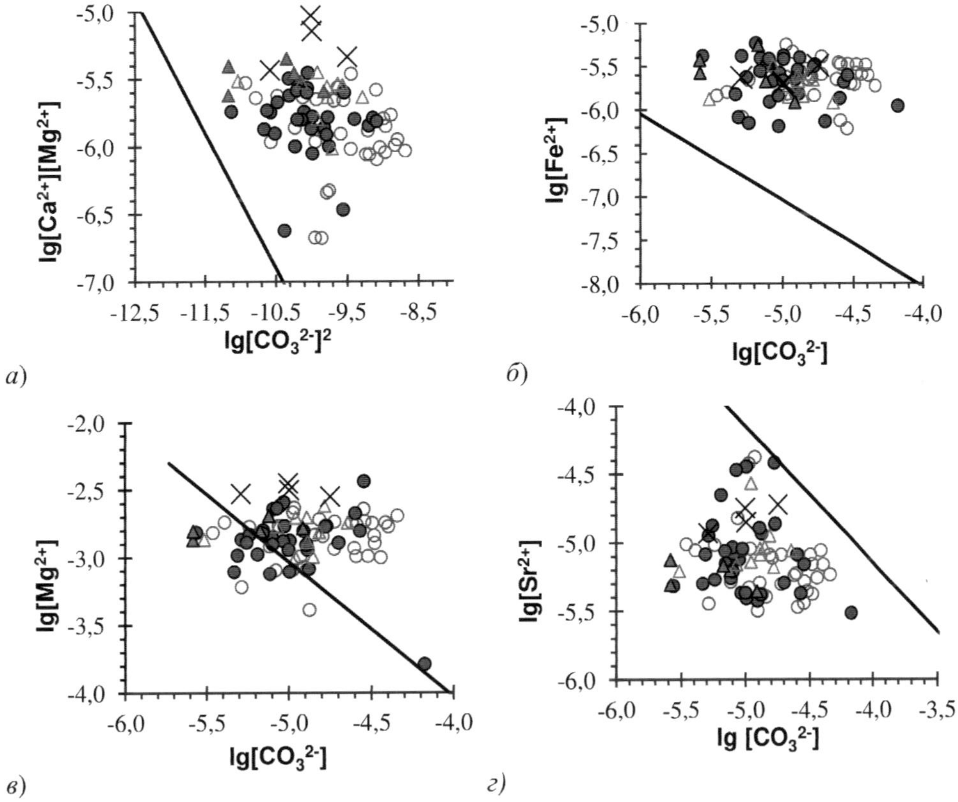


**Figure 3**. Diagram of equilibrium of groundwater in Kishinev with dolomite (*a*), siderite (*b*), magnesite (*c*) and strontianite (*d*) at 25°C (nomenclature see in figure 2)

**Acknowledgements**

Authors thank Prof. Dr. S L Shvartsev and Head of Laboratory of hydrogeology C E Moraru for their helpful advice and suggestions. The investigations were supported by The Russian Foundation for Basic Research (Project No 13-05-90919).

## 6. References

[1] Shvartsev S L 2013 *Procedia Earth and Planetary Science* **7** 810

[2] Shvartsev S L 2008 *Geoch. International* **46** (13) 1285

[3] Alekseev V A, Ryzhenko B N, Shvartsev S L, Zverev V P, Bukaty M B, Mironenko M V, Charykova M V and Chudaev O V 2005 *Geological evolution and water-rock self-organization V.1. Water-Rock System on the Earth Crust: interaction, kinetics, equilibrium, simulation* (Novosibirsk: Punlishing House SB RAS) p 244 [in Russian]

[4] *USSR Geology. Moldavian SSR. Moldavskaya SSR* 1969 **45** (Moscow: Nedra) p 456 [in Russian]

[5] Bukaty M B 2002 *Bulletin of Tomsk Polytechnic University* **305** (6) 348–65 [in Russian]

[6] Garrels R M and Christ Ch L 1965 *Solutions, Minerals and Equilibria* (New York: Harper & Row) p 450

[7] Moraru C E 1995 *Buletinul Academiei de Stiinte a Rep. Moldova.ser. Fizica si Tehnica* **1** 93–99

[8] Shvartsev S L 1998 *Hydrogeochemistry of hypergenesis zone* (Moscow: Nedra) p 366[in Russian]

[9] Perelman A I and Kasimov N S 1999 *Landscape geochemistry* (Moscow: Astreya-2000)

[10] Plyusnin A M, Zamana L V, Shvartsev S L, Tokarenko O G and Chernyavskii M K 2013 *Russian Geol. and Geoph.* **54** (5) 495–508

XVIII International Scientific Symposium in Honour of Academician M. A. Usov: PGON2014 IOP Publishing
IOP Conf. Series: Earth and Environmental Science **21** (2014) 012025 doi:10.1088/1755-1315/21/1/012025

# Change of trace elements content in sewage water under the influence of hydrophilic macrophytes

**K V Akhmed – Ogly[a] and O G Savichev[b]**
Tomsk Polytechnic University, Tomsk, Russia

E-mail: [a]algologichka@mail.ru, [b]osavichev@mail.ru

**Abstract**. According to the researches carried out by authors in 2013, the estimate of the effectiveness of domestic sewage treatment with the help of hydrophilic vegetation was received. It has been shown that if sewage is treated with the help of macrophytes, copper and lead concentration reduces. Thus, if the volume of sewage treated by reed mace is 500 ml and 1 l, lead concentration decreases 5 and 3,5 times, if sewage is treated by reed, lead concentration decreases 2,5 times in both cases; if sewage is treated by reed mace copper concentration decreases 0,9 and 1,8 times (if the volume of sewage is 500 ml and 1 l), if sewage is treated by reed, copper concentration decreases 1,4 and 1,5 times respectively. The conclusion has been drawn: in West Siberia it is possible to use the shallow reservoirs with natural aeration and hydrophilic vegetation for effective sewage treatment from such heavy metals as lead and copper.

## 1. Introduction

Pollution of water objects represents a global problem. Sustainable social and economic development in all regions of the world is impossible without solving this problem because resources of waters suitable for household-drinking use are getting reduced. In this regard it is necessary to provide the effective sewage treatment to the level providing steady functioning of water and land ecosystems. Achievement of this purpose can be connected with the construction of new treatment facilities and the expansion (reconstruction) of existing ones. But at the same time cost that limits the scope of environmental protection measures dramatically increases. To solve this problem is to use the artificial or natural reservoirs with natural aeration and hydrophilic vegetation.

There is operational experience of similar objects in many countries including the Russian Federation. In particular, reservoirs with hydrophilic vegetation are applied in Novosibirsk, Murmansk, Kursk regions [1,2]. There is such experience in Tomsk region, in Tomsk and Kargasok districts, and usually a basic purpose of treatment facilities is to decrease the concentration of organic and biogenic substances [3,4]. The possibilities of sewage treatment from metal under sufficiently severe climatic conditions of Siberia are less studied and it has determined the appropriate purpose of considered research, i.e. the effectiveness assessment of domestic wastewater treatment from Al, Zn, Cu, Pb, Cd with a help of local types of hydrophilic vegetation – reed and reed mace [5].

## 2. Methods and materials

The goal of research is to clear up two main questions: 1) how change of the metal content in the presence of a reed and reed mace in sewage happens; 2) at what ratio of sewage and vegetation the greatest efficiency of cleaning is reached. Taking into account these questions the research method has included:

1) sewage sampling from the receiver tank of sewage pumping station (SPS-6) of LLC Tomskvodokanal; it is a factory providing household-drinking and technical water supply, sewage treatment and sewage disposal of Tomsk, it is an administrative center of the Tomsk region of the Russian Federation located in southeast part of West Siberia; the chemical composition of sewage sample was defined in the accredited laboratory of LLC Tomskvodokanal;

2) vegetation sampling (reed mace and reed) from Peschanoe Lake, located 4 km to the southwest from Tomsk; the mass of each sample is 50 g and its linear size is 1-1,6 m;

3) placement of vegetation samples with root system in polyethylene containers with the capacity of 5 l with sewage; total quantity of containers – 12 pieces, including: a) 3 pieces with reed and volume of sewage of 1 l; b) 3 pieces with reed and volume of sewage of 0,5 l; c) 3 pieces with reed mace and volume of sewage of 1 l; d) 3 pieces with reed mace and volume of sewage of 0,5 l; all containers were placed on the sunny side with constant lighting during all light day in the open room with air temperature from 15°C at night and to 30°C during the day; the start of experiment – August 11, 2013;

4) measurement of Al, Zn, Cu, Pb, Cd concentrations in sewage in 5, 9, 15 days after the start of experiment (August 16, 20 and 26, 2013); chemical analysis are carried out in the accredited hydrogeochemical laboratory of Tomsk polytechnic university with a help of atomic absorption spectrometry and inversion voltamperometry methods;

Reed mace (Typha latifolia L.) and reed (Phragmites australis L.) were selected as test objects as they are widespread, have a wide range of environmental tolerance and dominate among other hydrophilic plants.

Peschanoe Lake has been chosen as a typical reservoir for the South taiga subzone of West Siberia. This lake can also be used for dumping of domestic sewage. Thus, the experiment allows to estimate efficiency of sewage treatment at the most rigid influence when 0,5 and 1 l of drains are the share of 50 g of a plant and it can be observed in low water level. The choice of hydrochemical indicators (concentration of Al, Zn, Cu, Pb, Cd) is due to the fact that these trace elements in accordance with their assessment of toxicity belong to hazard category III and II, and they are characterized as substances moderately dangerous (Cu, Zn) and highly hazardous (Al, Pb, Cd) nature. When maximum permissible concentration of these elements enters the human body, disruption of the functioning and the occurrence of development undesirable processes in an organism is observed. The loss of calcium from the bones depending on high contents of cadmium, paralysis of nerves of bending and unbending extremities, convulsions and coma, decrease of hemoglobin concentration in blood, being observed at poisoning with lead; growth stop, the low content of hemoglobin, abnormality in the liver, kidneys, brain tissue because of chronic excess of copper belong to such processes. Therefore it is so important to pay particular attention to these trace elements in the process of sewage treatment [6-8].

## 3. Results and discussion

Crude sewage of SPS-6 of LLC Tomskvodokanal according to O. A. Alekin's classification [9] is characterized as fresh with salinity, as to pH value it is alkalescent, it contains a significant amount of organic compounds, including hydrocarbons of the oil series, phenols, anionic surfactants (table 1). Sewage falls short of accepted standards of household-drinking water management in conformity with Fe, $PO_4^{3-}$, anionic surfactants, $NH_4^+$, oil products content, it falls short of accepted to standards of fishery water use according to $BOD_5$, $PO_4^{3-}$, $NH_4^+$ content.

**Table 1**.The average chemical composition of sewage of SPS-6 of LLC Tomskvodokanal, August 2013.

| Index | SPS- 6 | Index | SPS- 6 |
|---|---|---|---|
| pH | 7,67 | | |
| mg/dm$^3$ | | | |
| Suspended matters | 225 | Al | – |
| Dry residue | 541 | Zn | 0,0372 |
| $Cl^-$ | 62,7 | Cd | <0,0002 |
| $SO_4^{2-}$ | 40,5 | Pb | 0,0059 |
| $NO_3^-$ | 3,86 | Cu | 0,0165 |
| $NO_2^-$ | 1,25 | Oil products | 2,96 |
| $NH_4^+$ | 14,4 | Phenol | 0,0122 |
| $PO_4^{3-}$ | 5,07 | Anionic surfactants | 1,31 |
| Fe | 3,52 | $BOD_5$, $mgO_2$/дм$^3$ | 152 |
| Mn | 0,094 | COD, mgO/дм$^3$ | 333 |

The analysis of the received results showed that

1) at physiological level the increased concentration of considered trace elements can cause insufficiency of necessary elements, imbalance of plant food elements, change of permeability of membranes, causing their ruptures. At phenological level, adverse influence on plants is evident as marginal chlorosis, loss of their turgor, witherings and stooling delay.

2) the greatest efficiency of treatment is achieved by use reed mace and a ratio of 0,5 l of drains on 50 g of vegetation.

**Table 2.** The average chemical composition of sewage before and after treatment by hydrophilic vegetation.

| Index | SPS-6 | Sample volume 1 l | | | | | | Sample volume 0,5 l | | | | | |
|---|---|---|---|---|---|---|---|---|---|---|---|---|---|
| | | Reed | | | Reed mace | | | Reed | | | Reed mace | | |
| | 11.08.13 | 16.08.13 | 20.08.13 | 26.08.13 | 16.08.13 | 20.08.13 | 26.08.13 | 16.08.13 | 20.08.13 | 26.08.13 | 16.08.13 | 20.08.13 | 26.08.13 |
| mg/dm$^3$ | | | | | | | | | | | | | |
| Al | – | 0,068 | 0,096 | 0,151 | 0,069 | 0,074 | 0,072 | 0,124 | 0,124 | 0,165 | 0,085 | 0,034 | 0,037 |
| Zn | 0,0372 | 0,048 | 0,048 | 0,059 | 0,15 | 0,1 | 0,043 | 0,05 | 0,033 | 0,0345 | 0,16 | 0,047 | 0,077 |
| Cd | <0,0002 | <0,0002 | <0,0002 | <0,0002 | <0,0002 | <0,0002 | <0,0002 | <0,0002 | <0,0002 | <0,0002 | <0,0002 | <0,0002 | <0,0002 |
| Pb | 0,0059 | 0,0052 | 0,0024 | 0,0026 | 0,012 | 0,0038 | 0,0016 | 0,0018 | 0,0014 | 0,0022 | 0,0058 | 0,0076 | 0,0011 |
| Cu | 0,0165 | 0,017 | 0,0086 | 0,011 | 0,016 | 0,012 | 0,0092 | 0,013 | 0,01 | 0,012 | 0,028 | 0,018 | 0,017 |

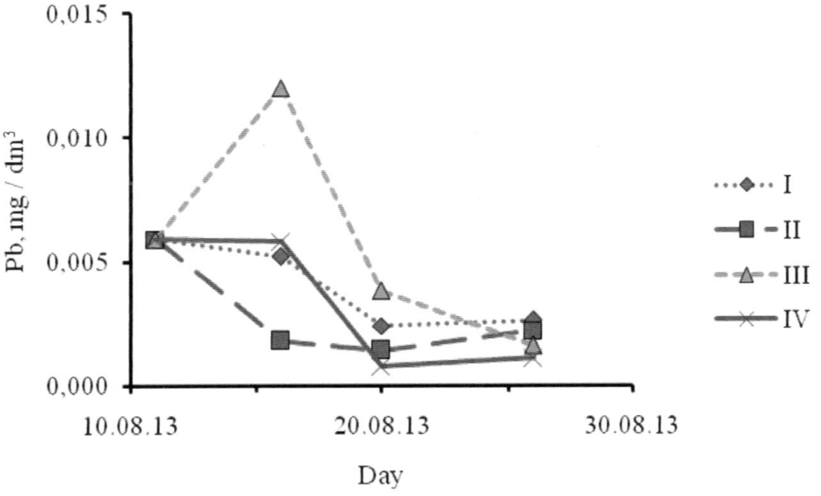


**Figure 1**. Change of Pb concentration in sewage according to conditions: I) vegetation is a reed, sample volume is 1 l; II) vegetation is reed, sample volume is 0,5 l; III) vegetation is reed mace, sample volume is 1 l; IV) vegetation is reed mace, sample volume is 0,5 l.

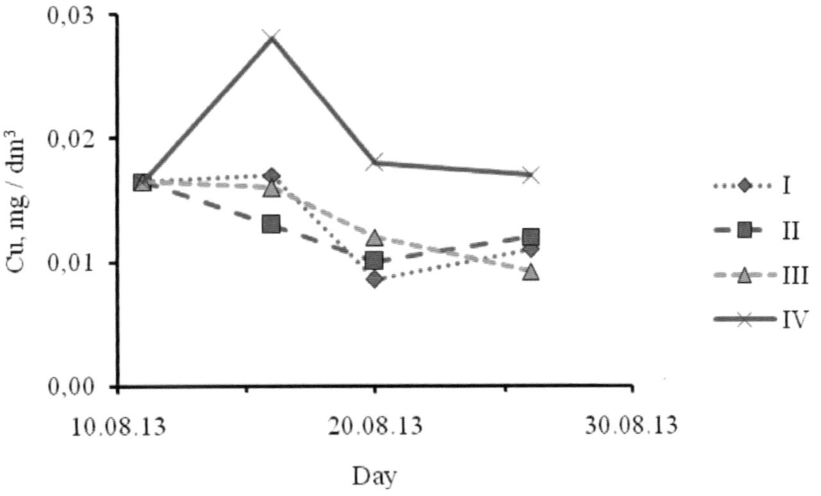


**Figure 2**. Change of Cu concentration in sewage according to conditions: I) vegetation is reed, sample volume is 1 l; II) vegetation is reed, sample volume is 0,5 l; III) vegetation is reed mace, sample volume is 1 l; IV) vegetation is reed mace, sample volume is 0,5 l.

## 4. Conclusion

The carried out investigation has shown that reed mace (*Typha latifolia L.*) and reed (*Phragmites australis L.*) can be applied to purify domestic wastewater from Cu and Pb in shallow reservoirs with natural aeration. The greatest effect is achieved by use of reed mace and a ratio of 0,5 l of drains on 50 g of the vegetation, the worst effect is achieved by use of reed and a ratio of 1 l of drains on 50 g of vegetation. Significant decrease of concentration is achieved on the average in 10 days.

Purification of domestic wastewater from Al, Zn in shallow reservoirs with natural aeration and hydrophilic vegetation isn't expedient. Probably, it is connected with that, on the one hand, absorbing ability of plants is effective till a certain moment (in case of experiment till 10th day). Then plants become unsuitable for further treatment or this biomass (50 g) isn't enough for effective removal of these elements from sewage. On the other hand, it can be assumed that such plants as reed mace and reed simply aren't suitable for removal of such trace elements as Al, Zn.

Work is performed with support RFBR 13-05-98045 r_sibir_a and TPU (project LRU_VAF_144_2014).

## References

[1] Sivkova E E and Semenov S Yu 2010 Ispolzovaniye tekhnologii «Constructed wetlands» dlya ochistki stochnykh vod malykh naselennykh punktov i predpriyatiy *Tomsk State University Journal of Biology* **4**(12) 123 – 130 [in Russian]

[2] Borzenkov A A, Cumani M V and Lukiyanchikov D I 2010 Primeneniye biologicheskikh prudov dlya doochistki stochnykh vod Kurskoy oblasti *The online academic journal of Kursk State University* **1**[in Russian]

[3] Savichev O G 2008 Biologicheskaya ochistka stochnykh vod s ispolzovaniyem bolotnykh biotsenozov *The Bulletin of the Tomsk Polytechnic University* **312** 69 – 74 [in Russian]

[4] Semenov S Yu and Shelepova L I 2008 Vodno-bolotnaya ochistka stochnykh vod *Bezopasnost zhiznedeyatelnosti* **1** 37-38 [in Russian]

[5] Alonso-Castro A J, Carranza-Álvarez C, Alfaro-De la Torre M C, Chávez-Guerrero L and García-De la Cruz R F 2009 Removal and Accumulation of Cadmium and Lead by *Typha latifolia* Exposed to Single and Mixed Metal Solutions *Archives of Environmental Contamination and Toxicology* **57** (4) 688-696

[6] Bogdanovsky G A 1994 *Khimicheskaya ekologiya* (Moscow: Moscow State University Press) p 237 [in Russian]

[7] 1989 *Spravochnik po gidrokhimii: uchebnoye posobiye* ed A M Nikanorov (Leningrad: Gidrometeoizdat) p 392 [in Russian]

[8] Molchanova J P, Zaika E A, Babkina E I and Surnin V A 2007 *Gidrokhimicheskiye pokazateli sostoyaniya okruzhayushchey sredy* (Moscow: Publishing House of the "Forum") p 192 [in Russian]

[9] Alekin O A 1953 *Osnovy gidrokhimii* (Leningrad: Gidrometeoizdat) p 296 [in Russian]

IOP Conf. Series: Earth and Environmental Science **21** (2014) 012026 doi:10.1088/1755-1315/21/1/012026

# Taxation indices of forest stand as the basis for cadastral valuation of forestlands

**V Kovyazin[1a], V Belyaev[1b], O Pasko[2c] and A Romanchikov[1d]**
[1]University of Mines, St. Petersburg, Russia
[2]Tomsk Polytechnic University, Tomsk, Russia

E-mail: [a]vfkedr@mail.ru, [b]belyaev_rosbi@list.ru, [c]helgapas@mail.ru, [d]romanchicov@inbox.ru

**Abstract.** Cadastral valuation of forestlands is one of the problems of the modern economy. Valuation procedures depend either on the profitability of timbering or forest areas are not differentiated according to value. The authors propose the procedure based on taxation indices of strata. The most important factors influencing the valuation are determined. The dependence that allows establishing the relative cost of a certain forest area is defined. Knowing the cadastral value of a model area, it is possible to determine the values of all other sites. The evaluation results correlate with the Faustman procedure with slight difference in the absolute value.

## Introduction

Forest has traditionally been one of the most important natural resources in many European countries: Russia [1, 2], Turkey [3], Italy [4], Greece [5], Romania [6], etc. Only Russian forest lands occupy 1.104 billion hectares or 64% of the total area of the country and are not available for private ownership. [7] Forestry planning is aimed at creating a reasonable balance between the real wood cutting and its reproduction [8], as well as its protecting [9]. Charges for forest use, which are obtained in the form of rent, or fees under the contract of sale must offset the costs of caring for and recovery of forest plantations [10, 11]. Planning tool, rationalization and economic development, environmental protection[12], as well as the basis for forest management is called cadastre [13]. Since the beginning of the 80s the forestland has been described in space and time in the framework of the cadastral systems adopted in different countries.

Cadastral valuation procedure defining the cadastral value of parcels includes the following:

1.Making a list of real estate items; 2. Gathering information about the market value; 3. Analysis of factors determining the value of property; 4. Processing graphic information; 5. Modelling; 6. Calculation of the cadastral value; 7. Economic analysis of cadastral valuation results. The information about the cadastral valuation is relevant at least for two years. Nowadays not only a visual inspection is used for cadastral valuation, but also the latest scientific technologies (digital maps based on aerial photographs and photogrammetric method. To calculate probable value of a parcel it is desirable to provide evaluation using different procedures. The comparative analysis of the procedures results, the analysis of accuracy and influence on the final value allows getting quite accurate results. The procedures planned for calculation should be practical and efficient.

In Russia, two procedures are mainly used for cadastral valuation of forest lands. The first procedure (2002) estimates the parcels in accordance with probable profits that can be earned from them, based on the capitalization of land rent. [7] It is convenient from economic point of view, but is applicable only when the use of a forest area can draw profit. After the crisis in Russia in 2008 logging became unprofitable. Calculations according to this method make the cadastral value negative or close to zero. There was no sense in this procedure. It was cancelled by the Government of the Russian Federation in 2010. A specific value of one hectare of forest land is used today as an alternative to the cancelled procedure. This value is unified for the whole country, which is certainly convenient for practical use. But we can hardly speak about any differentiation of forest lands according to the value.

The aim of this research is to develop an accurate cadastral valuation of forest lands on the basis of taxation indices of forest stand.

**Objects and methods of research**

The basis for calculating the cadastral value of forest lands are the taxation indices of forest stand which are determined by conventional forestry techniques. The diameters were measured by the Swedish calliper MontaxComputer, and the height of the trees was measured by the hypsometer made in Germany by Blume-Leyss. The age of trees was determined with the help of age defining auger. The age class was calculated according to the average age. The stand density was determined by the Bitterlich's angle gauge. The merchantability class was calculated according to the percentage of the merchantable timber output. The quality class was determined on the basis of M.M. Orlov's evaluation tables. The forest stand was calculated using the tables of timber volume depending on the height quality class. The storey composition was determined according to the percentage of each wood species in the foreststand. Usually the data for cadastral valuation are presented in a form that is uncomfortable for statistical processing. So, they were reorganized at the preparatory stage. Table 1 shows an example of an electronic form of forest management documentation (a sample of 100 plots at Kurortnyi forest park, Saint-Petersburg).

**Table 1.** Taxation indices of forest stand on stratum № 1 presented in the electronic form.

| Stratum № / S, ha / Year of inventory | Wood species, age | Storey / Average height, m | Height, m | Diameter, m | Age class / Age group | Merchantability class | Quality class | Density / Foresttype / Forest site type | Stand, $m^3$: per 1 ha / On stratum | Stand, $m^3$: Total / By species | Stock on stratum, $m^3$: Deadwood | Stock on stratum, $m^3$: Debris-strewn forest |
|---|---|---|---|---|---|---|---|---|---|---|---|---|
| 1 | 6E - 5 | 1 | 20 | 24 | 4 | 1 | 2 | 0,7 | 232 | 27 | 2 | 2 |
| 0,2 | 1C - 5 | -- | 19 | 24 | -- | 1 | - | KS | -- | 5 | --- | --- |
| 2013 | 3Б - 50 | 19 | 18 | 18 | 2 | 2 | - | S2 | 46 | 14 | - | 1 |

The following factors in table 1 were used as taxation indices: ***stratum number*** – serial number of the forest area in the compartment (it is convenient to use it for objects numbering); ***square of the land area*** (ha) (important information, however, when comparing objects some relative indices are required, square is not applicable here; in addition, the square size of a target site is considered in the index "Total stock of timber"); ***year of inventory*** *determines the relevance* of the given information, however, since all strata are recorded during one season, there is no need to use this index. If taxation indices were determined at different period of time, it is possible to make age corrections; *wood species and their age* – in this column NX indicates wood species, where N is a relative part of the specie in the forest stand in tens of percent, X is the abbreviated name of the specie. The average age of a particular species in the stratum is indicated through a hyphen. These indices are definitely among the most important ones because they help determine the composition of the forest stand, which due to the different cost of timber directly affects the final value of the forestland. Age also allows scheduling some economic activities in the forest stand (improvement thinning, final felling, sanitary felling, etc.); ***stand of timber*** is a key index to assess the economic attractiveness of this or that strata, values dependent on this index are also given in the column; ***height (m), diameter (cm)*** are the parameters describing the average shape and size of the tree trunks in the stratum. They play an important role in determining the specific purpose of wood - for example, whether to use it as a fireplace wood, or as a commercial wood; ***age class***, merchantability class, quality class are estimated values calculated on the basis of the values of other factors and statistically making no contribution to valuation; ***the amount of deadwood and debris-strewn forest*** $(m^3)$ comprises the non-merchantable wood. As a rule, it is quite small and does not affect the valuation significantly.

Thus, a set of factors for each stratum is presented in table 2. The data analysis shows that there are a lot of factors to be considered and their full use makes the model inconvenient and unsustainable. Then we considered the average indices of forest stand age and diameter. To be more precise, the obtained parameters were calculated as average values for each species (table 3).

**Table 2.** Factors influencing the cadastral valuation of the forestlands with different forest stand composition (fragment).

| Stratum № | Forest stand composition*, fractions | | | | Age, years | | | | Diameter, cm | | | | Storey height, m | Stand, m³/ha |
|---|---|---|---|---|---|---|---|---|---|---|---|---|---|---|
| | B | Al | F | P | B | Al | F | P | B | Al | F | P | | |
| 1 | 0,5 | 0,5 | 0 | 0 | 55 | 55 | - | - | 18 | 20 | - | - | 17 | 124 |
| 2 | 0,5 | 0,5 | 0 | 0 | 65 | 65 | - | - | 20 | 22 | - | - | 18 | 136 |
| 3 | 0 | 1 | 0 | 0 | - | 70 | - | - | | 24 | - | - | 18 | 136 |
| 4 | 0,3 | 0,7 | 0 | 0 | 55 | 35 | - | - | 16 | 18 | - | - | 16 | 93 |

*NB. Forest stand composition: B – birch, Al – alder, F – fir tree, P – pine tree, As – aspen

**Table 3.** Factors influencing the cadastral valuation of the forestlands (fragment).

| Stratum № | Forest stand composition*, fractions | | | | | Age, years | Diameter, | Storey height, m | Stand, m³/ha |
|---|---|---|---|---|---|---|---|---|---|
| | B | Al | F | P | As | | | | |
| 1 | 0,5 | 0,5 | 0 | 0 | 0 | 55 | 19 | 17 | 124 |
| 2 | 0,5 | 0,5 | 0 | 0 | 0 | 65 | 21 | 18 | 136 |
| 3 | 0 | 1 | 0 | 0 | 0 | 70 | 24 | 18 | 136 |
| 4 | 0,3 | 0,7 | 0 | 0 | 0 | 41 | 17,4 | 16 | 93 |

*NB. Forest stand composition: B – birch, Al – alder, F – fir tree, P – pine tree, As – aspen

After selecting the factors that can be used in a mathematical model for cadastral value calculation, it is necessary to determine the correlation ratio between them, since the correlation between the factors significantly reduces the accuracy of the results. The correlation matrix is presented in table 4.

The correlation matrix allows revealing a close connection ($|R| > 0,6$) between the factors. The correlation ratio between the fraction of birch and pine tree content in the forest stand is 0.74, which is caused by the competition between these tree species. This should be taken into account when developing the model. Such taxation indices as the average diameter, age, height and stand are mutually interconnected. Their correlation ratios range from 0.69 to 0.90. It is obvious that these 4 factors are interrelated – the maturity stands are higher and have a large stock of wood, than the young ones. Thus, it is relevant to exclude the average diameter and height of the stand from the model, as these parameters are closely related to the other two - the age and stand volume. It should be noted that the economic attractiveness of some forest area is determined by the stand index, which greatly depends on the factors mentioned.

**Table 4.** Correlation matrix

| Indices | Timber species* | | | | | Age | Dia-meter | Heigh t | Stand |
|---|---|---|---|---|---|---|---|---|---|
| | B | Al | F | P | As | | | | |
| B | 1 | | | | | | | | |
| Al | 0,020 | 1 | | | | | | | |
| F | -0,194 | -0,146 | 1 | | | | | | |
| P | -0,742 | -0,378 | -0,303 | 1 | | | | | |
| As | 0,402 | -0,077 | -0,116 | -0,451 | 1 | | | | |
| Age | -0,390 | 0,057 | 0,271 | 0,221 | -0,428 | 1 | | | |
| Diameter | -0,312 | 0,103 | 0,419 | 0,027 | -0,281 | 0,838 | 1 | | |
| Height | -0,155 | 0,019 | 0,369 | -0,073 | -0,093 | 0,763 | 0,903 | 1 | |
| Stand | -0,461 | -0,244 | 0,494 | 0,272 | -0,331 | 0,691 | 0,734 | 0,773 | 1 |

*NB. Forest stand composition: B – birch, Al – alder, F – fir tree, P – pine tree, As – aspen

The next stage of research is getting the factor or a system of factors that would reflect the composition of stands without overloading the model by excessive information. If the conifers prevail in the forest stand composition, it leads to the increase of the timber stock. If the broadleaved trees predominate, it reduces the timber stock. Therefore the whole forest stand can be divided into two corresponding groups. The unit of measurement will be the unit fraction. To use both fractions of conifers and broadleaved trees as two factors simultaneously is not reasonable because they are completely collinear (one can always be expressed in terms of the other). So, the fraction of the conifers may be used as the main index to reflect the structural composition of the forest stand on the stratum (table 5).

**Table 5.** The final set of factors influencing the cadastral valuation of the forest stand (fragment)

| Stratum № | Conifer fraction | Age, year | Stand, $m^3$/ha |
|---|---|---|---|
| 1 | 0 | 55 | 124 |
| 2 | 0 | 65 | 136 |
| 3 | 0 | 70 | 136 |

The basic data show that the vast majority of the stand has not reached the maturity stage. So the key parameter is not the age but the remaining time before reaching the age of maturity, i.e. the difference between maturity age and real age. For broadleaved trees the age of maturity is 60 years and for conifers it is 80 years. From the obtained data the weighted average time to start timbering was determined as the sum of productions of time before timbering the conifers and broadleaved trees and their fracture in the stand. To form to a unified system of measurements and correct recording of the distances between the factors the obtained data were standardized by the division of each value of the factor by the maximum value in the sample. The maximum stand of raw wood, taken as a unit, is found at the 34th stratum and equals 268 $m^3$/ha. Since the value dependence on time before timbering is inverse, it is turned into direct dependence for the convenience of calculations. To do this, all the standardized values obtained were subtracted from a unit and turned into reciprocal values. The index of the coniferous trees proportion in the stand composition essentially contributes to the cost of one cubic meter of wood. Thus, it can be taken into account only after all the basic calculations. According to the Federal State Statistics Service for 2013 one cubic meter of softwood round timber is 1.14 times more expensive than a cubic meter of hardwood. This difference is not spontaneous. It can be traced since 1998 on a regular basis. Then index that increases the value of wood for the broadleaf stand is 1, and 1.14 for the coniferous stand. For the mixed stand the formula will be the following:

$$k = 1 + f \cdot 1{,}14,$$

where:

$k$ – index that increases the value of the stand;

$j$ – conifer fraction in the forest stand.

Thus, after excluding the proportion of coniferous trees, there are still two taxation indices affecting the cadastral value. These are time before timbering and stand. The known values for each stratum can be projected onto a plane, so that the stand is marked on the Y-axis, and the time before timbering is marked on the X-axis (figure 1.1). The heights above the axis mark the cadastral value of the stratum. The direction of the regression line projection was determined at the next stage. It describes the dependence of the parcel value on two taxation factors. It should be noted that the maximum cost is common for the forestlands with little time left before timbering and maximum stand. So these strata are marked in the upper right part of the point cloud. It is obvious that the minimum cost is common for the forestlands where the stand tends to zero and the maximum time before timbering. Thus, we can draw the assumed regression line projection as the connection of the corresponding points (figure1.2).

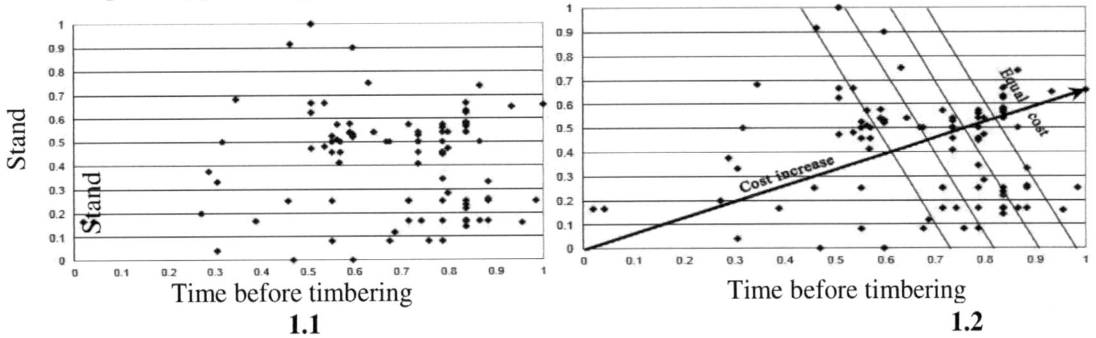


**Figure 1.1.** Point cloud on the plane. **1.2.** Cost line projection on the plane.

In this case, the segment length between the origin of coordinates and the point of intersection of the perpendicular dropped from the point with the values of the certain stratum factors with the cost line projection on the plane will be the geometrical expression of the forestland value (f).

The segment length is calculated according to the following formula

$$f = y \cdot \sin(\varphi) + x \cdot \cos(\varphi),$$ where:

$f$ – the segment length expressing the cost;

$x$ – on the X-axis, equals to the standardized time before timbering;

$y$ – on the Y-axis, equals to the standardized stand;

$\varphi$ – the inclination angle of the cost line projection.

The calculated value of $f$ for each stratum is multiplied by the value index $k$. Then these values are standardized by dividing them by the maximum value for forestlands. As a result the relative cost values (table 6) are calculated, which are fractions of the model stratum, the value of which is taken as a unit (in this case, it is stratum № 34).

In order to validate these results the same forestlands were evaluated using the procedure developed in 2002. Indices related to the economic activity of the enterprise were excluded from formulas, in particular the costs of transportation and timber processing, i.e. the value was determined only on the basis of income from timbering. The cost per one cubic meter of hardwood round timber was 1232 rubles, and the cost per one cubic meter of softwood round timber was 1407 rubles (according to the data of the Federal State Statistics Service in 2013). The capitalization ratio was set

IOP Conf. Series: Earth and Environmental Science **21** (2014) 012026 doi:10.1088/1755-1315/21/1/012026

at 0,014. The cadastral values obtained were standardized by dividing by the value of the most valuable stratum (stratum № 34). The values of the forestland value, obtained with the help of different procedures, correlate quite closely, R = 0,732.

**Table 6.** Standardization of the relative cost index.

| Stratum № | Relative cost index | Standardized relative cost index |
|---|---|---|
| 1 | 0.689354 | 0.577137 |
| 2 | 0.686628 | 0.574855 |
| 3 | 0.726742 | 0.608439 |
| 4 | 0.649571 | 0.543831 |

## Conclusion

Thus, the procedure described in the given paper characterizes the degree of change in value of the forestlands in the same way as the procedure developed in 2002. But the main difference is that the capitalization ratio is not used in the Principle component analyses. It is based purely on taxation indices of the forest stand for each forestland (stratum) in the forestry.

## References

[1] Karimov A State, Navy and the Origin of Petrine Forest Cadastre. Date Views 10.06.2014 http://ihst.ru/personal/krm/personal/petrine_forest_cadastre.htm

[2] Pasko O 2013 The forest resource of Tomsk Region and its use. Applied and fundamental Studies Proceedings of the 3rd International Academic Conference (St.Louis Missouri USA: "Science and Innovation Center ") 102-109

[3] Yalcin G 2012 Forest and Cadastre in Turkey and Sustainable Development Knowing to manage the territory, protect the environment, evaluate the cultural heritage Turkey (Rome: FIG Working Week ) 1-8

[4] Barliba L, C Barliba, G Eles et al 2010 Implementing the Forest Cadastre by Applying the Property Law in Order to Register a Forest Plot chapter 1/6 (Padova – Italy : Formec)

[5] Technical Chamber of Greece 2005 *Axes of Strategic and Operational Planning for National Cadastre Development* (*in Greek*) (Hellenic: Association of Rural and Surveying Engineers) p 120

[6] Aurel C, C Jenica, B Claudiu et al 2011 *Surveying of a land area partially covered withforest vegetation using GPS technology* (Craiova, Romania: University Publishing house) 289-293

[7] Kovyazin V, Martynov A and Melnikov E 2012 Basics of Forestry Management and Taxation (*SPb.*: Izdatelstvo "Lan")

[8] Drosos V 2007 *Forest opening up – shift and transportation of forest products* (Thessaloniki: Tziolas Publishing) p 124

[9] Amacher G, M Ollikainen and E Koskela 2009 Economics of Forest Resources. (Massachusetts: The MIT Press) p 148

[10] Giannoulas V 2001 Forest opening-upwith modern machinery. Technico-econmic, enviromental effects. Dissertassion, Aristotle University of Thessaloniki p 12

[11] Bettinger P, K Boston, J Siry et al 2009 *Forest management and planning* (New York: Academic Press) 331 13

[12] Zhivotyagina N 2011 Usage Specificity and Attitudes Adaptation to the Forest Resources Valution in the Present Day economic Situation *Scientific journal of Kuban State Agrarian University* **74** (10) 34-46

[13] Cadastral valuation of real estate in Russia.Date Views 10.06.2014 http://irt.su/wp-content/uploads/2012/01/Cadastral-valuation-of-real-estate-in-Russia.pdf

IOP Conf. Series: Earth and Environmental Science **21** (2014) 012027 doi:10.1088/1755-1315/21/1/012027

# New method for the onset point determination of the petroleum asphaltene aggregation

**A M Gorshkov[1], L V Shishmina[2] and A T Roslyak[3]**
Tomsk Polytechnic University, Tomsk, Russia

E-mail: [1]gorshkovam89@mail.ru, [2]shishmina@tpu.ru, [3]RoslyakAT@ignd.tpu.ru

**Abstract.** Work has been devoted to investigation of the process of asphaltenes aggregation in a model system asphaltene-toluene-heptane. The procedure of the onset point determination of the petroleum asphaltenes which is based on the method of photon correlation spectroscopy has been described. Two ways have been developed to determine the onset of asphaltenes aggregation by means of method of photon correlation spectroscopy. The first method allows defining the onset of asphaltenes aggregation of light scattering intensity depending on the quantity of *n*-alkane in the model system. The second method allows determining the onset point of asphaltenes aggregation parameter $k$ depending on the quantity of *n*-alkane. A comparative evaluation of onset points, obtained in different ways has been given. The influence of the model system composition on the process of asphaltenes aggregation has been studied. The characteristic diffusion time of asphaltenes has been calculated.

## 1. Introduction

Asphaltenes, the high molecular weight and the most aromatic component in the oil, are the subject many studies, due to the ever increasing share of the heavy oil in the world's hydrocarbon reserves. There is a change of thermobaric conditions, leading to the aggregation of asphaltenes colloidal particles in the process of fields exploitation. They can plug the pore medium of productive layers and lifting pipes column in the process of oil production and form deposits in the manifold during transport.

Currently asphaltenes considered as oil fraction which is insoluble in *n*-alkanes, and soluble in aromatic compounds such as toluene or benzene. The average composition of asphaltenes as a class is well known. Typical asphaltenes molecule consists of 3 – 10 condensed aromatic rings and the alkyl chains (about 3 – 5 carbons). Polycyclic nucleus contains a small amount of heteroatoms (N, O, S), and metals (Va, Ni, Fe) [1]. High polarity, aromaticity and heteroatom content in asphaltenes causes a strong tendency to self-association and the formation of molecular aggregates even in very dilute organic solvents. Asphaltenes aggregation in toluene solution at the concentration of 1.7 – 3.0 mg/L is accompanied by the nanoaggregates formation of 4 – 6 molecules with size from 0.5 to 4 nm [2]. Increasing the concentration of asphaltenes to 2 – 5 g/L results in the nanoclusters formation [3].

Asphaltenes can lose the sedimentation stability depending on the amount of the dispersed phase (asphaltenes) and the ratio in the dispersion medium precipitant/solvent the initial demonstration of them is onset of their aggregation. In modern publications the start of asphaltenes aggregation can be evaluated by using method of the point onset determination [4]. This procedure allows defining the parameter of start aggregation of asphaltenes by the deviation from the linearity dependence of the optical density of the asphaltenes solution from concentration *n*-alkane in accordance with the

Bouguer-Lambert-Beer law. The main difficulty of determining the onset point asphaltenes in this way is that the research results are strongly dependent on the laser wavelength [5]. We should look for new ways to determine the onset of the aggregation process due to the imperfection of this method.

We propose a new method for onset point determination of asphaltenes which is based on the method of photon correlation spectroscopy.

## 2. Experimental procedure

### *2.1. Materials*

The asphaltene samples were extracted from crude oil of Taymurzinskoe field by means of «cold» method Golde. The elemental composition of asphaltenes is presented in table 1.

**Table 1.** The elemental composition of asphaltenes.

| Content, % wt. | | | | | Degree of aromaticity H/C | Molecular weight, a.m.u. |
|---|---|---|---|---|---|---|
| C | H | N | S | O | | |
| 81.5 | 8.2 | 1.9 | 6.2 | 2.2 | 1.2 | 1900 |

The data of table 1 shows that the investigated asphaltenes have high molecular weight and value of degree of aromaticity. Sulfur predominates among the heteroatoms in asphaltenes.

Asphaltene aggregation was studied in a model system asphaltene-toluene-heptane which is the most common system for investigating the process of asphaltenes aggregation [6–9]. We used chromatographic grade toluene as a solvent and *n*-heptane of the same grade as a precipitant. All experiments were performed on freshly prepared solution of the asphaltenes in toluene. Concentration of asphaltenes in toluene was 0.4 and 0.6 g/L. Initiation of asphaltenes aggregation was performed by adding a certain amount of precipitant – *n*-heptane. After preparing investigated solutions the cell with sample was placed in an ultrasonic bath for 15 seconds to thoroughly mix the components together, prior to each experiment, and then measurements were conducted. A new cell has been chosen for each ratio *n*-heptanes/toluene. We used a round cell 15 mm diameter for all measurements of the aggregation process.

### *2.2. Method and equipment*

The method of photon correlation spectroscopy (PCS) or method of dynamic light scattering which allows determining the particle size in the dispersion system in real time during their growth is became widespread among the optical methods recently.

PCS method is based on determining of the diffusion coefficient $D$ of the colloidal particles by measuring the autocorrelation function of light scattering intensity fluctuations. If the particles take a spherical shape their radius $R$ can be calculated by Stokes-Einstein equation (1) [8]:

$$R = \frac{k_B \cdot T}{6 \cdot \pi \cdot \eta \cdot D} \tag{1}$$

where $k_B$ – the Boltzmann constant, $T$ – absolute temperature and $\eta$ – shearing viscosity of the medium.

We used PhotoCor Complex system to measure the size of asphaltenes associates and study the aggregation process. The device includes two parts: the optical-mechanical unit and control and data analysis unit [8].

A semiconductor laser beam with a wavelength $\lambda = 654$ nm was used as a radiation source. The main measuring tool of PhotoCor Complex device is photoelectron multiplier tube (PMT) operating in the photon counting mode. Light scattering intensity values increase amplify by a decade due to the greater sensitivity of the sensor through the use of an additional photodetector (Photocor-PC3), the main element of which is an avalanche photodiode. As a result, it becomes possible to track small changes in dispersion system. Thus in our work Photocor-PC3 was set at a 90° angle to determine the light scattering intensity. To study the asphaltenes aggregation and particle size determination second PMT (Photocor-PC1) is used. It is less sensitive but more effective for the study of the process. Photocor-PC1 was set at an angle of 40°. To avoid the influence of the «thermal lens» effect (occurrence of convection flows as a result of the liquid heating by laser beam) on the validity of the particle size determination the optimum neutral glass filter mounted on the laser for reduce the radiation power was selected for each experiment.

The main element of the control and data analysis unit in the PhotoCor Complex device is a programmable correlator which provides automatic control of the instrument operation according to the algorithm. It can also measure the correlation function and the integrated intensity of the light scattering. The autocorrelation function of the light scattering intensity fluctuations is recorded by cycles in our study. Duration of cycle is 1 minute, number of cycles is 1000.

## 3. Experiments and results

### *3.1. Onset point of asphaltenes*

The particle size growth occurs, which leads to an increase in the light scattering intensity in the process of asphaltenes aggregation. It is known that the scattering intensity for polarized incident light depends on the particle radius, ceteris paribus, as power function $I \sim R^6$ according to Rayleigh's theory [10]. As a result a sharp increase of the light scattering intensity is observed, when the particles go from molecular-dispersed to colloidal state. The process can be detected by optical photodetectors.

Figure 1 shows the dependence of the light scattering intensity on the amount of *n*-heptanes in a model system.

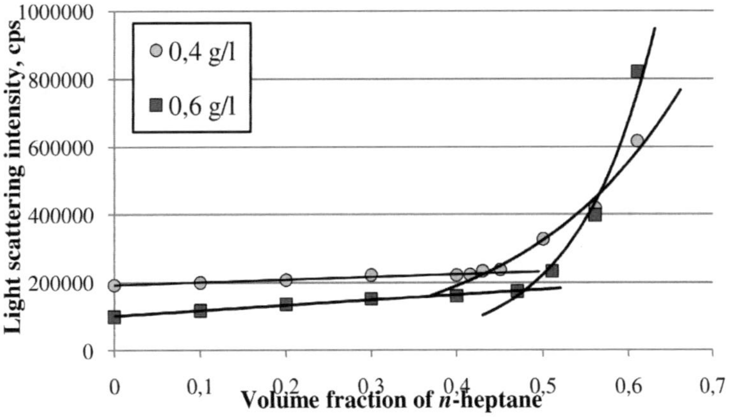


**Figure 1.** Dependence of the light scattering intensity on the volume fraction of *n*-heptanes in a model system.

A linear increase in the light scattering intensity from the amount of added *n*-alkanes for both asphaltenes concentrations in toluene (reliability approximation $\chi^2$>0.95) is observed with increasing volume fraction of *n*-heptanes to 0.4. For a given quantity of *n*-heptanes in solution the asphaltene aggregation does not occur but increasing transparency of the system asphaltenes-toluene-heptane takes place by the addition of *n*-alkanes. Moreover, the growth rate of scattering intensity for a solution with a asphaltenes concentration of 0.6 g/L is more than that of a solution with concentration 0.4 g/L, since it is more absorbing medium (figure 1).

There is a sharp growth in the light scattering intensity with increase volume fraction of *n*-heptanes over 0.4 for the asphaltenes solution with concentration of 0.4 g/L and over 0.45 for the solution with a concentration of asphaltenes of 0.6 g/L, that is evidence of the onset of aggregation. The particle size detection of asphaltenes order of several hundred nanometers using the method PCS confirms the aggregation when that value volume fractions of *n*-heptanes in a model system. In addition, this area of the experimental data is good approximated by an exponential dependence ($\chi^2$>0.98). The crossing point of linear and exponential functions will define the onset point of aggregation. Thus, for the asphaltenes concentrations in toluene 0.4 g/L the onset point is 0.43 volume fraction, while for a concentration of 0.6 g/L is 0.48 volume fraction of *n*-heptanes in a model system.

### *3.2. Process of asphaltenes aggregation*

Aggregation process, the variation with time of the particles number and the degree of their aggregation is described by Smoluchowski's equation. Solutions of the Smoluchowski's equation are known explicitly only for special cases, the appropriate modes of diffusion-limited aggregation (DLA) and reaction-limited aggregation (RLA). DLA mode dependence of the mean radius of aggregate $R$ from the time $t$ is given by [8]:

$$R = R_0\left(1+\frac{t}{\tau_D}\right)^{\frac{1}{d_f}}, \qquad (2)$$

where $d_f$ is fractal dimensionality, $R_0$ is an initial particle size, $\tau_D$ is the characteristic diffusion time. The value $d_f$ is in the range $1 \le d_f \le 3$ [8].

Figure 2 shows the results of a research of the aggregation process of asphaltenes with different concentrations in toluene by adding of *n*-heptanes. Solid curves are the processing result of the experimental data by least-squares method (LSM) on the power law in accordance with the expression (2).

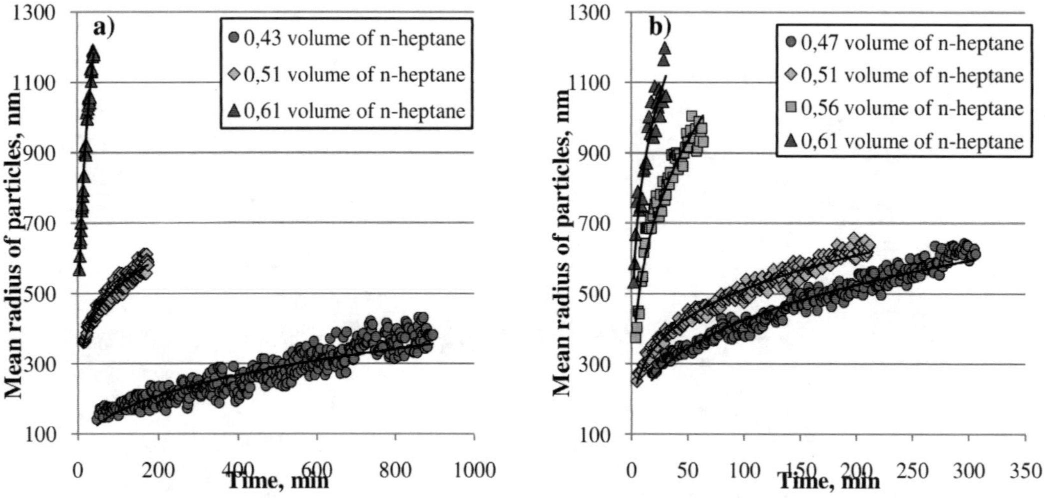


**Figure 2.** Process of asphaltenes aggregation upon adding of different *n*-heptanes quantity: a) – asphaltenes concentration is 0.4 g/L; b) – asphaltenes concentration is 0.6 g/L.

Figure 2 demonstrates the aggregation process is satisfactorily described by a power law ($\chi^2 \ge 0.93$). This suggests that the aggregation of asphaltenes occurs on the DLA mode. It should be also noted that the addition amount of *n*-heptanes is near to the threshold value the reliability approximation of the experimental data decreases. Probably, this is due to the fact that asphaltenes are a complex polyaggregative system in which molecules differed in physic-chemical properties, including the tendency to association are presented [11] When the *n*-heptane/toluene ratio is close to onset point, aggregation of different types asphaltene passes with different speed that leads to uneven growth process of the particles and, as a result, to a deviation from the DLA mode.

The characteristic diffusion time has been calculated experimentally for investigated asphaltenes concentrations in accordance with the decision of the Smoluchowski's equation for the DLA mode. In addition, the time of aggregation $t_{agr}$, in other words, the time during which there is growth of particles in solution to the critical size after which the aggregates start to precipitate and measurement becomes impossible, has been determined in this study. Results are presented in table 2.

**Table 2.** Parameters of petroleum asphaltenes aggregation.

| Concentration of asphaltenes in toluene, $gL^{-1}$ | Volume fraction of *n*-heptanes | Equation approximation on the method (LSM) | Reliability approximation $\chi^2$ | Parameters of asphaltenes aggregation | | |
|---|---|---|---|---|---|---|
| | | | | $t_{agr}$, *min* | $\tau_D$, *min** | *k*, – |
| 0.4 | 0.43 | $R=31.7t^{0.36}$ | 0.83 | 896 | $1.210^{-4}$ | 31.7 |
| | 0.51 | $R=207.3t^{0.35}$ | 0.95 | 166 | $4.4\cdot10^{-7}$ | 207.3 |
| | 0.61 | $R=310.2t^{0.36}$ | 0.99 | 39 | $2.1\cdot10^{-7}$ | 310.2 |
| 0.6 | 0.47 | $R=105.4t^{0.30}$ | 0.95 | 300 | $4.4\cdot10^{-7}$ | 105.4 |
| | 0.51 | $R=163.2t^{0.29}$ | 0.98 | 208 | $4.6\cdot10^{-8}$ | 163.2 |
| | 0.56 | $R=273.4t^{0.31}$ | 0.94 | 64 | $3.4\cdot10^{-8}$ | 273.4 |
| | 0.61 | $R=452.9t^{0.26}$ | 0.93 | 27 | $1.9\cdot10^{-10}$ | 452.9 |

*In processing the experimental data initial size was given $R_0$=1.25 nm.

Table 2 demonstrates the characteristic diffusion time of asphaltenes decreases with increasing concentration of asphaltenes in toluene and with increasing quantity of *n*-heptanes. Time of aggregation has a similar dependence.

Equation (2) for the DLA mode has no unique solution because it includes three indeterminate. Table 2 shows that the characteristic diffusion time is much shorter than the asphaltene aggregation $\tau_D \ll t_{agr}$, thus in the expression (2) can throw away a unit. The result is a power law dependence of the form $R=kt^{\alpha}$ which contains all the indeterminate in the parameter *k* [9].

$$k = R_0 \tau_D^{-\frac{1}{d_f}}, \ \alpha = \frac{1}{d_f}. \tag{3}$$

We construct the dependence of parameter *k* on the volume fraction of *n*-heptanes for different asphaltenes concentrations in toluene (figure 3). The parameter *k* values dependence on the quantity of *n*-alkanes is described by a linear function ($\chi^2 \geq 0.96$). The parameter *k* is equal 0 only if the characteristic diffusion time of asphaltenes equals infinity and this in turn is possible only in the absence of aggregation. The extrapolation of the linear function to the horizontal axis can define a threshold volume fraction of *n*-heptanes or the onset point.

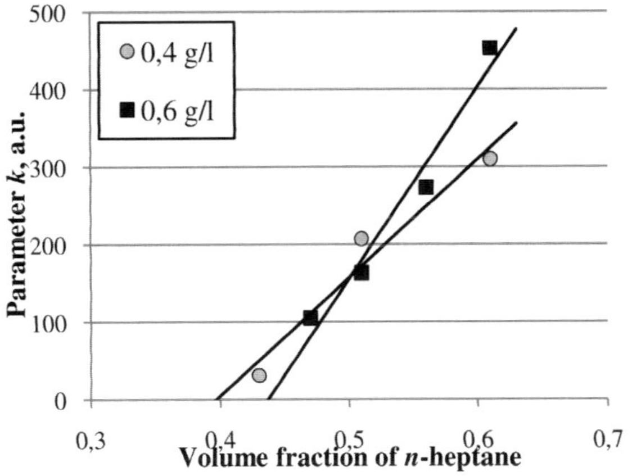


**Figure 3.** The parameter *k* dependence on volume fraction of *n*-heptanes for different asphaltenes concentration in toluene.

Thus, this method allows determining the onset of asphaltenes aggregation the parameter *k* of aggregation depending on the quantity of *n*-alkanes. The threshold volume fraction of *n*-heptanes for the

IOP Conf. Series: Earth and Environmental Science **21** (2014) 012027 doi:10.1088/1755-1315/21/1/012027

asphaltenes solution in toluene with a concentration of 0.4 g/L was 0.4, for the asphaltenes concentration of 0.6 g/ L was 0.44.

## 4. Conclusion

The study introduces a novel procedure of determining the onset point of asphaltenes which is based on the method of photon correlation spectroscopy. Two ways have been developed to determine the onset of asphaltenes aggregation. The first method allows defining the onset of asphaltenes aggregation of light scattering intensity depending on the volume fraction of *n*-alkanes. The second method allows determining the threshold amount of *n*-heptanes aggregation parameter $k$ depending on the volume fraction of *n*-alkanes.

The onset point values, calculated by the second method, is less than the values determined from the light scattering intensity. Perhaps this is due to the fact that asphaltenes are a complex polyaggregative system in which molecules with different tendency to association are presented. The light scattering intensity depends not only on the size of asphaltenes aggregates but also by their number. Because of non-simultaneity of the aggregation process of asphaltenes molecules of different types the onset of aggregation identified by the light scattering intensity is observed in more *n*-heptanes quantity. The amount of *n*-alkanes enhances the aggregation rate, which leads to a significant increase in the concentration of asphaltenes aggregates sufficient to fix the change in the light scattering intensity. Thus, the value of the onset point of asphaltenes can be previously estimated within 5% error in the light scattering intensity and by calculation of aggregation parameters $k$ can be updated an obtained value by means of photon correlation spectroscopy method.

## 5. References

[1] Sheu E Y 1996 Physics of asphaltene micelle and microemulsions – theory and experiment *Journal of Physics: Condensed matter.* **8** 125–41
[2] Zhang Y, Takanohashi T, Sato S, Kondo T, Saito I and Tanaka R 2003 Dissolution and dilution of asphaltenes in organic solvents *Energy and Fuels.* **17** 101–06
[3] Acevedo S, Castro A, Negrin J G and Fernandez A 2007 Relations between asphaltene structures and their physical and chemical properties: the rosary type structure *Energy and Fuels.* **21** 2165–75
[4] American Society for Testing and Materials (ASTM). *ASTM D 6703-01: Standard Test Method for Automated Heithaus Titrimetry*; ASTM: West Conshohocken, PA, 2001
[5] Safieva J O, Likhatsky V V, Filatov V M and Syunyaev R Z 2010 Composition of asphaltene solvate shell at precipitation onset conditions and estimation of average aggregate sizes in model oils *Energy and Fuels.* **24** 2266–74
[6] Mal'tseva E V, Gorshkov A M, Chekantseva L V, Shishmina L V and Yudina N V 2013 Influence of aggregation of asphaltenes on the rheological properties of oil *Russian Journal of Applied Chemistry.* **86** 1370–75
[7] Tukhvatullina A Z, Barskaya E E, Kouryakov V N, Ganeeva Yu M, Yusupova T N and Romanov G V 2013 Supramolecular structures of oil systems as the key to regulation of oil behavior *J Pet Environ Biotechnol.* **4** 1
[8] Mullins O C, Sheu E Y, Hammami A and Marshall A G 2007 *Asphaltenes, Heavy Oils, and Petroleomics* (New York: Springer) p 669
[9] Gorodetskii E E, Kuryakov V N, Yudin I K, Dechabo V A, Kosov V I and Yudin D I 2012 Light scattering study of heavy petroleum fractions stability and aggregation kinetic in oil systems *Georesources. Geoenergetics. Geopolitics.* **6** 1
[10] Fridrihsberg D A 1984 *Course of colloid chemistry. Textbook for institute of higher education* 2 (Leningrad: Chemistry) p 368
[11] Ganeeva Yu M, Yusupova T N and Romanov G V 2011 Asphaltene nano-aggregates: structure, phase transitions and effect on petroleum systems *Russian Chemical Reviews.* **80** 993–1008

# Turbulent flow of oil-water emulsions with polymer additives

**V N Manzhai, Monkam Monkam Clovis Le Grand and A V Abdousaliamov**
Tomsk Polytechnic University, Tomsk, Russia

**Abstract.** The article outlines direct and reverse oil-water emulsions. Microphotography study of these emulsions was carried out. The effect of water-soluble and oil soluble polymers on the emulsion structure and their turbulent flow velocity in cylindrical channel was investigated. It has been experimentally proven that if the fluid being transported is not homogeneous, but a two-phase oil-water emulsion, only the polymer that is compatible with dispersion medium and capable of dissolving in this medium can reduce the hydrodynamic resistance of the fluid flow. Thus, the resistance in direct emulsions can be reduced by water-soluble polyacrylamide, while oil-soluble polyhexene can be applied for reverse emulsions.

When a polymer is dissolved in a liquid, there occurs viscosity increase and a simultaneous acceleration of turbulent flow rate of the polymer solution in comparison with the flow rate of low-viscosity solvent. This paradoxical phenomenon is called the Toms effect and it has found a wide practical application in the oil pipeline transportation. When the Toms effect occurs (figure 1), it leads to reduction in the hydrodynamic resistance coefficient (λ), which is used in Darcy-Weisbach equation

$$\Delta P = \lambda \cdot \frac{\rho \cdot L}{4 \cdot \pi^2 \cdot R_w^5} \cdot Q^2$$

The λ value decrease is accompanied either by the increase of the volume flow rate (Q) at a constant predetermined pressure drop ΔP = const, or by the decrease of the pressure drop (ΔP) caused by friction at a constant flow rate Q = const.

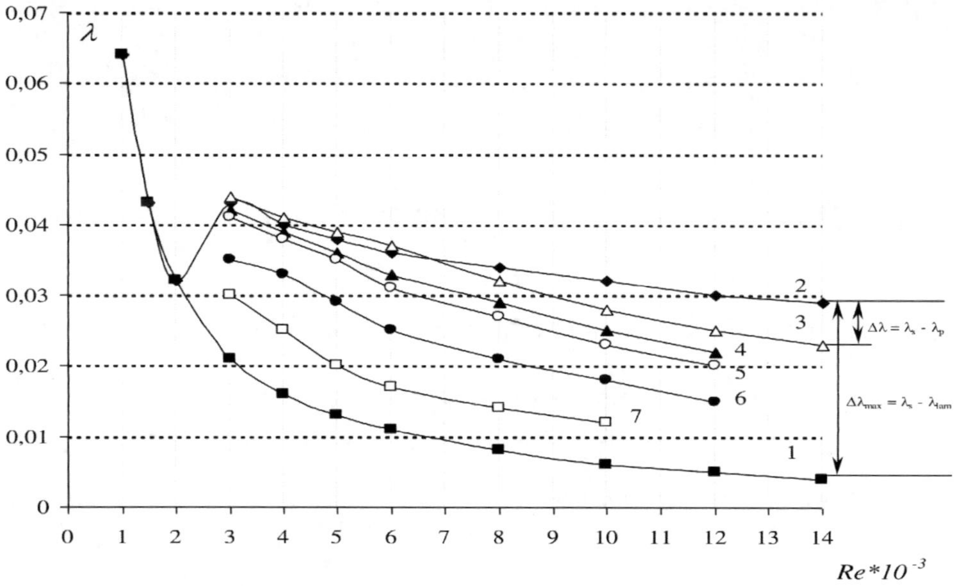


**Figure 1.** Dependence of hydrodynamic resistance on the Reynolds number for different polymer-solvent system [1].

1. Poiseuille theoretical curve for laminar flow $\lambda_{лам} = 64/Re$
2. Blasius empirical curve for turbulent flow $\lambda_{тур}=0,3164/Re^{0,25}$
3. Polyisoprene dissolved in toluene solution ($C$ = 0,05 кг/м$^3$; $M_r = 0,5 \cdot 10^6$);
4. Polyisoprene dissolved in toluene solution ($C$ = 0,1 кг/м$^3$; $M_r = 0,5 \cdot 10^6$)
5. Polybutadiene dissolved in toluene solution ($C$= 0,1 кг/м$^3$; $M_r = 0,6 \cdot 10^6$);
6. Polybutadiene dissolved in toluene solution ($C$ = 0,1 кг/м$^3$; $M_r$= $1,2 \cdot 10^6$);
7. Polyisoprene dissolved in oil ($C$ = 0,1кг/м$^3$; $M_r = 0,5 \cdot 10^6$).

The magnitude effect of reducing the hydrodynamic resistance (DR,%), which characterizes the energy yield obtained through pumping oil with polymer additives in comparison with pumping fluid without additives, is calculated by the formula

$$DR,\% = \frac{\lambda_S - \lambda_P}{\lambda_S} \cdot 100\% \quad (1)$$

Where $\lambda_S$ and $\lambda_P$ - hydraulic resistance coefficients of solvent and polymer solution, respectively.

Under actual operating conditions delivering oil through trunk pipeline by centrifugal pumps one can simultaneously observe decrease in the pressure drop in the pipe and increase of the volumetric flow rate (figure 2). When the pipeline is filled with oil-treated polymer additive, there is an increase in the Toms effect (figure 3).

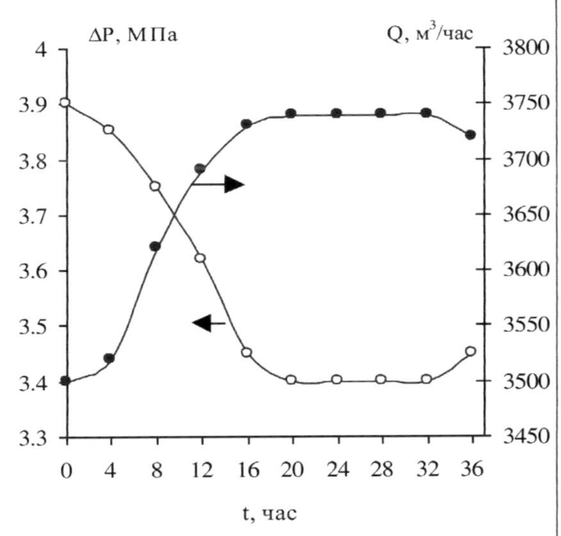


**Figure 2.** Changes in the pressure drop and flow rate observed in Tihoretsk - Novorossiysk pipeline when dosing pipe with the polymer additive [2].

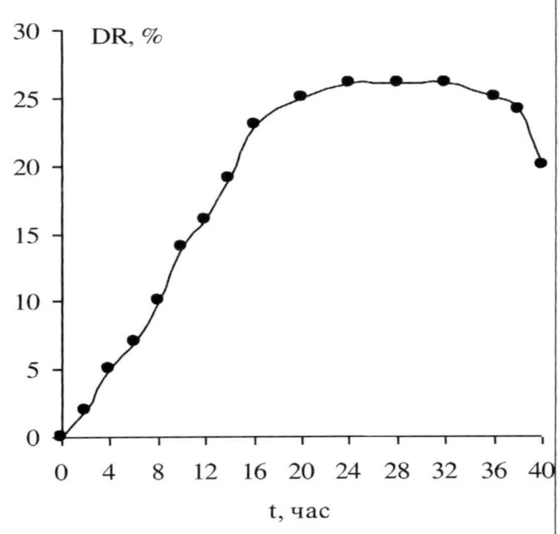


**Figure 3.** Changes in the magnitude of Toms effect in Tihoretsk - Novorossiysk pipeline when dosing pipe with the polymer additive [2].

In laboratory investigations, we conducted a pilot study of the effect of polymer additives on the flow rate of oil-water emulsion. During the in-situ filtration and delivery of oil-water mixture through gathering pipeline both "direct" emulsions (O/W) and "inverse" emulsions (W/O) can occur. Depending on the physicochemical nature of the dispersion medium, we used both a water-soluble polymer (polyacrylamide - PAA) and an oil-soluble polymer (polyhexene - PH).

The experimental part of the research was carried out using turbulent rheometer, which is structurally simple and similar to the capillary viscometer, but allows the study both in the laminar and turbulent flow regime.

The receiving chamber of the rheometer is filled with working fluids (solvents) and the polymer solutions produced from these solvents. The usual working fluids are water, ethanol, toluene, heptane, gasoline, oil and etc. By using the gas system operating on inert gas, in the chamber of the rheometer it is possible to produce different pressure drop effects ($\Delta P$) and, therefore, different flow rates of fluid through the tube with the radius ($R_w$) and the length (L), connected to the chamber. An electronic stopwatch records the elapsed time (t) required for a constant volume of fluid (V) to flow out of the chamber. Then we can calculate the volumetric flow rate Q = V / t, as well as the shear stress at the channel wall ($\tau_w$) and the Reynolds number (Re):

$$\tau_w = \Delta P \cdot R_w / 2L \qquad \mathrm{Re} = 2Q / \pi R_w \cdot \nu$$

More detailed information about the turbulent rheometer is given in the article [3].

At the initial stage of experiment we determined the effect of these polymers on the individual fluids (water and oil). The graphic illustration of the dependence of the effect (DR,%) on the concentration of PAA dissolved in water or PH dissolved in benzene is shown in Figures 4 and 5. The analysis of the results presented in the figures shows that these polymer samples are approximately equal in their anti-turbulent efficiency, but each of them can only be used in the dispersion medium of certain physical and chemical nature (water or oil).

Under hydrodynamic conditions typical for trunk pipeline operation (shear stress on the tube wall $\tau_W \approx 1 - 10$ Па) both of these samples are capable of reducing the resistance by more than 50% at the "optimal" concentration of $C_{opt.} \sim 10$ ppm (i.e. concentration at which the maximum magnitude of $DR_{MAX}$ effect is achieved).

However, the laboratory tests showed that if the transported liquid medium is not a homogeneous liquid (figure 4 and 5) but a two-phase oil-water emulsion, then its hydrodynamic resistance can be reduced only using the polymer that is compatible with the dispersion medium, i.e. which is soluble in that dispersion medium.

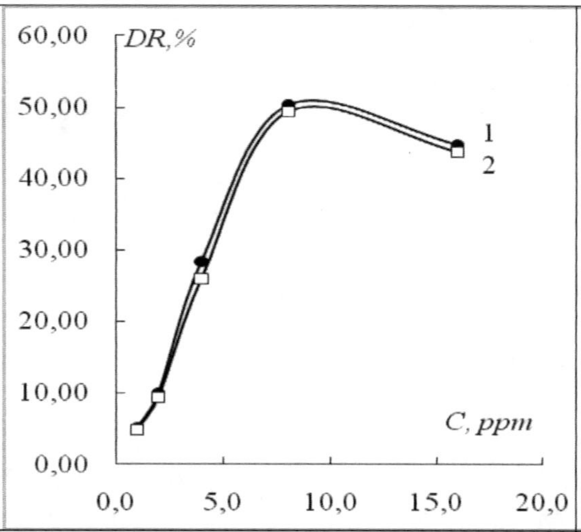


**Figure 4.** Dependence of the effect (DR,%) on the concentration (C, ppm) of polyacrylamide in water at different shear stresses: $1 - \tau_W = 8$ Па; $2 - \tau_W = 4$ Па.

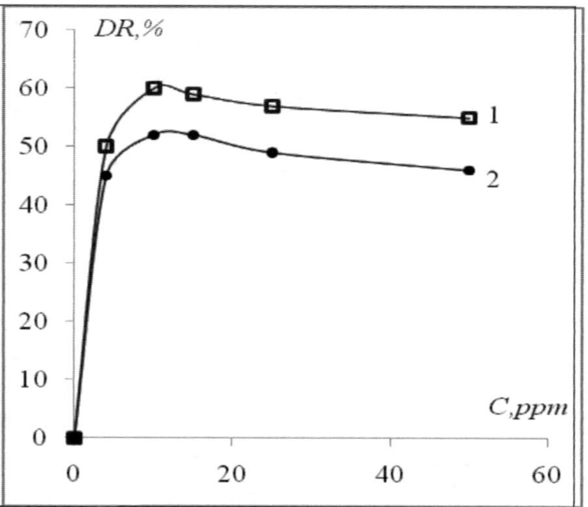


**Figure 5.** Dependence of the effect (DR,%) on the concentration (C, ppm) of polyhexene in benzene at different shear stresses: $1 - \tau_W = 11$ Па и $2 - \tau_W = 4$ Па.

Thus, the water-soluble PAA is capable to reduce the resistance in "direct" emulsions, while oil soluble PH can produce a similar effect in "reverse" emulsions. The experimental results of the effect of these polymers on the "direct" or "reverse" emulsions are shown in figure 6.

The figure shows that in the turbulent flow region at Re > 2300 hydrodynamic drag coefficients ($\lambda_S$) of water, oil and oil-water mixture without polymer additives, calculated by Darcy-Weisbach formula, can be naturally superimposed on the Blasius empirical curve. With the same polymer being added to oil-water emulsions, hydrodynamic drag coefficients ($\lambda_P$) decrease and enter $64/Re < \lambda_P < 0.3164/Re^{0.25}$ range, i.e. a zone between the Blasius curve and the hypothetical extension of the Poiseuille curve to a turbulent region (figure 6).

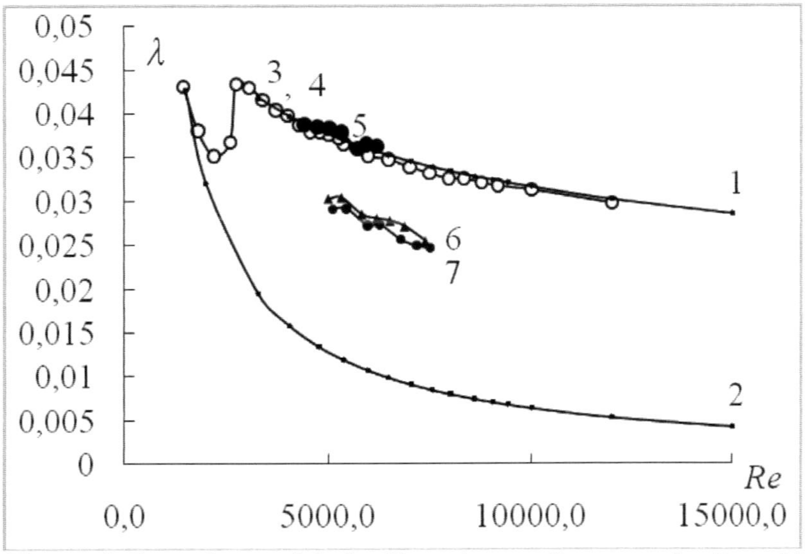


**Figure 6.** Dependence of hydraulic resistance factor on the Reynolds number for different systems:
1- Blasius empirical curve $\lambda = 0.3164/Re^{0.25}$;
2- Poiseuille curve and its hypothetical extension to a turbulent region $\lambda = 64/Re$;
3- Water;
4- Oil;
5- Oil-water emulsion;
6- 'Direct' oil-water emulsion containing PAA (C = 10ppm);
7- 'Reverse' oil-water emulsion containing PH (C = 50ppm).

After the experimental measurements with turbo-rheometer for given pressure drops ($\Delta P$) in volumetric flow rates of "direct" emulsions with additives of various concentrations of PAA ($Q_{PAA}$) and flow rates of "reverse" emulsions with additives of various concentrations of PH ($Q_{PH}$), hydrodynamic drag coefficients were calculated by Darcy-Weisbach equation. The further calculations were made by the formula (1) and determined hydrodynamic drag reduction values (DR,%). Their dependence on the concentration of polymer added to the oil-water emulsion is illustrated in figures 7 and 8. These figures show that for reducing hydrodynamic drag of turbulent flows of oil-water emulsions it is necessary to saturate these emulsions with polymer additives in amounts significantly above their "optimal" concentration for individual fluids (figure 4 and 5).

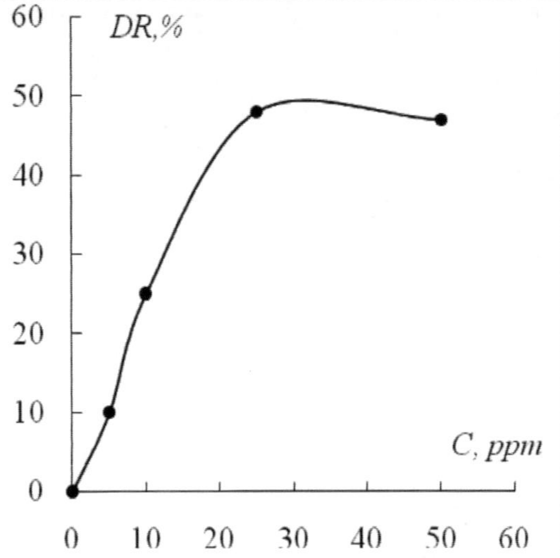


**Figure 7.** Dependence of effect DR (%) on the concentration of polyacrylamide in the "direct" emulsion.

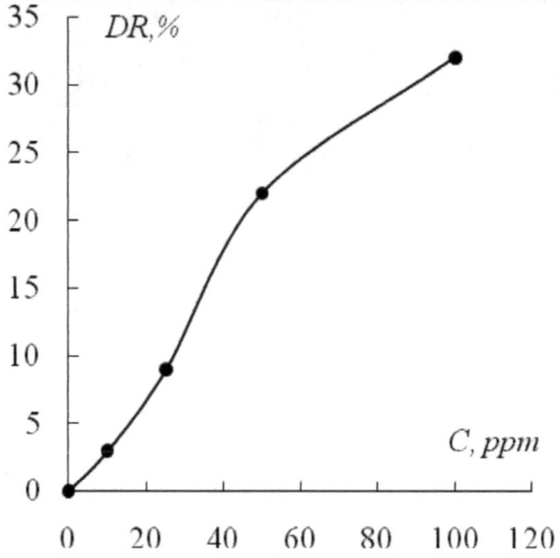


**Figure 8.** Dependence of effect DR (%) on the concentration of polyhexene in the" reverse" emulsion.

This substantial increase in the number of input anti-turbulent additives is required due to the fact that a significant portion of polymeric macromolecules with asphaltenes and resins from the water mixture is adsorbed at the interface, i.e. on solvent-adsorption layers of micro-emulsion droplets. This explanation is confirmed by photomicrographs which are taken with a modular biological microscope Olympus SKH41 and presented in figures 9 and 10.

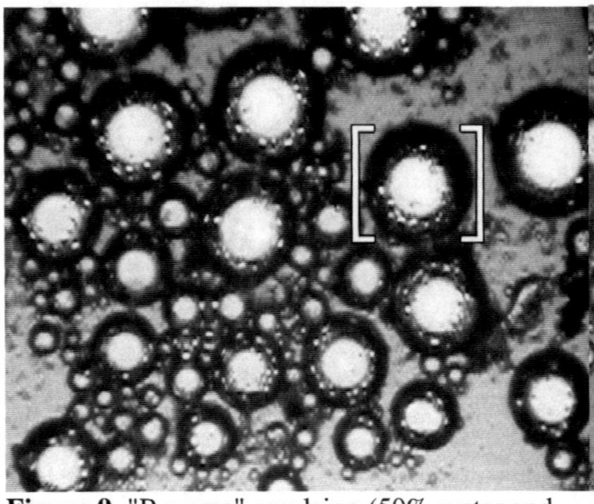

**Figure 9.** "Reverse" emulsion (50% water and 50% oil) with PAA.

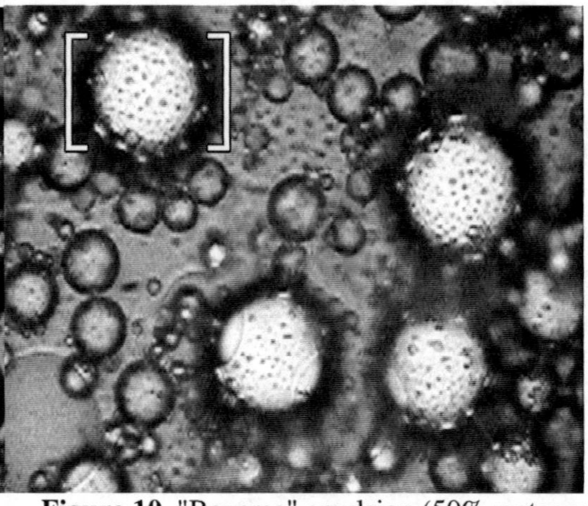

**Figure 10.** "Reverse" emulsion (50% water and 50% oil) with PH.

Summing up, transport of oil-water mixtures through pipeline requires a significantly greater amount of polymer additives than it does while delivering a single-phase liquid - either oil or water. This conclusion is, particularly, characteristic of emulsions of a "reverse" type.

## References

[1] Manzhai V N 1987 Influence of the concentration of polymer additives on reduction of hydrodynamic resistance of oil *Oil and gas. Proceedings of Higher Education* **1** 59-61

IOP Conf. Series: Earth and Environmental Science **21** (2014) 012028 doi:10.1088/1755-1315/21/1/012028

[2] Nesyn G V, Manzhai V N, Popov E A et al. 1993 An experiment to reduce hydrodynamic resistance of oil transported through Tihoretsk-Novorossiysk pipeline *Pipeline* **4** 28-30
[3] Nesyn G V, Manzhai V N and Souleymanova Y 2012 Mechanism of action, efficiency evaluation and peculiarities of polymer additive preparation for anti-turbulent transport of hydrocarbon liquids Polymer **54** (1) 65 - 72

# Conversion of straight-run gas-condensate benzenes into high-octane gasolines based on modified ZSM-5 zeolites

**V Erofeev[1], V Reschetilowski[2], A Tatarkina[3], I Khomajakov[4], L Egorova[5] and T Volgina[6]**
[1,2,3,4,5,6] Tomsk Polytechnic University, Tomsk, Russia
[3] Dresden University of Technology, Dresden, Germany

Email: [1]erofeevvi@mail.tomsknet.ru, [2]wladimir.reschetilowski@chemie.tu-dresden.de, [3]tatarkina.nastya@mail.ru, [4] khomyakov_i.s.@mail.ru, [5] egorova xf tsu@mail.ru, [6] volgina_t@mail.ru

**Abstract.** This paper describes the conversion of straight-run benzene of gas condensate into high-octane gasoline based on zeolite catalyst ZSM-5, modified in binary system oxide- based Sn (III) and Bi (III). It was defined that the introduction of the binary system oxide-based Sn(III) and Bi (III) into the basic zeolite results in the 2-fold increase of its catalytic activity.High-octane gasoline converted from straight-run benzene is characterized by a low benzol content in comparison to the high-octane benzenes produced during the catalytic reforming.

## 1. Introduction

Today one of the basic processs in petroleum refining is catalytic reforming , i.e. a technological procees where high-octanes benzenes are produced from straight-run gasoline.The catalytic reforming proceeds at 450-500 °C in hydrogen-containing gas environment based on alumino-platinum catalysts. Alumino-platinum catalyst reforming is rather sensitive to different micro-impurities, therefore, raw hydrocarbons should be preliminarily subjected to high purification from sulphur-, oxygen and nitrogen compounds. Produced high-octane benzene in catalytic reforming based on alumino-platinum catalysts includes up to 50-70 % aromatic hydrocarbon and up to 7-15% benzol, which , in this case, requires additional separation of arene and especially benzol so as to produce finished motor gasoline from liquid reforming products [1-2].

In this respect, the most practical process is zeoforming[3] on zeolite-containing catalysts-ZSM-5 type for production of high-octane benzenes with low aromatic hydrocarbon content. Zeolites have a unique property [4]: they have high activity and selectivity in such reactions as dehydration, cracking isomerization, oligomerization and dehydrocyclization of various hydrocarbon source groups due to their unique microporous structure (micropore size-0.5-0.8 nm) and molecular and size analysis properties.The application of zeolite-containing catalysts excludes the preliminarily raw material hydrofining.

The conversion process of straight-run gasoline based on zeolite catalysts is quite different from that of catalyst reforming as the first produces high-octane benzene with low benzol content (not more than 1-3%), total aromatic hydrocarbon content-not more than 30-35 % and sulphur- not more than 0.05–0.10 mass %. It should also be noted that preliminarily raw hydrocarbon hydrofiningis excluded due to the application of different modified zeolite-contaning catalysts [5-9].

The target of this research is to investigate the conversion process of straight-run benzene-gas condensate into high-octane gasoline based on zeolite catalysts, modified by binary systems on complex tin and bismuth oxides.

## 2. Experimental procedure amd applied methods

High-silica zeolites (HSZ-G) with high silic-module 50 was produced from alkaline alumosilicic gel at 175-180°C during 3-4 days applying organic structure-forming additive hexamethylenediamine as template. After crystallization synthezied zeolites are flushed out in distilled water several times, dried at 110°C and then calcinated at 550°C for 6 hours. Reactive HSZ-G zeolites were converted in 1m $NH_4NO_3$ water solution at 90°C for 2 hours and further drying-out at 110°C and calcinating at 550°C for 6 hours ($Na_20$ content in decationated zeolites is less than 0.01%). Modified HSZ-G is produced by liquid infiltration of 0. 1 M HCl solution, containing tin and bismuth oxides in the following element ratio: Sn/Bi = 10:1, to zeolite saturation capacity[10]. Complex tin and bismuth oxides are produced by coprecipitated hydroxide calcination. After this, zeolites with compound Sn and Bi oxide coating ( 1-5mass%)are dried at 110 $^0$C and calcinated at 550 $^0$C for 6 hours.

To identify the produced zeolite catalysts HSZ-G, infrared spectroscopy (infrared Fourier spectroscopy Nicolet 5700) and X-ray phase analysis (X-ray unit DRON-3, Mo-anode; Ni-filter) were applied. Infrared-spectrum was conducted within the range of 450-2000 $cm^{-1}$). The X-ray of synthesized samples showed lines with the following interplanar spacing (d, Å): 11.05, 10.19, 4.26, 4.07, 3.87, 3.83, 3.73, 3.66, typical for high- silica zeolites ZSM-5 type. Produced zeolite infrared -spectra showed absorption bands at 1000–1200, 795–800, 451 and 541 $cm^{-1}$, as in this range, the absorption bands (a.b.) of the major oscillations of $AlO_4$, $SiO_4$ tetrahedron skeleton are found. Intense absorption band of 1000-1200 $cm^{-1}$ is conditioned by antisymmetric stretch vibrations of $TO_4$ tetrahedron; band of 794 $cm^{-1}$ is related to the valence vibration in which $SiO_4$ tetrahedron is the basic one, whereas, the position of this band influences the silicate module ($SiO_2/Al_2O_3$) in zeolite scaffold. Absorption band of 541 $cm^{-1}$ is stipulated by five-membered rings in the zeolite skeleton and indicates the fact that this zeolite is of the ZSM group. All synthezied HSZ-G samples according to infrared spectroscopy and X-ray phase analysis could be related to zeolite group.

Conversion of straight-run benzene fraction of gas condensate at 70–170°C based on modified zeolite catalysts was performed in a circulating catalytic unit with a fixed-bed catalyst (reactor volume 8 $cm^3$) within the temperature interval of 350–425°C at feed space velocity of raw material2 $hrs.^{-1}$, atmosphere pressure and exposure for each fixed temperature process was 1 hour.

The analysis of gaseous hydrocarbons was performed in the stainless steel packed column (length-3m; inner diameter-3mm), filled up with 5% NaOH to $Al_2O_3$ (fraction- 0.25–0.50 mm), liquid hydrocarbons- in the quartz-glass capillary column (100 m. x 0.25 mm. x 0.25 mkm) with a fixed-bed of stationary phased ZB-1. Quantitative analysis of gaseous and liquid conversion products of straight-run benzene was performed by the following method: gas chromatography via hardware and software package of gas chromatographer "Chromatek-kristall 500", error-correcting by "Chromatek Analitik" program.Positional error of gaseous and liquid hydrocarbons by chromatography method is ±2.5%.

According to the hydrocarbon composition group, straight-run benzene fraction of 70 – 170°C gas condensate includes: 35 mass% n-paraffine,40 % isoalkane, 20 % naphthenes and 4 % arenes. The octane number of straight-run benzene fraction of 70 – 170°C gas condensate is 65 points in investigated method (IM). The catalytic activity unit was the number of produced arenes from straight-run benzene.

## 3. Results and discussion

The straight-run benzene conversion temperature effect on virgin zeolite catalyst HSZ-G activity showed that with a rise in process temperature from 375 to 425°C and straight-run benzene feed space velocity of 2 $hrs.^{-1}$ high-octane gasoline yield from straight-run benzene decreases from 63.2 to 54.8 % due to the increasing conversion intensity of raw hydrocarbons (table1, figure1). Firstly, gaseous product yield increases from 34.8 at 375°C to 45.1 % at 425°C, mainly paraffins $C_3$–$C_4$, while arene

$C_6$–$C_9$ content in liquid products increases from 23.4 to 26.7 %. Toluene and xylene are prodominate in the arene group, where, with a rise in process temperature from 1.5 and 1.2 at 375°C to 2.0 and 1.3 % at 425°C benzol and olefine $C_{5+}$ content increases, respectively; whereas, naphthene and n/isoparaffin hydrocarbon $C_{5+}$ content decreases with a rise in temperature (Table 1). Among the gaseous products of the straight-run benzene conversion process the prodominate products are propane and butan, where with a rise in temperature from 375 to 425°C, the propane concentration among other gaseous products increases from 57.6 to 59.3 %; while the total content of propane and butane among gaseous products is 93-97 % and methane and ethane content- not more than 3–7 %.

Introducing tin and bismuth oxides into HSZ-G (samples-1–3 % (Sn:Bi = 10:1)/99-97 % HSZ-G) significantly increases the content of arenes up to 32.9–43.8 % in straight-run benzene conversion process liquid products and the decreases the content of iso-paraffin and naphthene hydrocarbons from 27.4–34.9 % and 18.0–21.8 % ,respectively. High-octane benzene yield in the straight-run benzene conversion process with a rise in temperature from 375 to 425 $^0$C and feed space velocity of 2 hrs.$^{-1}$ ,based on these catalysts, changes from 62.6 to 49.3 %, while octane number increases with a rise in temperature from 93.1 to 96.3 units according to IM.

The most active of investigated samples in the conversion of straight-run benzene into high-octane gasoline was the catalyst 1 % (Sn:Bi = 10:1)/99 % HSZ-G. For example, based on the catalyst 1 % (Sn:Bi = 10:1)/99 % HSZ-G with a rise in temperature in straight-run benzene conversion process from 375 to 425 $^0$C liquid product yield decreases from 60.5 to 52.4 %, while octane number of these liquid products increases from 94.6 to 95.3 units according to IM. With a rise in temperature arene and benzol content in liquid products of straight-run benzene conversion process increases from 35.0 to 43.8 % and from 2.1 to 3.7 %, respectively.

**Table 1.** Composition of conversion products of straight-run benzene gas condensate based on catalysts : **1** - HSZ-G, **2** – 1% (Sn:Bi=10:1)/99 % HSZ-G, **3** – 3 % (Sn:Bi=10:1)/97 % HSZ-G, **4** – 5 % (Sn:Bi=10:1)/95 % HSZ-G for 2 hrs.$^{-1}$

| Item | Catalyst type | | | | | | | | | | |
|---|---|---|---|---|---|---|---|---|---|---|---|
| | 1 | | | 2 | | | 3 | | | 4 | |
| Temperature, $^\circ$C | 375 | 400 | 425 | 375 | 400 | 425 | 375 | 400 | 425 | 375 | 425 |
| Gas phase, mass % | 36.8 | 43.0 | 45.1 | 39.5 | 45.9 | 47.6 | 37.4 | 45.5 | 50.7 | 13.2 | 18.7 |
| Liquid phase, mass % | 63.2 | 57.1 | 54.8 | 60.5 | 54.1 | 52.4 | 62.6 | 54.5 | 49.3 | 86.8 | 81.3 |
| Gas phase composition, mass% | | | | | | | | | | | |
| Methane | 0.5 | 0.9 | 1.6 | 0.3 | 0.5 | 1.0 | 0.3 | 0.5 | 1.0 | 0.2 | 0.5 |
| Ethane | 1.5 | 2.3 | 3.5 | 1.3 | 2.1 | 3.3 | 1.3 | 2.1 | 3.3 | 1.1 | 1.9 |
| Ethylene | 0.6 | 0.9 | 1.3 | 0.4 | 0.7 | 0.9 | 0.4 | 0.6 | 1.0 | 1.8 | 4.4 |
| Propane | 57.6 | 58.5 | 59.3 | 57.4 | 58.9 | 61.4 | 56.0 | 59.5 | 61.7 | 52.6 | 47.0 |
| Propylene | 1.5 | 2.0 | 2.7 | 0.8 | 1.5 | 1.8 | 0.7 | 1.2 | 1.8 | 3.9 | 10.3 |
| Iso-butan | 19.6 | 18.2 | 16.1 | 22.5 | 20.6 | 17.9 | 23.0 | 20.6 | 17.8 | 19.7 | 15.1 |
| n-butan | 17.2 | 15.2 | 13.3 | 17.3 | 15.5 | 13.3 | 18.1 | 15.3 | 13.0 | 17.1 | 13.6 |
| Iso-butene | 1.3 | 1.6 | 1.8 | 0.1 | 0.3 | 0.4 | 0.3 | 0.3 | 0.4 | 3.5 | 7.1 |

| Liquid phase composition mass % | | | | | | | | | | | |
|---|---|---|---|---|---|---|---|---|---|---|---|
| Arenes | 23.4 | 26.7 | 26.7 | 35.0 | 40.0 | 43.8 | 33.8 | 32.9 | 43.0 | 13.0 | 12.6 |
| Benzol | 1.5 | 2.0 | 2.0 | 2.1 | 3.2 | 3.7 | 1.9 | 1.9 | 4.2 | 0.4 | 0.6 |
| Isoalkanes | 41.8 | 40.7 | 40.7 | 34.0 | 31.6 | 28.5 | 34.9 | 34.6 | 27.4 | 38.6 | 37.3 |
| Naphthenes | 19.1 | 18.0 | 17.9 | 21.0 | 18.0 | 18.0 | 21.8 | 21.2 | 18.5 | 30.4 | 28.7 |
| Alkanes | 14.5 | 13.3 | 13.4 | 8.6 | 9.0 | 8.2 | 8.4 | 10.0 | 8.2 | 15.2 | 16.3 |
| Olefines | 1.2 | 1.3 | 1.3 | 1.4 | 1.3 | 1.5 | 1.2 | 1.3 | 3.0 | 2.8 | 5.0 |
| Octane number (IM), units | 89.7 | 91.7 | 90.5 | 94.6 | 94.3 | 95.3 | 93.1 | 93.3 | 96.3 | 79.7 | 80.8 |

It should be noted that the produced high-octane gasoline from straight-run benzene based on zeolite catalysts modified 1–3 % (Sn:Bi = 10:1)/99-97 % HSZ-G fully complies with the corresponding motor gasoline "Evro-4 and 5". Further increase of tin and bismuth content (catalyst 5 % (Sn: Bi = 10:1)/95 % HSZ-G ) significantly decreases the arene content from 12.6–15.0 % in liquid products of the straight-run benzene conversion, while arene formation selectivity is not more than 30 %.

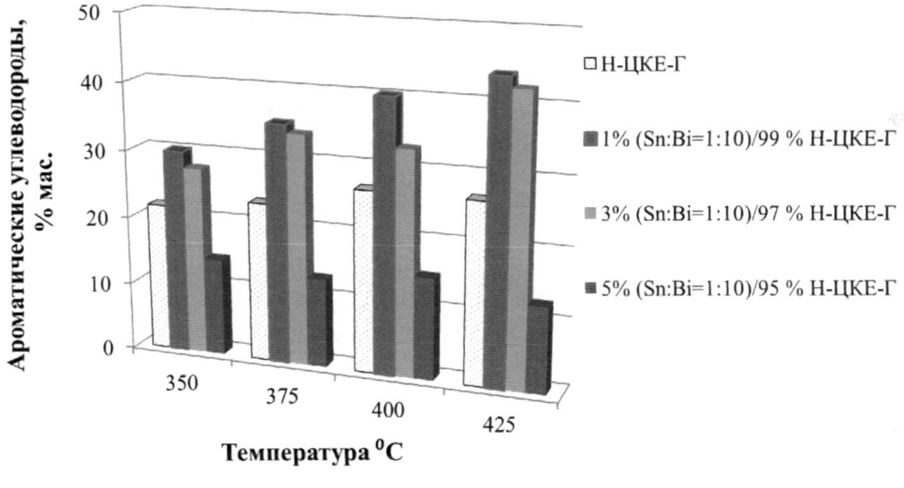


**Figure 1.** Yield dependence of arene from process temperature

Temperature $C^0$ aromatic hydrocarbons HSZ-G

Thus, introducing tin and bismuth oxide mixture as 1–3 % (catalysts: 1–3 % (Sn:Bi = 10:1)/99-97 % HSZ-G) into zeolite ZSM-5 significantly increases the catalyst activity in the conversion process of straight-run benzene into high-octane gasoline. In this case, produced zeolites modified by complex tin and bismuth oxide mixture already show a high aromatic activity at the temperature from 375–400 °C. at the temperature reaction of 375 °C for sample 1 - 1 % (Sn:Bi = 10:1)/99 % HSZ-G , the aromatic hydrocarbon yield is 35 %, which is 1.5 times higher than in the case of catalyst HSZ-G at the same temperature. This proves the perspective application of zeolite catalysts, modified by complex tin and bismuth oxide mixture in the conversion process of straight-run benzene into high-octane gasoline.

## 4. Conclusion

Thus, introducing modified complex additive based on tin and bismuth oxide mixture into zeolite ZSM-5 significantly increases the catalyst activity in the conversion process of straight-run benzene into high-octane gasoline, in comparison to HSZ-G feed. The most perspective are HSZ-G catalysts, modified 1–3 % complex tin and bismuth oxide mixture.

## References

[1] Gureyev A A, Zhorov Yu M and Smidovich Ye V 1981 *Proizvodstvo vysokooktanovykh benzinov* (Moskva: Khimiya) s 224 [in Russian]

[2] Maslyanskiy G N and Shapiro R N 1985 *Kataliticheskiy riforming benzinov* (Leningrad: Khimiya) s 224 [in Russian]

[3] Stepanov V G and Ione K G 2005 Proizvodstvo motornykh topliv iz pryamogonnykh fraktsiy neftey i gazovykh kondensatov s primeneniyem protsessa «Tseoforming» *Khimiya v interesakh ustoychivogo razvitiya* 13 809-822 [in Russian]

[4] Christopher J Rhodes 2007 Zeolites: physical aspects and environmental *Applications. Annu. Rep. Prog. Chem., Sect. C.* **103** 287–325

[5] Vosmerikov A V, Velichkina L M, Korobitsyna L L, Antonova N V, Vagin A I and Yerofeyev V I 1977 Prevrashcheniye uglevodorodnykh fraktsiy gazovogo kondensata na tseolitsoderzhashchikh katalizatorakh *Neftepererabotka i neftekhimiya* 2 16 – 19 [in Russian]

[6] Barbashin Ya E, Ryabov Yu V, Vosmerikov A V, Velichkina L M, Korobitsyna L L and Yerofeyev V I 1998 Dezaktivatsiya tseolitnykh katalizatorov v protsessakh prevrashcheniya metanola, geksana i benzinovoy fraktsii gazovogo kondensata *Neftepererabotka i neftekhimiya* 8 17 – 21 [in Russian]

[7] Yerofeyev V I, Medvedev A S, Koval L M, Khomyakov I S, Yerofeyev M V and Tarasenko V F 2011 Vliyaniye UF aktivatsii na kislotnyye i kataliticheskiye svoystva tseolitsoderzhashchikh katalizatorov v protsesse prevrashcheniya pryamogonnykh benzinov gazovogo kondensata v vysokooktanovyye benziny *Russian Journal of Applied Chemistry* **84** (10) 1668–1674 [in Russian]

[8] Yerofeyev V I, Medvedev A S, Khomyakov I S and Yerofeyeva Ye V 2013 Prevrashcheniye pryamogonnykh benzinov gazovogo kondensata v vysokooktanovyye benziny na tseolitnykh katalizatorakh, modifitsirovannykh nanoporoshkami metallov *Russian Journal of Applied Chemistry* **86** (7) 1044–1051 [in Russian]

[9] Yerofeyev V I, Khomyakov I S and Yegorova L A 2014 Polucheniye vysokooktanovykh benzinov iz pryamogonnykh benzinov na modifitsirovannykh tseolitakh ZSM-5 *Theoretical Foundations of Chemical Engineering* **48** (1) 77 – 82 [in Russian]

[10] Pat. 2313488, RF, MPKS 01V 39/48 Sinteticheskiy tseolit i sposob yego polucheniya Yerofeyev V I, Koval L M – Zayavl. 24.04.2006 Opubl. 27.12.2007 [in Russian]

# Mathematical model of straight run diesel catalytic hydroisomerization

**N Belinskaya[a], E Ivanchina[b], E Ivashkina[c], E Frantsina[d] and G Silko[e]**
Tomsk Polytechnic University, Tomsk, Russia

E-mail: [a]belinskaya@tpu.ru, [b]ied@tpu.ru, [c]ivashkinaen@tpu.ru, [d]evf@tpu.ru, [e]silko_gy@mail.ru

**Abstract.** Straight run diesel catalytic hydroisomerization mathematical model has been developed. There are the following steps for sequential accomplishment: assessment of thermodynamic possibility of spontaneous reactions under the existing technological conditions, development and formalization of hydrocarbon reaction network scheme, kinetic parameters evaluation, industrial verification of the model. The temperature impact within 345-405 °C, the pressure impact in the range of 5-8 MPa and hydrogen bearing gas consumption impact within 3000-51000 $m^3$/h on higher molecular n-alkanes conversion and i-alkanes yield as the basic components which determine diesel fuels low-temperature characteristics to the fullest extent, have been investigated using the developed model. It has been revealed that hydroisomerization process ought to be carried out under such temperature and pressure which do not lead to catalyst exploitation properties come-down in order to receive the product with the required low temperature characteristics. They are 355 °C and 6,7 MPa. The optimal hydrogen bearing gas consumption has been adopted as 39000 $m^3$/h when the feed flow rate is 301 $m^3$/h.

## 1. Introduction

At present the most important problems of petroleum refining industry development are the following:

1. Involvement in processing "heavy" oil feedstock, which is characterized by high-boiling oil fractions, sulfur residuals, resins and high metal content
2. Increase of oil conversion ratio
3. Toughening of ecological requirements for fuels quality
4. Growing demand for high quality motor fuel.

In addition, the climatic conditions of northern regions in the Russian Federation stipulate the demand for fuels with improved low temperature characteristics.

One of the latest technologies which are used in producing high quality diesel fuels is straight run diesel hydroisomerization. The hydroisomerization unit allows the processing of atmospheric gas oil into straight run diesel in order to produce ecologically friendly summer and winter diesel fuel components with ultra-low sulfur and polyaromatic compounds, as well as improved low temperature characteristics, which, in its turn, correspond to the European Standards for motor fuels [1,2].

During the technological process, sulfur-, nitrogen- and oxygen containing compounds and aromatic hydrocarbon hydrogenation reaction occurs based on modern Co-Mo and Ni-Mo catalysts, as well as alkanes hydroisomerization, in order to improve the low temperature properties of diesel fuels.

IOP Conf. Series: Earth and Environmental Science **21** (2014) 012030 doi:10.1088/1755-1315/21/1/012030

The purpose of this research is to improve hydroisomerization unit work efficiency by technological schedule optimization using mathematical modeling method.

## 2. Materials and methods

The full-scale experimental data for different days, which have been obtained from the industrial unit working in the regular technological schedule, are initial data for model development. Namely, they are material balances, technological conditions, and raw material and product compositions.

Mathematical modeling method as an effective tool for solving complex multifactorial problems within petroleum refining and petrochemical industries has been used in the current research. The reactor model which is based on physical-chemical laws provides a means of industrial process prognosis in dynamics without any interference in unit operation [3-7].

The main stages of mathematical model development include estimation of spontaneous reactions under the existing technological conditions, hydrocarbon reactions network scheme development and formalization, kinetic parameters evaluation, industrial verification of the model.

## 3. Model formulation of catalytic hydroisomerization

### *3.1. Thermodynamic analysis and reaction network scheme of hydroisomerization*

The experimental data analysis has showed that the product composition most significantly changes due to the decrease in high molecular n-alkanes concentration and increase in i-alkanes and cycloalkanes concentration. The alkenes content in raw material differs from that in the product negligible, because of n-alkanes isomerization reaction mechanism, which includes the intermediate stage of alkenes formation. The monoaromatic compounds (MAC) and polyaromatic compounds (PAC) concentrations practically do not change during the process.

Thermodynamic possibility of spontaneous reaction has been proved by changing Gibbs free energy values which have been calculated under an average technological conditions (temperature 353 °C and pressure 6,9 MPa) using quantum-chemical methods and Gaussian-98 software, (table 1).

**Table 1.** Change in free Gibbs energy during the hydroisomerization process reactions.

| № | Reaction | $\Delta G$, [kJ/mol] |
|---|---|---|
| 1 | Hydrocracking of n-alkanes $C_{10}$–$C_{27}$ | –85,16 |
| 2 | Dehydrogenation of n-alkanes $C_5$–$C_9$ | –52,22 |
| 3 | Hydrogenation of alkenes with i-alkanes formation | –67,75 |
| 4 | Cyclization of i-alkanes | –4,98 |
| 5 | Hydrogenation of MAC | 32,52 |
| 6 | Hydrogenation of PAC | –65,14 |

According to the reaction reversibility condition ($-70 \leq \Delta G \leq +70$ kJ/mol [8]), low molecular n-alkanes isomerization, i-alkanes cyclization and alkenes, MAC, PAC hydrogenation are reversible reactions, while high molecular n-alkanes hydrocracking is irreversible one.

According to thermodynamic analysis results, the hydroisomerization process reaction network scheme has been drafted (figure 1).

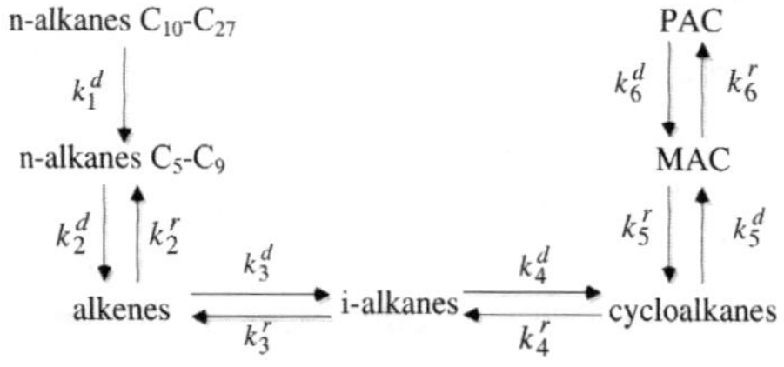


**Figure 1.** The hydroisomerization process reaction network scheme

where, $k_j^d$ is rate constant of $j^{th}$ direct reaction, $k_j^r$ is rate constant of $j^{th}$ reverse reaction.

### *3.2. Hydroisomerization process kinetic model*

On the basis of drafted reaction network scheme the hydroisomerization process kinetic model has been developed. It can be written as

$$\frac{dC_i}{d\tau} = \sum_{j=1}^{6} W_{fj} - \sum_{j=1}^{6} W_{ej} . \quad (1)$$

Initial conditions are the following: $\tau=0$, $C_i=C_{i0}$.

here, $C_i$ is $i^{th}$ components group concentration (alkanes $C_{10}$–$C_{27}$, alkanes $C_5$–$C_9$, etc.), mol/l
$C_{i0}$ is $i^{th}$ components group concentration at the initial time, mol/l
$\tau$ is residence time, s
$W_{fj}$ is $i^{th}$ components group formation reaction rate
$W_e$ is $i^{th}$ components group consumption reaction rate

Reactions rates have been written according to the law of mass action.

The residence time is calculated as

$$\tau = \frac{V_{cat}}{G_f + G_{HG}} . \quad (2)$$

where, $V_{cat}$ is the volume of the catalyst, $m^3$
$G_f$ is flow rate of the feed, $m^3/h$
$G_{HG}$ is flow rate of the hydrogen bearing gas, $m^3/h$

### *3.3. Kinetic parameters evaluation*

Preexponencial factors in the Arhenius equation for all reactions have been derermined by calculated and experimental values divergence minimization. Direct reactions rate constants have been calcutated by the Arhenius equation. Reverse reactions rate constants have been calculated thought equlibrium constants.

**Table 2.** Reactions kinetic parameters (under temperature 355 °C).

| № | Reaction | $k_j^d$ | $k_j^r$ |
|---|---|---|---|
| 1 | Hydrocracking of n-alkanes $C_{10}$–$C_{27}$ | 1,410 | – |
| 2 | Dehydrogenation of n-alkanes $C_5$–$C_9$ | 1,933 | $4{,}448 \cdot 10^{-6}$ |
| 3 | Hydrogenation of alkenes with i-alkanes formation | 3,305 | $1{,}491 \cdot 10^{-4}$ |
| 4 | Cyclization of i-alkanes | 0,016 | $6{,}162 \cdot 10^{-3}$ |
| 5 | Hydrogenation of MAC | 0,141 | $2{,}773 \cdot 10^{-4}$ |
| 6 | Hydrogenation of PAC | 0,139 | $5{,}274 \cdot 10^{-7}$ |

The determined kinetic parameters values correlate to the theoretical and thermodynamic regularities. Namely, primary reactions, which are high molecular n-alkanes hydrocracking, low molecular n-alkanes dehydrogenation and alkenes hydrogenation with i-alkanes formation pass with top speed.

### *3.4. Model verification*

The model verification to the real industrial process has been carried out by means of calculated and experimental components group concentrations in the product comparison (figure 2).

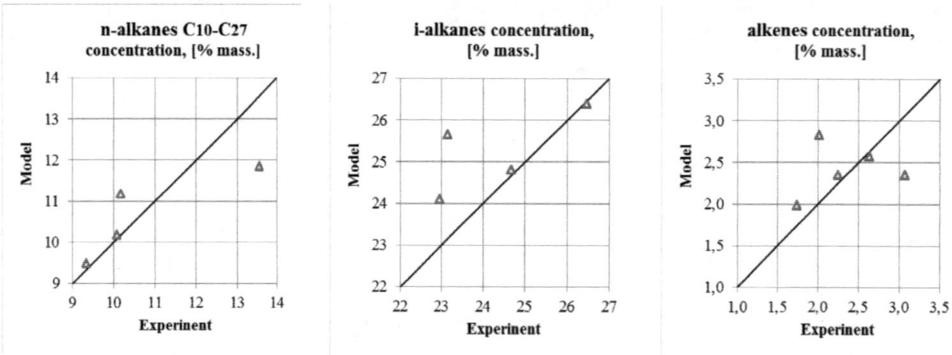


**Figure 2.** Model verification.

## 4. Results and discussion

### *4.1. The temperature impact on high molecular n-alkanes conversion and i-alkanes yield*

High molecular n-alkanes and i-alkanes are of the most importance for diesel fuels low temperature characteristics values [9,10].

The temperature in the hydroisomerization reactor impact on high molecular n-alkanes conversion and i-alkanes yield has been studied within the temperature range of 345-405 °C.

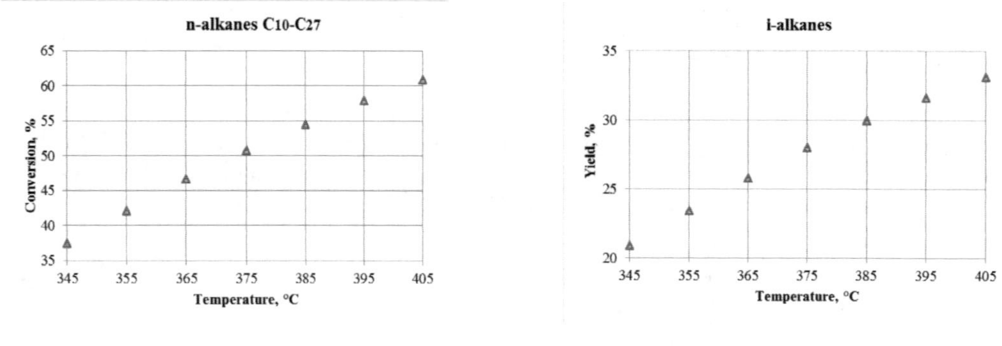


**Figure 3.** The temperature in the hydroisomerization reactor impact on high molecular n-alkanes conversion and i-alkanes yield

Figure 3 shows, that when the temperature in the hydroisomerization reactor raises from 345 to 405 °C n-alkanes $C_{10}$–$C_{27}$ conversion increases by 23 % from 37 % to 60 %. I-alkanes yield goes up by 12 % from 21 % to 33 %.

### *4.2. The pressure impact on high molecular n-alkanes conversion and i-alkanes yield*

The pressure in the hydroisomerization reactor impact on high molecular n-alkanes conversion and i-alkanes yield has been studied in the range of 5-8 MPa (figure 4).

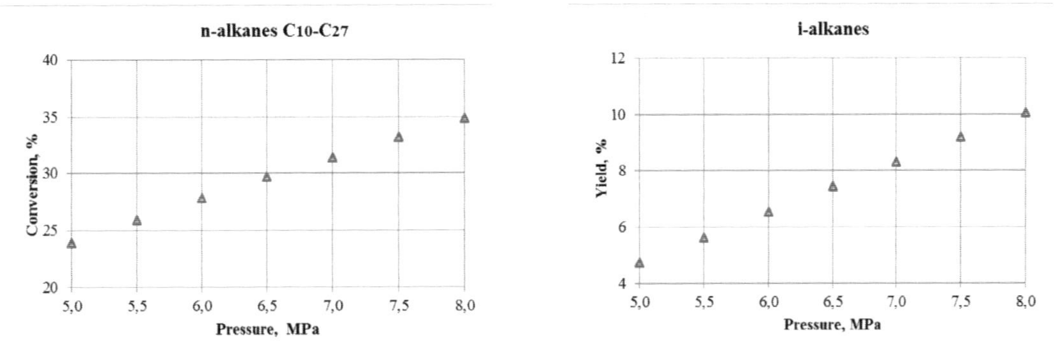


**Figure 4.** The pressure in the hydroisomerization reactor impact on high molecular n-alkanes conversion and i-alkanes yield.

Figure 4 shows that if the pressure is from 5 to 8 MPa n-alkanes $C_{10}$–$C_{27}$ conversion increases up to 10 %, i.e. from 24 % to 34 %. I-alkanes yield rises up to 5 %, i.e. from 5 % to 10 %.

Thus, increase in temperature and pressure in the hydroisomerization reactor furthers the production of diesel fuels with lower pour point, maximum filtration temperature and cloud point due to an increase in alkanes $C_{10}$–$C_{27}$ conversion, as well as i-alkanes yield.

### *4.3. The hydrogen bearing gas consumption impact on high molecular n-alkanes conversion and i-alkanes yield*

The hydrogen bearing gas consumption impact on high molecular n-alkanes conversion and i-alkanes yield has been studied within the limits of 3000-51000 $m^3/h$ (figure 5). The feed flow rate has been adopted as 301 $m^3/h$.

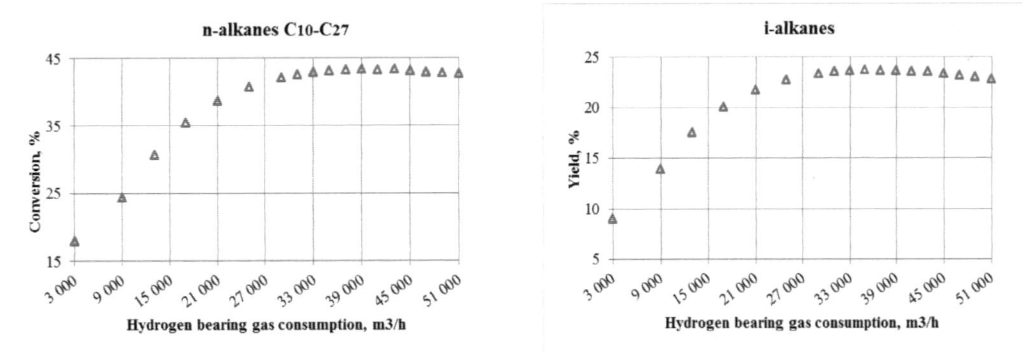


**Figure 5.** The hydrogen bearing gas consumption impact on high molecular n-alkanes conversion and i-alkanes yield.

It follows from Fig. 5 that the increase in hydrogen bearing gas consumption from 3000 to 39000 $m^3/h$ leads to n-alkanes $C_{10}$–$C_{27}$ conversion raise of 25 %, i.e. from 18 % to 43 %. I-alkanes yield increases up to 14 %,i.e. from 9 % to 23 %. There is no point in further increase in hydrogen bearing gas consumption because it does not lead to the raise in n-alkanes $C_{10}$–$C_{27}$ conversion and i-alkanes yield.

## 5. Conclusion

Straight run diesel catalytic hydroisomerization mathematical model has been developed by the following steps for sequential accomplishment: assessment of thermodynamic possibility of spontaneous reactions under the existing technological conditions, development and formalization of hydrocarbon reaction network scheme, kinetic parameters evaluation, industrial verification of the model.

The temperature impact within the limits of 345-405 °C, the pressure impact in range 5-8 MPa and hydrogen bearing gas consumption impact within range 3000-51000 $m^3$/h on higher molecular n-alkanes conversion and i-alkanes yield as the basic components which determine diesel fuels low temperature characteristics to the fullest extent, have been investigated using the developed model.

Thus, hydroisomerization process ought to be carried out under such temperature and pressure which do not lead to catalyst exploitation properties come-down in order to receive the product with the required low temperature characteristics. They are 355 °C and 6,7 MPa. The optimal hydrogen bearing gas consumption has been adopted as 39000 $m^3$/h when the feed flow rate is 301 $m^3$/h.

## Acknowledgements

The reported study has been carried out in the frame of the State Task “Science” 1.1404.2014.

## References

[1] Yacine Rezgui and Miloud Guemini 2010 Hydroisomerization of n-decane over Ni-Pt-W supported on amorphous silica–alumina catalysts, Applied Catalysis A: General. **374** 31-40

[2] Konnov S V, Ivanova I I, Ponomareva O A and Zaikovskii V I 2012 Hydroisomerization of n-alkanes over Pt-modified micro/mesoporous materials obtained by mordenite recrystallization *Microporous and Mesoporous Materials* **164** 222–231

[3] Chekantsev N V, Gyngazova M S and Ivanchina E D 2014 Mathematical modeling of light naphtha (C5, C6) isomerization process *Chemical Engineering Journal* **238** 120-128

[4] Chsherbakova Y, Dolganova I and Belinskaya N 2012 Benzene alkylation with ethylene process mathematical modeling *7th International Forum on Strategic Technology* **12** 50-53

[5] Karaba A, Zamostny P, Lederer J and Belohlav Z 2013 Generalized model of hydrocarbons pyrolysis using automated reactions network generation Industrial *and Engineering Chemistry Research* **52** (44) 15407-15416

[6] Frantsina E V, Ivashkina E N, Ivanchina E D and Romanovskii R V 2014 Developing of the mathematical model for controlling the operation of alkane dehydrogenation catalyst in production of linear alkyl benzene *Chemical Engineering Journal* 238 129-139

[7] Luís Pereira de Oliveiraa, Jan J Verstraetea and Max Kolb 2014 Simulating vacuum residue hydroconversion by means of Monte-Carlo techniques *Catalysis Today* **220–222** 208– 220

[8] Peter Sykes 1986 *A guidebook to mechanisms in organic chemistry* (London: Longman Scientific & Technical)

[9] Ovchinnikova A V, Boldinov V A, Esipko E A and Prozorova I S 2005 Effect of n-paraffins on the low-temperature properties of aviation diesel fuels *Chemistry and Technology of Fuels and Oils* **41** (6) 462-467

[10] Siddharth S Ray, Naval K Pandey and Alok K Chatterjee 2009 Effect of aromatics and iso-alkanes on the pour point of different types of lube oils *Fuel* **88** 1629-1633

IOP Conf. Series: Earth and Environmental Science **21** (2014) 012031 doi:10.1088/1755-1315/21/1/012031

# Determination of platinum in mineral raw materials by switching chronoamperometry

**E Pakrieva[1], Y Oskina[2] and E Ustinova[3]**
Tomsk Polytechnic University, Tomsk, Russia

E-mail: [1]epakrieva@mail.ru, [2]oskinaina@tpu.ru, [3]emg87@mail.ru

**Abstract.** The technique of platinum (IV) determination in mineral raw materials with the application of switching chronoamperometry has been offered. The graphite electrode impregnated with polyethylene was used as the working electrode. The hydrolytic precipitation method with 3% NaOH solution has been developed to separate platinum from the sample matrix. The use of switching chronoamperometry applied to the assessment of the platinum content in geological objects has been demonstrated.

## 1. Introduction

Nowadays the majority of conventional precious metal deposits have been exhausted. Therefore, it is necessary to search for alternative composite sources with poor ores. However, the interest to such deposits in recent years has revealed serious difficulties in the analytical base in the process of the platinum group metals content assessment in this type of mineral raw materials. It is connected, first of all, with the deficiency of high sensitivity and good reproducibility techniques of analyses. Besides, large scattering and low content of precious metals in minerals and rocks are also concerned [1].

Several ways of platinum determination have been suggested by means of using a highly sensitive method of stripping voltammetry [2, 3, 4]. However, this method is complex and multistage, because the determination of platinum (II or IV) in solutions with anodic stripping voltammetry is feasible only after the platinum electrodeposition in an alloy with a metal-activator on the surface of the graphite electrode with the following selective electrooxidation from the electronegative component alloy.

Until recently the application of the simplest express way of platinum determination obtained in chloride complexes by switching chronoamperometry has been complicated by its low sensitivity.

The purpose of the study is to develop the platinum determination technique in mineral raw materials with the application of switching chronoamperometry by means of the limiting impulse current testing for platinum (IV) in the solution.

## 2. Experimental part

*Reagents and chemicals*

All the chemicals used were of analytical or pharmaceutical grade and the solutions were prepared with deionized water. The standard platinum solutions were supplied by Merck. The hydrochloric, hydrofluoric and nitric acids were provided by Sigma. The control was carried out by Pt content measuring in the following reference standards (RS) from «SIBTZVETMETNIIPROEKT», Krasnoyarsk: RS 8770-2006 (SOP 107.1 RMK-4), RS 8771-2006 (SOP RMO-5 107.2), RS 519-88P (ST-2), RS 927-76 (VP-2), RS 8772-2006 (SOP 108.1).

*Electrochemical measurements*

Chronoamperometrical measurements were carried out with the universal research analyzer "MultiLab" (Tomsk). A three-electrode system and special quartz beakers of $V$=20 ml were used. A polyethylene-impregnated graphite electrode produced according to [5] was used as the working electrode. The determination of platinum was carried out by means of using 0.1 M HCL as the supporting electrolyte. Saturated silver chloride electrodes served as the reference and the auxiliary electrodes. Deaeration of solutions was not conducted.

## 3. Results and discussion

The principle of switching chronoamperometry or pulsed potentiostatic method of the platinum determination in mineral raw materials is the following: the potential of the working electrode being set and constant, the current contour of the electrochemical cell with the determined solution is periodically commuted [6]. Thus there occurs the determined ion recharging: the transition of platinum (IV) in platinum (II) in the solution near the electrode surface. The recorded maximum impulse current is proportional to the concentration of the input element.

The determination of platinum ions (IV) in the study was carried out close to the surface of the graphite electrode in the solution being stirred in real time with the controlled potential electrolysis -0.24 V. Selecting of the working electrode and the electrolysis potential value was due to the reproducibility and the maximum value of the analytical signal under these conditions. As the platinum was introduced in the form of chloride complexes $[PtCl_6]^{-2}$, the measurements were carried out using 0.1 M HCl supporting electrolyte with the following chronoamperometrical curves registration. Platinum (IV) ions concentration was determined with the signal response in relation to the saturated silver/silver chloride electrode (figure 1).

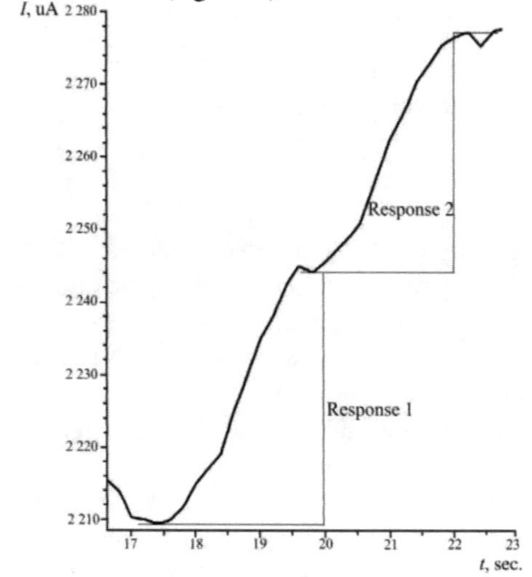


**Figure 1.** Chronoamperogramm of the platinum (IV) determination on the surface of the graphite electrode in the solution. The experimental conditions: supporting electrolyte - 0.1 M HCl; Ee = -0.24 V. Response 1 ($\Delta I$ = 34 mA) – the sample containing 0.1 Pt (IV) and Response 2 ($\Delta I$ = 33.5 mA) – the additive containing 0.1 mg $L^{-1}$ standard solution of Pt (IV).

The platinum concentration was determined at the limited current of the standard additive according to the following equation:

$$C = \frac{I_{sp} C_{ad} V_{ad}}{(I_{ad} - I_{sp}) V_{sp}}.$$

$C$ – the concentration of platinum in the solution, mg $L^{-1}$;

$I_{sp}$ , $I_{ad}$ – the current limit of the sample and the additive of platinum standard solution respectively, uA;

$V_{ad,}$ $V_{sp}$ – the volume of the sample and additive, L;
$C_{ad}$ – the concentration of the platinum standard solution additive, mg $L^{-1}$.

The platinum (IV) determination results in the supporting electrolyte indicate that the maximum error in the measurements in the experiment is equal to 20 % (table 1). The calculation of the detectable platinum concentrations was carried out according to the standard addition method.

**Table 1.** The results of platinum (IV) chronoamperometrical determination in 0.1 M HCl supporting electrolyte (n = 9; P = 0.95).

| Added of Pt, g $t^{-1}$ | Found of Pt, g $t^{-1}$ | Relative accuracy ± δ, % |
|---|---|---|
| 0.025 | 0.020 | 20 |
| 0.050 | 0.058 | 16 |
| 0.100 | 0.087 | 13 |
| 0.500 | 0.438 | 12 |
| 0.900 | 1.000 | 11 |

The linear dependence of the platinum (IV) current electroreduction on its concentration in the solution is observed (figure 2).

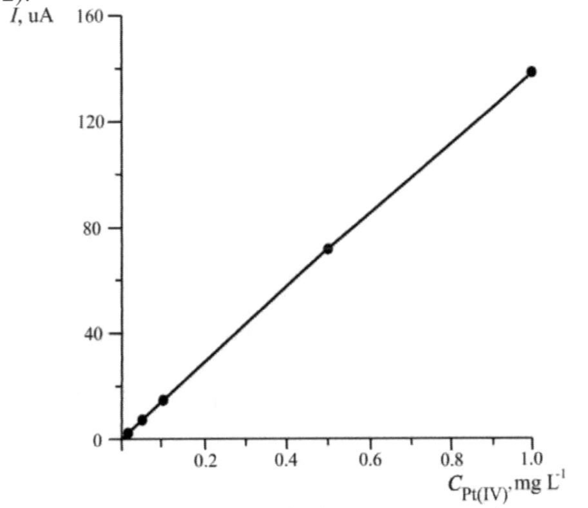


**Figure 2.** The dependence of the platinum (IV) limited current on its concentration in solution. Straight line equation: Y = 138.13·X + 0.08. The experimental conditions: supporting electrolyte - 0.1 M HCl; Ee = -0.24 V.

The detection limit was calculated to determine the platinum content by means of measurement limiting pulse-current with the application of switching chronoamperometry.

The detection limit was calculated on the 3-δ criterion with the help of the equation below:

$$C_{\min}, p = \frac{3\delta}{tg\alpha} = \frac{3 \cdot 0.01}{138.13} = 0.00022 \text{gt}^{-1}.$$

Where δ is the minimum signal of Pt (IV) in the supporting electrolyte; tg α is the angle of inclination from the calibration curve (figure 2).

*Sample pretreatment for platinum determination in geological objects*

Preconcentration of metals is required for reliability improvement of their low content determination with the simultaneous separation of metals from the bulk of macro components.

At the beginning of the sample preparation, dry ashing of the sample was carried out in the furnace at t = 500 °C for 45 minutes. Then the sample was cooled down. $H_2O$, S, $CO_2$, As, Sb, Hg, etc.

volatilize under these conditions. The matrix sample was treated with concentrated HF for the decomposition of silicates. The obtained mixture was heated at t = 170 °C until white fumes associated with the formation of volatile $SiF_4$. The solution was evaporated to dryness on the hotplate. The residue in the crucible was treated with aqua regia by means of heating (t = 170 °C). There Pt and Pd dissolved and formed $H_2[PtCl_6]$ and $H_2[PdCl_6]$.Then, after the concentrated HCl with heating (t = 170 $^0$C) being added, the complex $(NO)_2[PtCl_6]$ was decomposed and it was converted into $H_2[PtCl_6]$ [10]. The mixture was filtered, and then was washed with the double amount of distilled water. The filtrate was evaporated to wet salts and then diluted with a small amount of distilled water. The use of hydrolytic precipitation method with 3% NaOH solution makes the platinum transform into another complex – hexachloroplatinate sodium $Na_2[PtCl_6]$ and remains in the solution. The flasks with the filtrate were evaporated to dryness. Further HCl was added to neutralize the excess of NaOH, and to transit chloride complexes of platinum: $[PtCl_6]^{2-}$. The obtained mixture was evaporated on the plate to a dry residue. The residue was diluted with water. The platinum content was then determined by means of switching chronoamperometry (table 2).

**Table 2.** The results of switching chronoamperometry determination of the platinum (IV) content in the reference standards of copper-nickel ore (n = 9; P = 0.95).

| Ref. standard | Content of Pt in ref. sample, g t$^{-1}$ | Result of determination of Pt, g t$^{-1}$ | Relative accuracy ±δ, % |
|---|---|---|---|
| SOP 107.1 | 4.10 | 3.00 | 34 |
| VP-2 | 2.64 | 3.18 | 36 |
| SOP 108.1 | 2.50 | 1.87 | 38 |
| SOP 107.2 | 2.10 | 2.51 | 39 |
| ST-2 | $2.50 \cdot 10^{-2}$ | $1.98 \cdot 10^{-2}$ | 40 |

As table 2 shows, the systematic error is not significant in the chronoamperometrical method of the platinum determination, after the separation of interfering impurities by the 3% NaOH solution hydrolytic precipitation.

The suggested technique for the platinum (IV) determination was tested on real objects of mineral raw materials. The research objects were ores of different formation types from Siberia, the Urals, Tuva, Yakutia and Kazakhstan. The choice of these objects is due to the fact that the Innovative Scientific and Educational Center "Gold and Platinum" of Tomsk Polytechnic University revealed low concentrations of platinum by means of methods of the stripping voltammetry and atomic absorption spectroscopy in these deposits (table 3).

**Table 3.** The comparative analyses of platinum determination in the mineral raw materials by switching chronoamperometry (SCA) and atomic absorption spectroscopy (AAS), (n = 9, P = 0.95).

| Type and name of deposits | Sample no | Content of Pt, g t$^{-1}$ | |
|---|---|---|---|
| | | SCA | AAS |
| Maisko-Lebedskoe gold-skarn | PR-3 | 0.003±0.00061 | 0.0009±0.00018 |
| | PR-10 | 0.004±0.0008 | 0.0035±0.0006 |
| | ML-4 EPG | 0.007±0.0020 | 0.0063±0.0015 |
| | CAF-1 EPG | 0.004±0.0007 | 0.003±0.00065 |
| Bakchar iron ore | PR-3 | 0.003±0.00061 | 0.0009±0.00018 |
| | PR-10 | 0.004±0.0008 | 0.0035±0.0006 |
| | ML-4 EPG | 0.007±0.0020 | 0.0063±0.0015 |
| | CAF-1 EPG | 0.004±0.0007 | 0.003±0.00065 |
| Aksu copper-molybdenum | K-79 | 0.004±0.0010 | 0.0035±0.0009 |
| | K-90 D | 0.006±0.0013 | 0.0067±0.0013 |
| | K-92 G | 0.007±0.0016 | 0.003±0.0006 |
| | K-97 ZH | 0.005±0.0011 | 0.0017±0.00034 |

As table 3 shows, the results of the comparative analysis of the methods of switching chronoamperometry with AAS demonstrate a good data convergence of platinum content determination in the ore material.

Thus, the discussed method for the platinum determination in mineral raw materials with SCA application by means of measurement of limited pulse current can be used to control the platinum content in geological objects.

## 4. Conclusion

1. The technique of the platinum determination in mineral raw materials with the application of switching chronoamperometry by means of the limited impulse current of platinum (IV) in the solution has been proposed.

2. The platinum detection limit in the solution has been calculated with the 3σ-criterion application and is equal to 0.22 mg $t^{-1}$.

3. The data convergence obtained by means of the atomic absorption spectroscopy and switching chronoamperometry in the process of the platinum content determination in ore materials has shown the reliability of the developed technique.

## Acknowledgments

The study was supported by the government project "Science" (no. 1.1312.2014).

The authors gratefully acknowledge A. Pshenichkin, PhD, Head of the Research Centre "Gold and Platinum". The authors are also thankful to the Research Center "Gold and Platinum" TPU for the equipment, reagents and sample materials providing.

## References

[1] Aksenov S A, Bavlov V N, Mikhailov V K and Nekrasov A I 2013 The results of the mineral resources base development of solid minerals Russia from 2005 to 2012 and their direction until 2020 *Journal of Exploration and Protection of Natural Resources* **4** 5–11

[2] Svancara I, Galik M and Vytras K 2007 Stripping voltammetric determination of platinum metals at a carbon paste electrode modified with cationic surfactants *Talanta* **72** (2) 512–518

[3] Shekhovtsova N S, Glyzina T S, Romanenko S V and Kolpakova N A 2012 Estimation of the composition of electrolytically prepared intermetallic bismuth–platinum deposits *Journal of Solid Electrochemistry* **16 (7)** 2419–2423

[4] Gabdurakhmanova (Ustinova) E M, Gorchakov E V and Kolpakova N A 2012 Anodic stripping determination of Pt(IV) based on the anodic oxidation of In from electrochemically deposited Pt–In alloy phases *Journal of Solid Electrochemistry* **16** (7) 2455–2458

[5] Braynina X Z and Neyman E Ya 1982 *Solid state-reactions in electroanalytical chemistry* (Moscow: Chemistry) p 294

[6] Moshkin V V and Khustenko L A 2010 Switching chronoamperometry way Patent RF N. 2382356

[7] Barefoot R R and Van Loon J C 1999 Recent advances in the determination of the platinum group elements and gold *Talanta* **49** (1) 1–14

[8] Balcerzak M 2002 Sample digestion methods for the determination of traces of precious metals by spectrometric techniques *Analytical Sciences* **18** 737–750

[9] Hu Z 2014 Sample digestion methods *Reference Module in Earth Systems and Environmental Sciences. Treatise on Geochemistry sec. ed.* **15** 87–109

[10] Chugaev L V 1987 *Metallurgy of precious metals second ed* (Moscow: Metallurgiya) p 432

# Aromatization of Propane over Element-Alumosilicate Catalysts with ZSM-5 Structure

**L N Vosmerikova[1,a], A N Volynkina[2,b] and A V Vosmerikov[1,c]**
[1] Institute of Petroleum Chemistry, Siberian Branch, Russian Academy of Sciences, Tomsk, Russia
[2] Tomsk State University, Tomsk, Russia

Email : [a]lkplu@ipc.tsc.ru, [b]a.volynkina@inbox.ru, [c]pika@ipc.tsc.ru

**Abstract.** A method of hydrothermal crystallization of alkaline alumosilicagels is used to manufacture element-alumosilicates with ZSM-5 structure. Their physicochemical and acid properties are investigated and their catalytic activity in the course of propane conversion to aromatic hydrocarbons is determined. The Ga-alumosilicate is found to be the most efficient zeolite catalyst for propane aromatization.

## 1. Introduction

High-silica zeolites are widely used as catalysts in oil refining and petrochemistry. A combination of their unique structural and acidic properties with a dehydrogenation ability of the promoting additives makes it possible to manufacture effective catalysts for aromatization of light hydrocarbons, in particular, propane. To increase the selectivity of aromatization, a chemical or thermal modification of high-silica ZSM-5 zeolites is used. Typically, zeolites are modified by the methods of ion exchange and impregnation with solutions of metal salts, mechanical mixing with metals or their oxides, and isomorphous substitution of aluminum in the crystal lattice for other elements [1-2, 4]. A partial or complete substitution of aluminum ions in the aluminosilicate lattice for cations of other metals is an effective way of controlling the properties of zeolites aimed at changing their activity and selectivity. The element-alumosilicates (EAS) formed in this case preserve the structure of the respective zeolite; they however possess a number of individual features due to the origin of ions constituting the zeolite skeleton [3].

The aim of this work is to study the mechanisms of propane aromatization over an ZSM-5 element-alumosilicates.

## 2. Experimental

An aluminosilicate (AS) with a silica modulus (M) 40 was synthesized under hydrothermal conditions from the alkali alumina silica gels using hexamethylenediamine (HMDA) as an organic additive. The liquid glass composed of 29% $SiO_2$, 9% $Na_2O$, and 62% $H_2O$ was the source of silicon and $Al(NO_3)_3$ 9 $H_2O$ was the source of aluminum.

The AS was synthesized using HMDA by the following procedure: aqueous solutions of HMDA and aluminum salts were added to the liquid silica glass under vigorous stirring. The pH of the reaction mixture was controlled by adding a 0.1 N solution of the nitric acid.

The resulting reaction mass was crystallized in steel autoclaves at 170 °C for 4-5 days. The crystallization completed, the solid phase was separated from the liquid phase by filtration, washed with distilled water until the wash water pH $\leq 9$ was reached, and then air-dried at 110 °C and calcined at 540 °C for 16 hrs.

Gallium- and indiumalumosilicate (Ga-AS and In-AS) were synthesized by the same method via partial substitution of aluminum with gallium or indium in the initial ($SiO_2/Al_2O_3 + Me_2O_3 = 40$) aluminum silica gel. The content of gallium and indium oxides was 1.85 and 1.82 wt%, respectively.

To convert the zeolites to the active H-form, the samples were treated with a 25% aqueous $NH_4Cl$ solution at 90 °C for 2 hours. After removal of the sodium cations, the zeolite samples in the $NH_4$-form were dried at 110 °C and air-calcined at 540 °C for 6 hrs.

The quality of the resulting samples was controlled by IR and XRD. The IR spectra were recorded from the zeolite pellets using a FT-IR Nicolet 5700 spectrometer within the range of 2000-400 $cm^{-1}$ in the presence of KBr (1.2 mg of high silica zeolites per 300 mg of KBr) in air. An XRD analysis of samples was performed in a DRON-3M X-ray apparatus (Cu-anode, Ni-filter).

The acidic properties of the catalysts were tested by the method of temperature-programmed desorption (TPD) of ammonia. The TPD method allows the energy spectrum of the catalyst to be recorded from adsorption of the probing substance and the concentrations of active sites to be quantified for different desorption types. The TPD experimental procedure was described in [5]. The concentration of the acid sites in the investigated samples was determined by the amount of ammonia desorbed at the moment of identification of desorption peaks and expressed as micromoles per 1 g of the catalyst. Ammonia was selected as an adsorbate for two reasons: its high basicity that allows determining not only the strong acid sites but also weakly acid ones and its small molecular size and simple dosing.

The activity of catalysts studied was investigated in the course of propane conversion into aromatic hydrocarbons in a flow-type reactor with a fixed-bed catalyst ($V_{cat.} = 3$ $cm^3$) at atmospheric pressure at the reaction temperature 400-600 °C and the feed space velocity 500 $h^{-1}$. The reaction products were analyzed by GLC using a Khromatek-Kristall 5000.2 chromatograph. To determine the composition of the liquid phase, use was made of a BP-1 PONA capillary column (100 m x 0.25 mm x 0.5 mm) and the gas-phase compositions were determined in the GS-Gas-Pro (60 m x 0.32 mm) capillary and Carbosieve S-II (4 m × 2 mm, 60/80 mesh) packed columns. The catalytic activity of the samples was estimated from the propane conversion data. The yield and formation selectivity of the gaseous and liquid products were also calculated.

## 3. Results and discussion

The data on the composition and degree of crystallinity of elementoalumosilicate synthesized are presented in table 1.

**Table 1.** Composition and crystallinity of elementoalumosilicate.

| Sample | Concentration, % | | Molar ratio $SiO_2/Al_2O_3 + Me_2O_3$ | Degree of crystallinity, % |
|---|---|---|---|---|
| | $Me_2O_3$ | $Al_2O_3$ | | |
| AS | – | 4.07 | 40 | 100 |
| Ga-AS | 1.85 | 3.01 | 40 | 97 |
| In-AS | 1.82 | 3.34 | 40 | 95 |

The results of X-ray studies showed that all samples synthesized correspond to the structure of ZSM-5 zeolite and belong to the rhombic system.

Figure 1 shows the IR spectra for the resulting element-alumosilicates, indicating the presence of absorption bands characteristic of high-silica zeolites. All samples exhibit absorption bands in the region 550-560 $cm^{-1}$ attributed to oscillations of the tetrahedra, [$SiO_4$] and [$AlO_4$], over the external constraints of the skeleton due to the presence of four-, five- and six-member double rings, which determines their structure. The zeolites synthesized are characterized by a high degree of crystallinity that is somewhat reduced by an addition of gallium and indium, which is probably due to the slowing crystallization of silicate ions in the presence of gallium or indium ions (table 1).

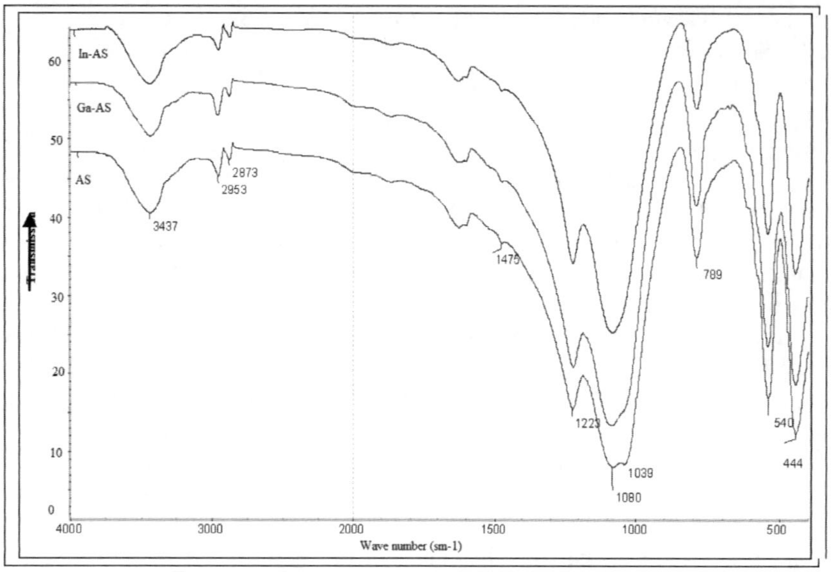


**Figure 1.** IR spectra for element-alumosilicates.

An important characteristic of zeolite catalysts, determining their activity, are their composition-controlled acid-base properties. The data on acid property studies are presented in table 2. An aluminosilicate possesses acid sites of two types as evidenced by the presence of two ammonia desorption types in the thermal desorption spectrum: weak acid sites with the temperature of peak maximum $T_{max}$ = 205 °C and strongly acidic with $T_{max}$ = 425 °C. The concentration of weakly acid sites is 676 μmol/g and that of strong acid sites – 251 μmol/g (table 2).

An introduction of gallium into the zeolite in the course of hydrothermal synthesis results in a change in its acidic properties, primarily, an increase in the strength and concentration of weak acid sites. The reason is that the new stronger active sites with a high energy of ammonia retention are formed by the introduction of gallium into the zeolite. The formation of strong aprotonic acid sites that contain gallium atoms results in an increased activity of galloalumosilicates in reactions of aromatization of lower alkanes, since their formation is accompanied by a sharp increase in the rate of the limiting stage of the reaction of dehydrogenation of lower saturated hydrocarbons, while oligomerization of the resulting olefins follows the protonic mechanism; both protonic and aprotonic acid sites of the zeolite participate in a subsequent cyclization of oligomers and their unsaturated fragments.

An indiumalumosilicate exhibits a higher acidity compared with the aluminosilicate but yields to the galloalumosilicate in its acidic characteristics.

Thus, the investigation of acidic properties of the catalysts has shown that element-alumosilicates are characterized by different distributions of acid sites in their strength and concentration. The most significant differences are observed in the concentrations of low-temperature acid sites and in the strength of high-temperature acid sites. The differences in the ratio of weak to strong acid sites of element-alumosilicates affect their catalytic properties manifested in the course of propane conversion, since different active centers are responsible for different chemical reactions.

**Table 2.** Acidic properties of zeolite catalysts.

| | $T_{max}$, °C | | Concentration, μmol/g | | |
|---|---|---|---|---|---|
| Catalyst | $T_I$ | $T_{II}$ | $C_I$ | $C_{II}$ | $C_\Sigma$ |
| AS | 205 | 425 | 676 | 251 | 927 |
| Ga-AS | 210 | 450 | 837 | 281 | 1118 |
| In-AS | 205 | 430 | 795 | 287 | 1082 |

Note. $T_I$, $T_{II}$ are the temperatures of maxima of low- and high-temperature peaks on the thermal desorption curves;
$C_I$, $C_{II}$ and $C_\Sigma$ are the concentrations of weak and strong acid sites and their sums, respectively.

The investigation of the catalytic properties of element-alumosilicates has shown that the catalysts differ from each other in their total (measured by propane conversion degree) and aromatizing reactivities. The data of Table 3 suggest that conversion of propane over aluminosilicate begins at the reaction temperature 400 °C, while the desired products – aromatic hydrocarbons consisting mainly of benzene, toluene and xylenes, are formed at at 500 °C or higher; note that $C_{9+}$ alkylbenzenes, naphthalene and alkylnaphthalenes are also formed in small quantities. The byproducts are represented by gaseous hydrocarbons – methane, ethane and minor amounts of hydrogen and $C_2$-$C_3$ olefins.

**Table 3**. Characteristic of the products of propane conversion over zeolite catalysts.

| Catalyst | T, °C | X, % | $S_1$, % | $S_2$, % | $S_3$, % | $S_4$, % | $Y_{ArH}$, % |
|---|---|---|---|---|---|---|---|
| | 400 | 14 | 0.4 | 25.9 | 5.2 | 3.6 | 0.5 |
| | 450 | 30 | 1.3 | 55.7 | 8.5 | 7.7 | 2.3 |
| AS | 500 | 59 | 2.1 | 66.4 | 8.9 | 17.1 | 10.1 |
| | 550 | 88 | 2.5 | 59.9 | 7.6 | 29.1 | 25.6 |
| | 600 | 99 | 3.7 | 60.9 | 6.0 | 28.7 | 28.4 |
| | 400 | 12 | 4.9 | 36.9 | 4.6 | 12.5 | 1.5 |
| | 450 | 31 | 6.6 | 54.8 | 5.2 | 22.9 | 7.1 |
| Ga-AS | 500 | 70 | 6.1 | 49.7 | 4.1 | 38.3 | 26.8 |
| | 550 | 95 | 5.7 | 47.0 | 2.7 | 43.9 | 41.7 |
| | 600 | 99 | 6.3 | 47.6 | 2.3 | 43.3 | 42.9 |
| | 400 | 6 | 0.4 | 20.6 | 9.1 | 3.3 | 0.2 |
| | 450 | 17 | 0.9 | 39.1 | 13.8 | 3.5 | 0.6 |
| In-AS | 500 | 37 | 1.7 | 60.5 | 17.8 | 5.7 | 2.1 |
| | 550 | 67 | 2.5 | 66.0 | 18.1 | 10.7 | 7.2 |
| | 600 | 91 | 3.3 | 60.5 | 14.6 | 21.3 | 19.4 |

Note. X is the propane conversion; $S_1$, $S_2$, $S_3$, and $S_4$ is the selectivity towards formation of hydrogen, $C_1$-$C_2$ alkanes, $C_2$-$C_4$ alkenes and aromatic hydrocarbons, respectively; $Y_{ArH}$ is the yield of aromatic hydrocarbons.

The conversion of feedstock and the selectivity towards formation of aromatics increase with increasing process temperature. The conversion and the selectivity towards formation of aromatic hydrocarbons over the AS sample reach 98.5 and 28.8% respectively, at the reaction temperature 600 °C. Thus, the main pathway of conversion of propane over aluminosilicates is its cracking and the selectivity towards formation of arenes is not higher than 30%.

We have found out that an introduction of gallium into the zeolite in the stage of hydrothermal synthesis results in a significant increase in its catalytic reactivity in the course of propane conversion into aromatic hydrocarbons and its aromatizing activity is exhibited even at a temperature of 450 °C. For the 1.85% Ga-AS sample, the yield of aromatic hydrocarbons at the reaction temperature 600 °C is ~ 43%, which is 1.5 times higher than that for the aluminosilicate at the same reaction temperature. This indicates a promising application of galloalumosilicates as catalysts for aromatization of lower alkanes.

Indiumalumosilicate exhibits the lowest activity in propane aromatization as compared to all the catalysts investigated. An abundant formation of aromatic hydrocarbons was observed only at 550 °C. An increase in the temperature to 600 °C does not result in any significant increase in selectivity towards formation of aromatic hydrocarbons, while the conversion increases to 91%. The reaction products formed in reactions in the presence of this catalyst contain a large number of lower $C_2$-$C_4$ olefins.

Figure 2 shows the plots of the selectivity towards formation of aromatic hydrocarbons over element-alumosilicates on the process temperature. It is evident that all samples demonstrate an increased selectivity towards formation of aromatic hydrocarbons with increasing process temperature. The highest aromatizing activity within whole investigated temperature range compared to other catalysts was exhibited by the Ga-AS sample. The monotonic increase in selectivity towards formation of aromatic hydrocarbons with increasing process temperature is characteristic of the In-aluminosilicate. The sharp increase in aromatizing catalytic activity with the process temperature increasing from 450 to 550 °C and the insignificant change in selectivity towards formation of aromatic hydrocarbons during further temperature increase to 600 °C were observed for this aluminosilicate.

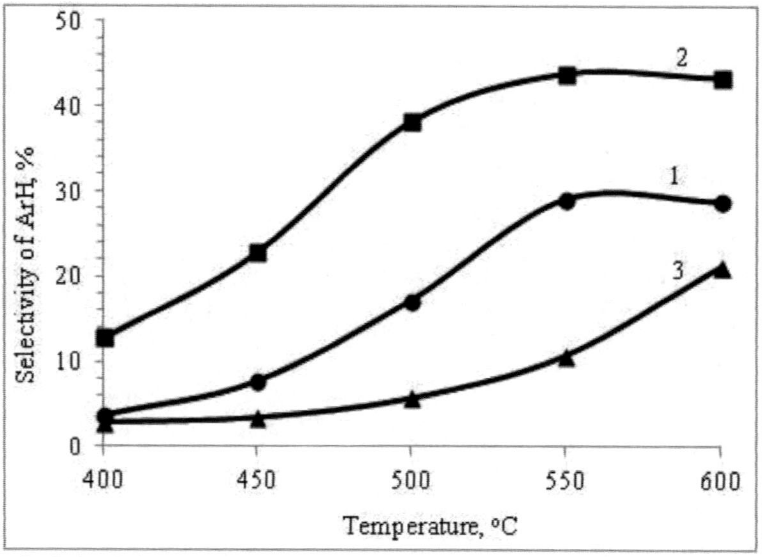


**Figure 2.** Selectivity towards formation of aromatic hydrocarbons as a function of temperature of propane conversion over the following catalysts: 1 – AS, 2 – Ga-AS; 3 – In-AS.

Thus, an introduction of various modifying additives into the zeolite in the course of hydrothermal synthesis results in samples with different catalytic properties. A zeolite system with an addition of gallium represents the most active and selective catalyst.

The bar chart in Figure 3 shows the effect of the nature of the element introduced into the zeolite on the conversion and yield of aromatic hydrocarbons that are formed on the catalysts at the reaction temperature 600 °C. It is evident that element-alumosilicates exhibit a high catalytic activity in the course of propane conversion but the highest activity is exhibited by the Ga-alumosilicate. A 99%

propane conversion over this catalyst results in a 43% yield of arenes at 600 °C for 500 $hr^{-1}$. In-AS exhibits the lowest activity during propane conversion and the yield of aromatic hydrocarbons under these process conditions is 19.4%.

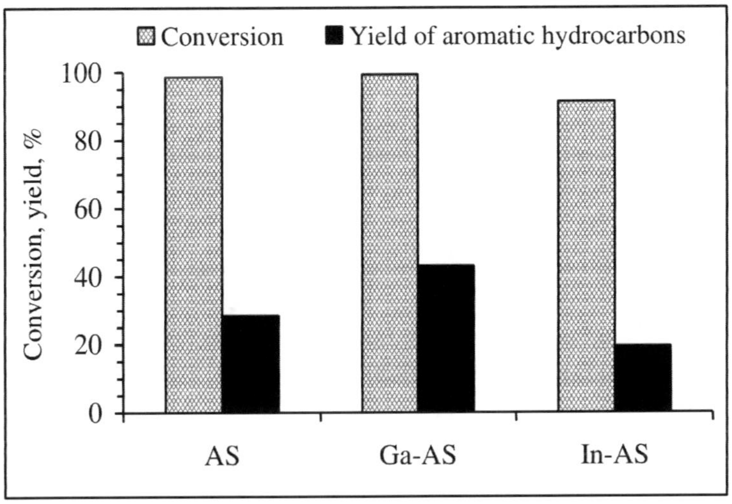


**Figure 3.** Propane conversion and yield in aromatic hydrocarbons formed over elementoalumosilicate catalysts (T = 600 °C).

## 4. Conclusion

Thus, we can conclude that an introduction of various modifying additives into the zeolite in the stage of its hydrothermal synthesis results in catalysts that differ in their functional properties. A zeolite, whose crystal lattice contains atoms of gallium along with those of silicon and aluminum, is the most promising for propane aromatization.

## 5. References

[1] Vosmerikova L N, Ryabova N V and Vosmerikov A V 2008 Synthesis, acid and catalytic properties of element-alumosilicates in the course of lower alkane aromatization *Neftepererabotka i neftekhimia* 2 35 - 39 [in Russian]

[2] Dergachyov A A and Lapidus A L 2008 Catalytic aromatization of light alkanes *Journal of D.I. Mendeleyev Russian society* **LII** (4) 15 - 21 [in Russian]

[3] Ione K G and Vostrikova A A 1987 The isomorphism and catalytic properties of silicates with the zeolite structure *Uspekhi khimii* **56** (3) 393 - 427 [in Russian]

[4] Lapidus A L, Dergachyov A A, Kostin V A and Silakova A A 2008 Ethane aromatization on Ga-Pt- pentasil zeolites *Petroleum Chemistry* **2** 2 83 - 86 [in Russian]

[5] Vosmerikov A V and Erofeev V I 1995 Effect of mechanical treatment on the catalytic properties of zeolite catalysts for aromatization of lower alkanes *Journal of Physical Chemistry* **69** (5) 787 - 790 [in Russian]

XVIII International Scientific Symposium in Honour of Academician M. A. Usov: PGON2014 IOP Publishing
IOP Conf. Series: Earth and Environmental Science **21** (2014) 012033 doi:10.1088/1755-1315/21/1/012033

# Research of inverted emulsions properties on the base of new emulsifiers

**K Minaev[1,a], A Epikhin[1], D Novoseltsev[1], M Andropov[2,b], V Yanovsky[2,b] and O Ulyanova[1,c]**
[1]Tomsk Polytechnic University, Tomsk, Russia
[2]Tomsk State University, Tomsk, Russia

E-mail: [a]minaevkm@bk.ru, [b]hudojnick@yandex.ru, [c]shamaim_7@mail.ru

**Abstract.** Emulsifiers on the base of tallol and ethanolamines derived acids have been researched in the paper. Electrical stability of emulsions drilling muds has been investigated. It is proved that synthesized emulsifiers according to emulsion stability can be divided into two groups. The first group is emulsifiers with high initial electrical stability but low emulsion stability under long-term storing, and the second group is emulsifiers with low electrical stability but with high emulsion stability. Emulsions flow characteristics have been researched. It is established that emulsifier on the base of ethanolamine provides better emulsion characteristics for drilling muds emulsions.

## 1. Introduction.

Oil and gas field depletion causes modification of drilling course and horizontal drilling with a large horizontal length, construction of multi-lateral wells and downhole splitters, including drilling in continental shelf waters. Under these conditions, traditional drilling muds are of little use for drilling. Emulsion drilling muds provide required drilling quantity. They permit to keep bottom-hole zone permeability, eliminate the risk of drill string drag and sidewall sticking, and provide wellbore stability as well as cutting transport [1-3].

In drilling, two types of emulsions are used: water-in-oil emulsions and oil-in-water emulsions. According to the drilling experience water-in-oil emulsions (inverted emulsions, oil-based emulsions, etc.) are good for drilling with zen more than 70°.

Oil-in-water emulsions (some types of clayless drilling mud and thin clay drilling mud) are good for drilling well with zen less than 70° [4 - 5].

These groups of emulsions have both advantages and disadvantages. The strong points of hydrocarbon emulsions are low water content in filtrate and a good lubricating property. Due to these properties, hydrocarbon emulsions minimally affect productive formation, and inhibit terrigenous deposits, as well as involve drilling wells with large zen and distance.

However, hydrocarbon emulsions have a number of drawbacks, such as ecological unfriendliness to environment, instability to pollution by produced water and a great quantity of solids, dispersed phase thickening, as well as high hydraulic characteristic compared to water-based emulsions.

High content of hostile SSAS (synthetic surface active substances) in emulsion negatively affects formation fluids in some cases. Some types of oil can be thickened in a contact with SSAS. Low conductivity of hydrocarbon emulsions negatively influences on electric coring [6-9]. Thus, the urgent purpose is the formulation of emulsion muds without the drawbacks mentioned.

The paper studies inverted emulsions properties on the base of new emulsifiers and evaluation of their availability.

## 2. Research methodology

The authors have invented new emulsifiers on the base of tallol and ethanolamines derived acids [10]. The emulsifiers are complex mixtures of amide and ester of fatty and resin acids and some appropriate ethanolamines. Formulae of the chemical compounds according to the main components are presented in figure 1.

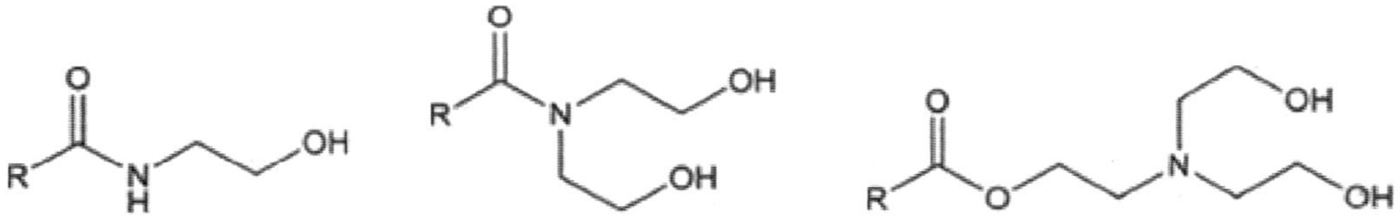

DcM DcD DcT

monoethylamine diethanolamide monoester of trietanolamine

**Figure 1.** Conventional formulae of chemical compounds (R – residuum of fatty or tallol resin acid).

The main characteristic of the emulsifiers is that they have firstly been synthesized under the condition of azeotropic water distilling off with acid catalyst. The emulsifiers have been synthesized of tallol residual matter and some appropriate ethanolamines. Azeotropiser (toluene) is used to eliminate water. After distillation, toluene is put back into the mother substances. One more advantage of this method is that azeotropiser facilitates to carry out experiment under low temperature – 95 – 100 °C. Concentrated sulphuric acid is used as azeotropiser.

Emulsions include 3 components: diesel fuel, experimental emulsifier and distilled water. Initially, emulsifier was dissolved in a required amount of diesel fuel, after which small portions of water were added, and finally, this substance was being mixed by mixer Hamilton beach (13000 r/min) for 5 minutes.

Rheological properties of the emulsions were studied by viscosimeter OFITE, modification 1100. Emulsions electrical stability was measured by OFITE according to API 13B-2 [11].

## 3. Discussion

The purpose of the paper is to study properties of inverted emulsion on the base of mentioned emulsifier as a future base for a drilling mud. The objective of the paper is to compare the electrical stability and rheological properties of the emulsifiers and Emultal.

Emultal as a comparative object is not chosen randomly. The emulsifier is a mixture of compound ester of fatty and resin acids and trietanolamine as the closest chemical product in nature. However, existing Emultal industrial technology does not infer to apply the catalyst and absorb additionally formed water.

Electrical stability is one of the important characteristics of a drilling mud. It is a complex parameter of drilling mud stability to the phase separation. Electrical stability facilitates to estimate aggregative stability of drilling mud quickly. The dependence of inverted emulsions electrical stability on the base of emulsifiers DcM, DcD, DcT and industrial Emultal model on emulsifier concentration C is studied. Electrical stability was measured immediately after emulsion obtaining. The results are shown in figure 2.

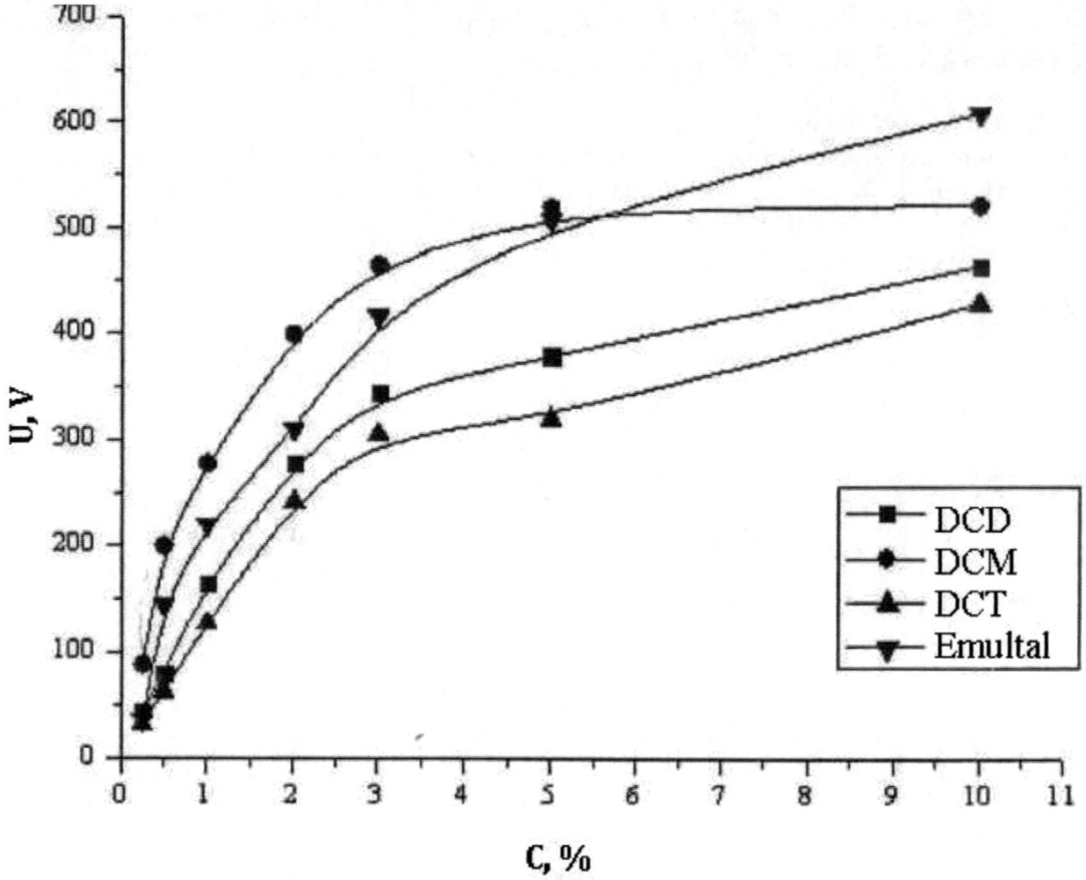


**Figure 2.** Inverted emulsions electrical stability dependence on emulsifier concentration, C.

The results show that all emulsifiers are characterized by a sharp increase in inverted emulsions electrical stability with ascending emulsifier concentration up to 3%. It's obvious that at the given concentration range the phase boundaries are saturated with emulsifier molecules, it increases aggregative emulsion stability. Under further emulsifier concentration increase, when interphase boundary is completely filled with SSAS molecules, there is a weak electrical stability rise. In this case, odd molecules of emulsifier split in the dispersion phase up as a molecular dispersion either in a form of micelles or in more complicated nematic phases, it increases dispersion phase viscosity (gel firms) as well as emulsion stability. To sum up, similar dependences are typical for the majority of emulsifiers.

The data given in figure 2, show that the researched emulsifiers in the range of standard concentrations 2 – 5% form emulsions with electrical stability of more than 220 V which makes them potentially applicable for a drilling mud. Among the three studied emulsifiers, only DcM has higher electrical stability than Emultal (if the concentration is less than 5 – 5.5%). If the concentration of DcM is more than 5 %, there is a pressure regulation of mud breakdown.

The important feature of drilling mud is emulsion stability with time. However, this parameter is rarely studied in the process of mud formulation which can lead to some difficulties in the well drilling process. Thus, the dependences of emulsions electrical stability on the concentration and time were calculated. The results are given in figure 3 (a-d).

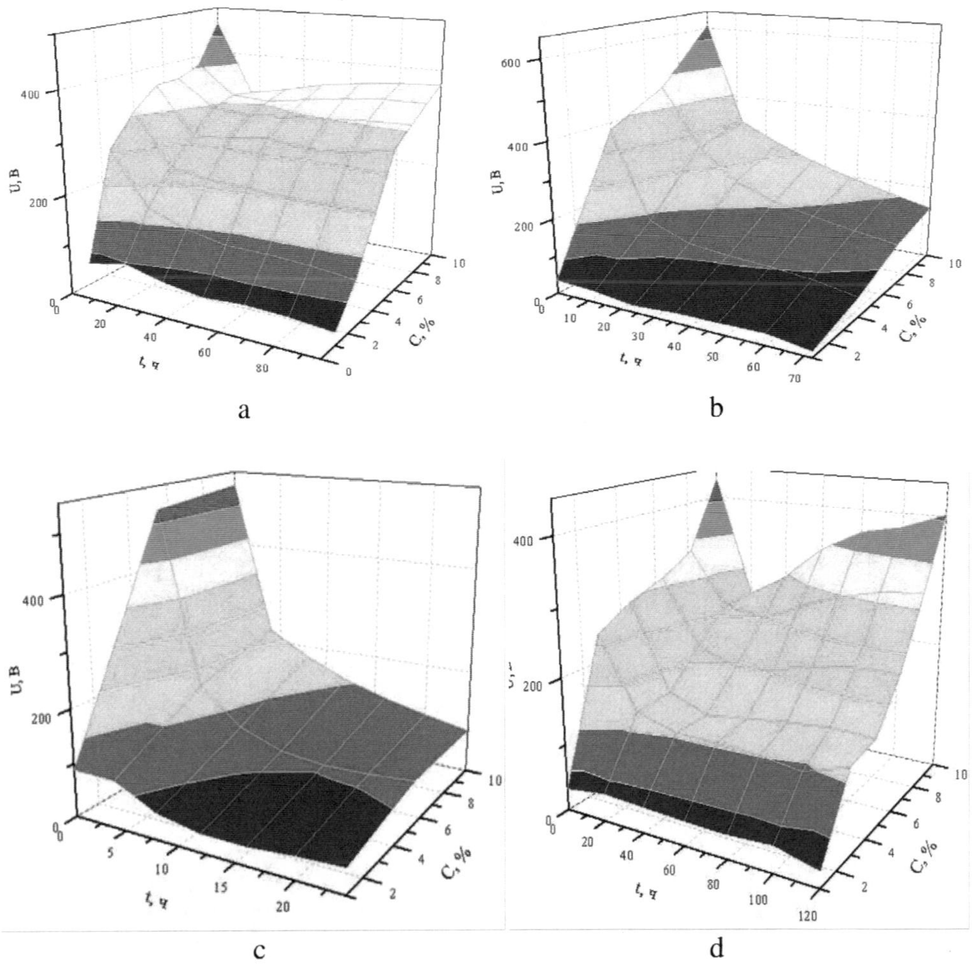


**Figure 3.** Dependence of emulsion electrical stability on the emulsifier concentration and time a – Emultal, b – DcT, e – DcM, d – DcD.

The analysis of the dependences involves making some important conclusions: according to the data in picture 3, the studied emulsifiers can be divided into 2 groups: 1) the emulsifiers with high initial electrical stability but its fast decrease with time; 2) the emulsifiers with low initial and high time electrical stability.

The emulsifiers of the first group (DcM and Emultal) are characterized by increased electrical stability of freshly-mixed emulsions which comes to 400-600 V under 3-10% emulsifier concentration. When stored, the emulsions are broken-up quickly and their electrical stability decreases sharply.

DcM emulsifier has the fastest decrease in electrical stability with time: in 24 hours it does not exceed 90 V. The higher concentration of the emulsifier, the faster decrease in its emulsion stability is. For instance, for 0.5% emulsifier concentration, electrical stability decreases 3.3 times in 24 hours; for 3-10% emulsifier concentration, it decreases as many as 6.2- 6.3 times.

Emultal has the similar features, with only the differences that the electrical stability parameters lower than 100 V are reached in 72 hours. The maximum relative decrease in electrical stability is observed in emulsions with medium emulsifier concentration (2 – 3 %).

Electrical stability of freshly-mixed emulsions for emulsifiers of the second group (DcD, DcT) is in the range of 300 – 450 V under the same concentrations of 3 – 10% emulsifiers. Unlike emulsifiers of the first group, emulsifiers of the second group according to picture 3, are characterized by high timing stability. Relative decrease in electrical stability for them is in as many as 1.3 – 3.6 times in 96 hours, whereas maximum decrease equals to average emulsifier concentration parameters (1 – 3 %). Thus, inverted emulsions based on emulsifiers DcD or DcT have rather high aggregative stability even in 96 – 120 hours.

A special feature of emulsifiers of the second group is the dependence character of electrical stability on time. In the sphere of low emulsifier concentrations, within the first hours after the emulsion formulation, its electrical stability slightly increases on average to 10-20 V. In 24 hours electrical stability of these emulsions starts slowly decreasing as mentioned above. But for emulsions with high content of emulsifier (2 – 10%), electrical stability of emulsions sharply decreases within the first hours. For medium emulsifier concentrations, electrical stability of emulsions is slightly decreasing, whereas for high emulsifier concentrations, electrical stability is slowly increasing.

Such strong changes of emulsions properties within the first hours are connected with the processes in the emulsions system structure. Firstly, there is regrouping emulsifier's molecules in the adsorbed layer. Secondly, there is some rearrangement of the dispersion phase among globes, as a consequence of emulsion polydispersity level changes. The main process of this type is Ostwald ripening, the core of which is transportation of the dispersion phase from smaller to larger globes. Thus, the number of small globes is decreasing, whereas large globes are becoming greater in size. These processes are especially noted for emulsions with high content of emulsifier, because of emulsifier excess in the dispersion phase (due to associates or micelles), the solvability of the dispersion phase (water) increases. It is quite an unexpected fact that DcT emulsifier and Emutal have completely different dependence character between electrical stability and time. As mentioned above, Emultal refers to the first group of emulsifiers, whereas DcT – to the second group. Such a radical difference in properties seems strange, because both emulsifiers have the same chemical nature, i.e they are a tallol ester and trietanolamine mixture. It means they must be differed only by residual carboxylic acids. This problem requires further thorough study.

The research of emulsions rheological properties shows that the character of viscosity changes from emulsifier concentration is hardly depended on emulsifier properties. In all the cases the maximum of dynamic shear stress and static shear stress for DcM and DcT is 1%, for DcD – 0.5%. Decrease in drilling mud viscosity on maximum is connected with high content of emulsifiers that causes hydrocarbon phase increasing.

DcM emulsifiers produce the largest emulsions gelling, dynamic shear stress and static shear stress are about 2 times as high as DcD and DcT emulsifiers. This drilling mud has the lowest plastic viscosity that can provide a high drilling speed and a very high cutting transportation [12]. In addition, in this case drilling mud requires less emulsifier concentration. It cuts drilling well cost.

## 4. Summary and conclusions.

To sum up, some emulsifier properties were comparatively researched in the paper. These emulsifiers are Emultal and derived tallol and ethanolamines acids obtained under the condition of azeotropic water distilling off. Dependences between electrical stability, emulsifier concentration and their lifetime, as well as rheological properties are studied here. According to electrical stability and rheological properties, DcM emulsifier has better results in both electrical stability and rheologic properties. This mud has the least plastic viscosity that can provide a high drilling speed and a very high cutting transportation. Also, due to less concentration of emulsifier it helps to cut cost for drilling wells.

## 5. References

[1] Petrov N A 2008 *Èmulsionnye rastvory v neftegazovyh processah* (Moskva: Himia) p 439 [in Russian]

[2] Glushhenko V N 2008 *Obratnye ehmul'sii i suspenzii v neftegazovoj promyshlennosti* (Moskva: Interkontakt Nauka) p 725 [in Russian]

[3] Ustkachkintsev E N, Husaenov R Ya, Makarov N V, Minaev K M, Rastegaev R A and Mavliev A R 2013 Effect of cementing mounting quality on selection of wells for sidetrack construction *Oil Industry* (2) 78 – 81

[4] Wei Wang, Pengyu Wang, Kai Li, Jimiao Duan, Kunyi Wu and Jing Gong 2013 Prediction of the apparent viscosity of non-Newtonian water-in-crude oil emulsions *Petroleum Exploration and Development* **40** (1) 130–133

[5] Fefelov Yu V, Karasev D V, Nekrasova I L and Natsepinskaya A M 2009 Osobennosti invertno-ehmul'sionnykh burovykh rastvorov pri burenii pologikh i gorizontalnykh skvazhin na mestorozhdeniyakh OOO «Lukojl-Perm» *Neft. Gas. Novatzii* 10 45 – 48 [in Russian]

[6] Ilyasov S E, Popov S G, Okromelidze G V, Garshina O V and Natsepinskaya A M 2011 Emulsionnye burovye rastvory - tendentsii razvitiya tekhnologii *Territoriya neftegaz* 11 14-17 [in Russian]

[7] Hubbatov A A and other authors 2010 Formation of properties of oil fluids *Construction of Oil and Gas Wells on-Land and off-Shore* 8 28–31

[8] Khismetov T V, Bershtejn A M, Silin M A, Magadova L A, Efimov N N, Khalzov A A, Orekhov V V, Gorobets E A and Grishchenko A S 2009 Squeeze cementing in oil wells with use of hydrocarbon based grouting mortars *Oil industry* 6 50-54

[9] Basma Yaghi 2003 Rheology of oil-in-water emulsions containing fine particles *Journal of Petroleum Science and Engineering* **40** (3–4) 103–110

[10] Yanovsky V A, Churkin R A, Andropov M O and Kosova N I 2013 Synthesis and study of properties of emulsifiers for reverse emulsions based on derivatives of acids of tall oil distillate and ethanolamines *Tomsk State University Journal* 370 194–199

[11] *Recommended Practice for Laboratory Testing of Drilling Fluids* 2008 (Eighth Edition) ANSI/API 13I/ISO 10416

[12] Medentsev S V 2010 Stabilizatsiya reologicheskogo profilya burovykh rastvorov na uglevodorodnoj osnove *Territoriya neftegaz* 10 28–31 [in Russian]

# Lightweight Cement Slurries based on vermiculite

**K Minaev[1], V Gorbenko and O Ulyanova[2]**
Tomsk Polytechnic University, Tomsk, Russia

E-mail: [1]minaevkm@bk.ru, [2]shamaim_7@mail.ru

**Abstract.** The main purpose of the research is to study the lightweight cement slurry based on vermiculite and its parameters in accordance with GOST 1581-96 requirements as well as improvement of its formulation by polymer additives. Analysis of vermiculite-containing mixture providing the lowest density while maintaining other required parameters was conducted. As a cement base, cement PTscT-I-G-CC-1, cement PTscT - 100 and vermiculite M200 and M150 were used. Vermiculite content varied from 10 to 15 %; and water-to-cement-ratio ranged from 0.65 to 0.8. To sum up, despite the fact that lightweight cement slurry based on vermiculite satisfies GOST 1581-96 requirements under laboratory conditions, field studies are necessary in order to make a conclusion about applicability of this slurry for well cementing.

## 1. Introduction

Well cementing is a hard technological process where proper selection of the cement slurries affects the successful realization. The main requirements for cement slurries used for casing cementing are to ensure sealing of annular space and to exclude cross-flows of fluids between casings. That is why cement slurries and cement stone must comply with necessary rheological and structural mechanical properties depending on hydrogeological conditions of well construction. Construction of wells in certain fields is complicated by high-permeability formations, and formations with low pore pressure. The most efficient process solution, which allows significant reducing of cost on casing cementing under these conditions, is the use of lightweight cement slurry with specified density. It can increase well productivity and enhance quality of plugging [1-4].

Often the reduction of cement slurry density is done by increasing. However, the increasing of cement-water ratio results in the necessity of adding water-retaining agents, otherwise cement slurry segregation could occur as well as the increasing of cement slurry filtration. Increasing of water content in Portland cement slurry, results in the decrease of sedimentation stability. When cement-water ratio is about 0.55, the dehydration is observed. If this parameter is 0.6, over-dehydration may occur. Filtration rate of fluid through the suspension could be reduced by increasing fluid viscosity and dispersity of solid phase, as well as air-entraining agents. Most of all, viscosity of cement is increased by clay, but in this case hydrated cement has lower compressive strength. As air-entraining agents potassium alumen hollow microspheres (PAHM) are used, which are waste products of heat electro power stations or other industries. This cement mixtures are well studied, have predictable and easily controlled properties. However, the PAHM is a scarce product as it is not produced in sufficient amounts which results in the increase of its cost. Foamed vermiculite might be used as an alternative to PAHM.

Vermiculite, a hydromica group mineral, is a product of secondary biotite conversion (figure 1). It has heat-and-sound-insulating properties and high absorbency. It absorbs to 500% of liquid from its self-weight. The mineral is not subjected to biological decomposition under acid and alkalis action. It should be noted that vermiculite is a non-toxic environmentally friendly material.

According to the opinion of different researchers [5 – 8], efficiency of lightweight cement slurry based on vermiculite is connected with vermiculite and cement physicochemical interaction. New hydrated phase formation reinforces composite material structure. In comparison with gel-cement slurry, vermiculite cement slurry has better plugging ability that increases the level of cement in fractured rocks. Moreover, this cement slurry has lower thermal conductivity providing better facilities for cement hardening, and increases frost resistance and fracture strength.

**Figure 1.** Vermiculite brand M200, 0.5mm fraction (Zooming is 40 times)

Despite the first application of vermiculite as an air-entraining additive was studied in early 1980s [6, 9] vermiculite is rarely used in well construction. It is connected with lack of vermiculite production in oil extraction regions, besides the transportation of vermiculite at a distance more than 400 km is unprofitable.

The purposes of the article - developing of lightweight vermiculite-cement slurry in accordance with GOST 1581-96 "Oil-well Portland cement. TS"; as well as improvement of cement slurry by polymers.

## 2. Research methodology

Distilled water with necessary water-solid ratio is used for cement gauging. Density and flowability of cement slurry have been measured. Water separation and thickening of cement were measured by consistometer OFITE 80 and graduated cylinder. Flexural strength tests of cement were carried out by hydraulic press MATEST E160. Volumetric shrinkage of experimental cement, cement PTscT-I-G-CC-1 and expansive backfill materials PTM-75 was carried out by OFITE 120-54 under constant pressure equals to 0.7MPa.

## 3. Discussion

At the first stage, lightweight cement slurry with the lowest density has been chosen. It keeps flowability of cement slurry in accordance with GOST 1581-96 "Oil-well Portland cement. TS". Vermiculites M200 with bulk density about 200 kg/m$^3$ (0.5 mm fraction) and PTscT-I-G-CC-1 are used as additives. Experimental results are shown in table 1.

**Table 1.** Density and flowability of cement slurry

| Parameters | Cement I 90% cement, 10% vermiculite | | | Cement II 87.5% cement, 12.5% vermiculite | | | Cement III 85% cement, 15% vermiculite | | | GOST 1581-96 |
|---|---|---|---|---|---|---|---|---|---|---|
| | Water-solid ratio | | | | | | | | | |
| | 0.65 | 0.75 | 0.80 | 0.65 | 0.75 | 0.80 | 0.65 | 0.75 | 0.80 | |
| Density, g/cm$^3$ | 1.60 | 1.55 | 1.50 | 1.57 | 1.53 | 1.48 | - | 1.46 | 1.44 | - |
| Flowability, mm | 145 | 170 | 250 | 120 | 150 | 210 | - | <90 | 165 | >180 |

IOP Conf. Series: Earth and Environmental Science **21** (2014) 012034 doi:10.1088/1755-1315/21/1/012034

Decrease in water-solid ratio and increase of vermiculite content in cement, lead to the decrease in density and flowability of cement slurry. Thus, for further experiments cements I and II were chosen (water-solid ratio 0.8).

Experimental results are shown in table 2.

**Table 2.** Test results of cements with water-solid ratio 0.8

| Parameters | Cement I<br>90% cement,<br>10% vermiculite | Cement II<br>87.5% cement,<br>12.5% vermiculite | GOST 1581-96 |
|---|---|---|---|
| Density, g/cm$^3$ | 1.5 | 1.48 | - |
| Flowability, mm | 250 | 210 | > 180 |
| Thickening time, min - 30Bc (75°C) | 110 | 75 | > 90 |
| Thickening time, min (room temperature) | >430 | 155 | Uncontrolled |
| Flexural strength after two days, MPa (30°C) | 1.7 | 1.45 | 1.0 |
| Flexural strength after two days, MPa (60°C) | 1.65 | 1.50 | Uncontrolled |
| Dehydration, cm$^3$ | 3.5 | 4.2 | <7.5 |

Table 2 shows that both cements comply with GOST 1581-96 requirements. It should be noted that the greater vermiculite content is, the lesser is the time of cement slurry thickening. At the second stage physical-mechanical properties of experimental lightweight vermiculite cement slurries were tested in accordance with GOST 1581-96 requirements to III-type cements.

During flexural strength tests cement stone expansion has been observed. Figure 2 shows the comparison of vermiculite cement stone and cement PTscT-I-G.

Volumetric shrinkage of experimental cement, cement PTscT-I-G-CC-1 and expansive backfill materials PTM-75 was carried out under constant pressure equals to 0.7MPa. Results are given in figure 3.

**Figure 2.** Cement stone surface during tests:

1 – Vermiculite cement stone

2 – Cement PTscT-I-G.

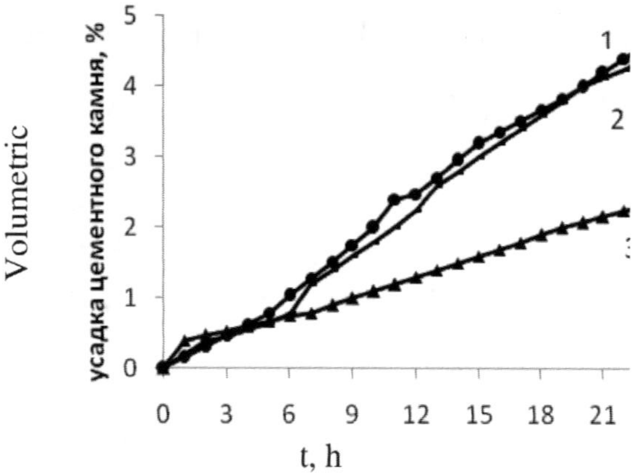


**Figure 3**. Volumetric shrinkage vs time:
1 – cement PTscT-I-G-CC-1;
2 – backfill materials PTM-75;
3 – experimental cement I

It is obvious, that experimental cement I has the smallest volumetric shrinkage. As this effect can be observed under low pressures and low temperatures, it could be assumed that the expansion of vermiculite cement stone is caused by replacement of air contained in vermiculite pores by water and its further expansion in cement slurry.

Recently, TyumenNIIgiprogaz did well cementing with vermiculite cement slurry using direct and reverse cementing of production wells. This type of cement slurry is patented by authors [10]. Calcium chloride was used as a setting rate regulator. It should be noted that only systems with water-solid ratio from 0.8 to 0.9 and density about 1500 kg/m$^3$ have optimal technological parameters. Nowadays cement slurry with density equals to 1350-1450 kg/m$^3$ is required to be used for single-stage cementing of well production. Thus, vermiculate cement slurry using has been reduced; it is used only for collar string cementing in permafrost rocks [11].

At present, vermiculite cement slurry with density ranges from 1350 to 1400 kg/m$^3$ is researched by us for single-stage well cementing. Moreover, to cut cementation cost, cement PTscT-I-G-CC-1 was replaced by cheap cement PTscT-I-100. In order to reduce cement slurry density, in comparison with previous ones, vermiculite with average bulk density (about 150 kg/m$^3$) and 1.0 mm fraction is used. This vermiculite provides greater air-entrainment that leads to major reduction of cement slurry density with the same proportions as in previous ones. However, such cements are unstable. Low stability as well as high filtration rate results in cross-flows and oil and gas casing blowouts during cement-slurry thickening time and formation of permeable cement stone. It decreases formation isolation quality and reduces workover period. Therefore, additional research was done to study the influence of bentonite clay and polymer additives on viscosity of vermiculite cement slurry and cement stone. Experimental results are shown in table 3.

In all cases polymer additives (such as carboxymethylcellulose, methylcellulose and starch) decrease cement slurry dehydration, stabilize cement slurry density, and reduce the probability of vermiculite formations. Moreover, it increases thickening of cement time, while cement stone flexural strength remains approximately at the same level. For commercial application methylcellulose could be recommended. This chemical agent provides the lowest density of cement slurry, dissolves well, and increases thickening time greatly, as well as reduces water separation.

**Table 3.** Test results of cements with polymer additives (water-solid ratio 0.8)

| Parameters | Additive | | | | |
|---|---|---|---|---|---|
| | - | carboxymethylcellulose 0.12% | methylcellulose 0.15 % | methylcellulose 0.075% bentonite 1.5% | 0.06% starch |
| Density, g/cm$^3$ | 1.35-1.40 | 1.37 | 1.35 | 1.40 | 1.40 |
| Flowability, mm | >250 | 250 | 220 | 215 | 205-210 |
| thickening time, min - 30Bc (75°C) | 110 | 148 | 145 | 105 | 152 |
| Flexural strength after two days, MPa (30°C) | 1.35 | 1.40 | 1.35 | 1.30 | 1.46 |
| Dehydration, cm$^3$ | 40 | 2.4 | 0 | 0 | 1.0 |

## 5. Summary and conclusions

To sum up, despite the fact that lightweight cement slurry based on vermiculite satisfies GOST 1581-96 requirements under laboratory conditions, field studies are necessary in order to make a conclusion about applicability of this slurry for well cementing. Some researchers [12] note the difficulty during vermiculite cement slurry use. For example, it's impossible to pump it into the annular space after short break.

Conducted experiments showed that reason for this is water separation, vermiculite formations and water belt formations, as well as uncontrolled cement setting time. The researchers draw a conclusion that it is impossible to use vermiculite cement slurry for well cementing.

There are main advantages of vermiculite as lightening additives are its unlimited quantity, possibility to control bulk density and fraction. In case of correct cement slurry usage, in accordance with the operating conditions it is possible to eliminate the drawbacks.

This work was financially supported by Foundation for Assistance to Small Innovative Enterprises in Science and Technology (Program "Participant of youth scientific innovation contest").

## 6. References

[1] Sharafutdinov Z Z, Chegodaev F A, Sharafutdinova R Z 2007 *Burovyye i tamponazhnyye rastvory. Teoriya i praktika: spravochnik* (SPb: Professional) p 416 [in Russian]

[2] Melekhin A A, Dorovskikh I V and Devyatkova N V and other authors 2012 Modelling of lost circulation formation isolating using expansive cement blend *Oil Industry* 8 114-116

[3] Ustkachkintsev E N, Husaenov R Ya, Makarov N V and other authors 2013 Effect of cementing mounting quality on selection of wells for sidetrack construction *Oil Industry* (2) 78-81

[4] Nguyen D, Kagan M and Rahman S 1992 Evaluation of drilling fluid removal by cement slurry from horizontal wells with the use of an accurate mathematical model *Journal of Petroleum Science and Engineering* **8** (3) 191-204

[5] Klusov A A 1993 Razrabotka i issledovaniye tsementnykh tamponazhnykh kompozitsiy, tverdeyushchikh pri ponizhennykh temperaturakh *PhD Thesis* 560 [in Russian]

[6] Patent 884367 USSR, MKIZ E 21 B 33/138 Lightweight cement slurry for low useable temperature wells / A.A. Klusov (USSR) - № 2977437/ 22-03

[7] Patent 1339233 USSR, MKIZ E 21 B 33/138 *Cement slurry* (A A Klusov, T V Kuznesova, M M Shalyapin, N A Danukin, E M Nanivsky and Yu F Zakharov )(USSR) - № 4056952/22-03. Received 16.04.86.

[8] Klusov A A 1985 K effektivnosti ispolzovaniya tamponazhnykh rastvorov ponizhennoy plotnosti *EI Geologiya, bureniye i razrabotka gazovykh i morskikh neftyanykh mestorozhdeniy* 10 9-11 [in Russian]

[9] Gorskiy A T, Batalov D M and Shvezov V D 1983 Primeneniye vermikulitotsementnykh rastvorov dlya tsementirovaniya skvazhin *Tr. in-ta ZapSibNIGNI* 66 54-59 [in Russian]

[10 Oblegchennaya tamponazhnaya smes'. Ippolitov V V, Podshibyakin V V, Belej I I, Konovalov V S and Vyalov V V № RU2243358C1. Received 29.04.2003 [in Russian]

[11] Beley I I, Shtol V F and Sherbich N E 2005 Tipy primenyayemykh oblegchennykh tamponazhnykh rastvorov dlya krepleniya skvazhin na gazokondensatnykh mestorozhdeniyakh severa Tyumenskoy oblasti *Stroitelstvovo neftyanykh i gazovykh skvazhin na sushe i na more* 3 30-32 [in Russian]

[12] Danushevsky V S, Aliev R M and Tolstykh I F 1987 *Spravochnoye rukovodstvo po tamponazhnym materiala* (Moscow: Nedra) p 327

IOP Conf. Series: Earth and Environmental Science **21** (2014) 012035 doi:10.1088/1755-1315/21/1/012035

# Influence of Primary Glyoxal on Properties of Tested Drilling Mud

**K Minaev[1,a], D Martynova[2,b], A Knyazev[2], A Zaharov[1] and I Shenderova[1,c]**
[1]Tomsk Polytechnic University, Tomsk, Russia
[2] Tomsk State University, Tomsk, Russia

E-mail: [a]minaevkm@bk.ru, [b]martynova1847@gmail.com, [c]Inna-shenderova@yandex.ru

**Abstract.** The research was done to observe the influence of primary glyoxal on properties of tested drilling mud. Glyoxal was used as a reagent to reduce viscosity and stabilize shale. Drilling mud processing parameters were estimated before and after affecting by glyoxal. Suggested is the method of polysaccharide modification by glyoxal for improving its resistance to microorganisms; carried out the research of bio- and thermal resistance of starchy and xanthan reagents.

## 1. Introduction

According to the present knowledge, it has been found that microorganisms negatively affect the process of oil production starting from well drilling to commercial oil transport.

At drilling process, the activity of cellulose-decomposing and other bacteria leads to a rapid deterioration of processing properties of drilling mud based on polysaccharide (starch, xanthan gum, various types of cellulose). This results in changing of the flow properties and the increasing of fluid loss indicator [1. № 2. P.23]. Loss in flow and processing properties of mud, resulted from biodegradation, leads to the necessity of extra processing with costly chemical reagents and thus increases the cost of drilling operations. Therefore, critical task is to prevent biodegradation of polysaccharide.

The most rational and promising way to suppress bacteria is to use special antimicrobial agents - bactericides. In this case, bactericides should meet the following requirements: to have broad spectrum of activity, to be of high activity at low concentration, non-corrosive, safe for humans and environmentally friendly, do not affect processing properties of drilling mud, to have available resource base and low cost [2 . P.15].

High biodegradability of polysaccharides affected by bacteria gives the ground to carry out the study aimed to determine the effect of bactericidal agents on the consistency of bio-polymer mud.
Phenol and formaldehyde (paraformaldehyde) have antibacterial effect, but their usage for drilling mud treatment is limited by hygienic and ecological restrictions. Formaldehyde has found wide application as an antibacterial additive to the drilling mud and a corrosion inhibitor. However, this reagent has several drawbacks, in particular, low efficiency and duration of antibacterial effect, as well as high toxicity index (2$^{nd}$ hazard class according to GOST 12.1.007-76) [3. C.1]. Thus, the most optimal bactericide agent is primary glyoxal.

Glyoxal is more than twice as active as formaldehyde and greatly exceeds in environmental characteristics. One molecule of glyoxal is able to fix four sulfur-containing molecules (hydrogen sulphide, mercaptans, etc.), hence it is possible to use it for prevention of H2S corrosion [4. P.34]. Biocidal properties of glyoxal are widely used to extend the service life of drilling mud and for oil decontamination, in particular, to prevent the growth of sulfate-reducing bacteria. High cost of glyoxal reduced its application. Currently "Novokhim" company, firstly in Russia, launched the production of glyoxal by using unique technology, hence the products became competitive. Thus, the aim of present research was to study the flow and processing properties of tested drilling mud modified by glyoxal.

## 2. Experiment

*2.1 Preparation*

Drilling mud was prepared by dissolution of bentonite clay (PBMA) in distilled water, than the required amount of glyoxal was added.

*2.2 Characterization*

Study of flow properties was performed on a rotor viscometer OFITE 1100. Glyoxal showed the highest chemical reactivity in a temperature interval of 50-60 ° C, and therefore the study of drilling mud rheology was carried out at temperature of 20 ° C and 55 ° C. Filtration characteristics of drilling mud was determined by pressure filter OFITE. Effect of drilling mud on clay swelling was studied by longitudinal swelling tester OFITE in dynamic regime.

Antibacterial activity was evaluated by the number of colony-forming bacteria and by observing the structural change in flowing properties of 1% biopolymer aqueous solution during 7 days.

## 3. Results and discussion

*Research of flow and processing properties of drilling mud containing glyoxal.*

During the study of drilling mud rheology it was defined that processing of bentonite slurry by glyoxal leads to the considerable reduction of drilling mud funnel viscosity (table 1). With the increase of glyoxal concentration the effect becomes more obvious, thus when using glyoxal in clay mud, thin properties of the reagent should be taken into account. Despite the fact that viscosity of drilling mud decreases when adding glyoxal, the filtration index is slightly changing during glyoxal concentration increasing.

**Table 1.** Flow and processing properties of tested glyoxal and bentonite drilling mud.

| Mud | ρ, g/cm$^3$ | Funnel viscosity, sec. | Filtration, cm$^3$/30 min. | SGS, pound/100ft$^2$, | | PV, cPs | | DSS, pound/100ft$^2$ | |
|---|---|---|---|---|---|---|---|---|---|
| | | | | 20 °C | 55 °C | 20 °C | 55 °C | 20 ° C | 55 ° C |
| Bentonite 5% | 1,035 | 28,5 | 19,5 | 4,0/6,0 | 6,0/8,0 | 3,3 | 3 | 3,3 | 4,1 |
| Bentonite 5%+ glyoxal 0,25% | 1,035 | 22,0 | 19,4 | 2,0/3,0 | 3,0/4,0 | 2,9 | 2,3 | 1,8 | 2,3 |
| Bentonite 5%+ glyoxal 0,5 % | 1,035 | 20,0 | 19,3 | 2,0/2,0 | 2,0/2,0 | 2,4 | 1,9 | 1,6 | 1,8 |
| Bentonite 5%+ glyoxal 1,0 % | 1,035 | 18,0 | 19,2 | 1,0/2,0 | 2,0/2,0 | 2,5 | 1,9 | 1 | 1,3 |
| Bentonite 5%+ glyoxal 2,0 % | 1,035 | 18,0 | 19,0 | 1,0/2,0 | 2,0/2,0 | 2,2 | 2,2 | 1,3 | 1,3 |

Static gel stress (SGS), plastic viscosity (PV) and dynamic shear stress (DSS) decrease with the increasing of glyoxal concentration in mud

Glyoxal has the ability to fix polymer reagents containing hydroxyl groups. This can result in viscosity increaing and mud shearling. Therefore, flow properties study of tested starch and glyoxal-containing drilling mud (5% bentonite and 0.25% starch) was made. Measurement of rheology was done one day after mud preparation.

**Table 2.** Flow and processing properties of tested drilling mud containing glyoxal and starchy additives (5% bentonite, 0.25% starch).

| Glyoxal concentration, % | ρ, g/cm$^3$ | Funnel viscosity, sec. | Filtration, cm$^3$/30min | SGS, pound/100ft$^2$, | PV, cPs | DSS, pound/100ft$^2$ |
|---|---|---|---|---|---|---|
| | | | | 20 °C | 20 °C | 20 °C |
| 0,025 | 1,035 | 23.8 | 14 | 4,0/6,0 | 10,2 | 3,5 |
| 0,075 | 1,035 | 25 | 14 | 4,0/7,0 | 10,4 | 4,8 |
| 0,10 | 1,035 | 25.2 | 14,4 | 5,0/8,0 | 10,7 | 4,9 |
| 0,25 | 1,035 | 27 | 13 | 6,0/9,0 | 11,4 | 5,6 |
| 0,50 | 1,035 | 24.5 | 15 | 5,0/8,0 | 10.3 | 4,4 |
| 0,75 | 1,035 | 23.5 | 15 | 4,0/7,0 | 9,8 | 3,4 |
| 1,0 | 1,035 | 23 | 14.6 | 4,0/6,0 | 8,9 | 3,7 |

From the results of drilling mud flow properties measurements, shown in table 2, it can be seen that increasing of glyoxal concentration to 0.25% results in the increase of the tested mud viscosity. This may be the result of glyoxal and starch interaction, at which polysaccharide polymer units are fixing and shearling is increased. Further increasing of glyoxal concentration in drilling mud leads to the decrease in viscosity. Thus, it can be assumed that excess of glyoxal reacts with bentonite clay developing thin effect as it was shown above (table 1).

Effects of glyoxal on colloidal clay mud in water can be used to reduce the swelling of shale clay. The swelling process is accompanied by increase in humidity, rock volume and swelling pressure. As a result, during the process of drilling, challenges associated with sidewall sticking, cavings and rock falls may emerge due to water-based drilling mud application [5. P.42].

Studies (figure 1) show the interaction between bentonite and sodium during 70 hours of analysis. The graph clearly displays a discrepancy of four fluids. Water demonstrates the highest degree of swelling, exceeding approximately in 4 times the axial size of bentonite sample.

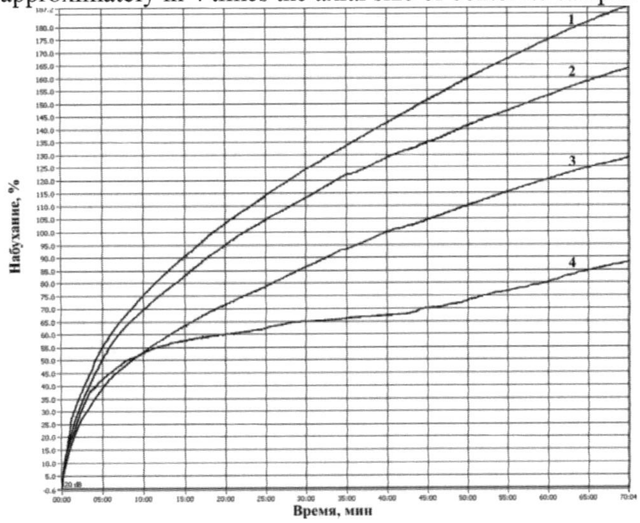


**Figure 1.** Bentonite clay swelling in drilling mud (1 - water: 2 - 5% bentonite + 0.25% Bur-S (modified starch); 3 - 5% of bentonite + 0.25% starch; 4 - 5% bentonite + 0.25% starch + glyoxal 1%)

Clay mud containing Bur-S and starch demonstrate shale stabilizing ability, although it inferiors to the drilling mud modified by glyoxal. The glyoxal add into drilling mud containing starch, results in significant inhibition of the swelling process of compacted bentonite clay samples. This effect starts to emerge only after 9 hours of drilling mud affecting on clay sample. Apparently, the highest swelling property of mud containing glyoxal, before that period, is determined by lower viscosity, which in its turn resulted in high diffusion of water into the bentonite. Reduction of samples linear swelling placed in starch drilling mud modified by glyoxal reaches 40% as compared to non-glyoxal mud.

*Study of antibacterial properties.*

Addition of bactericidal reagents allows reducing the number of microorganisms and extends the drilling mud life. In course of research the attempt to increase polysaccharides enzyme resistance was made. In this case, drilling mud was glyoxal free. Glyoxal was used as a modifier of biocidal properties to starch reagent.

**Figure 2**. Visual determination of microorganisms colonies.

**Table 3.** Number of colony-forming units of starch drilling mud samples.

| Composition | $1^{st}$ day | $2^{nd}$ day |
|---|---|---|
| Natural starch | $1{,}2\cdot10^8$ CFU/ml | $8{,}3\cdot10^{10}$ CFU/ml |
| Starch with bactericide additive MI CIDE | $7{,}8\cdot10^5$ CFU/ml | – |
| Glyoxal-modified starch | $3{,}6\cdot10^5$ CFU/ml | $2{,}3\cdot10^6$ CFU/ml |

As criteria to measure the dynamics of structural and rheological characteristics changing, dynamic shear stress and static gel stress was selected. From the results shown in table 4, it can be seen that modified starch solution saves primary flow properties for a long period of time. While the source starch reactant, under the influence of bacteria, completely degraded after 7 days. Thus, the use of modified starch allows refusing from bactericide additives in drilling mud, this in its turn will reduce the total costs.

**Table 4**. Rheological stability of starch reagent (1 - natural starch reagent, 2 – glyoxal-modified starch).

| Flow properties | 1st day | | 3rd day | | 7th day | |
|---|---|---|---|---|---|---|
| | 1 | 2 | 1 | 2 | 1 | 2 |
| Plastic Viscosity, cPs. | 40,0 | 45,0 | 29,6 | 41,7 | 6,09 | 39,9 |
| DSS, pound/100ft$^2$ | 19,7 | 20,2 | 11,93 | 21,2 | 0 | 20,5 |
| SGS pound/100ft$^2$ | 4/5 | 7/8 | 2/2 | 7/8 | 0/0 | 6/7 |

*Research of bactericide influence on flow properties of xanthan gum based drilling mud.*
For qualitative processing of biopolymer 0.1% of potassium hydroxide was added by stirring in a triaxial mixer till homogeneous substance for alkaline condition, while xanthan gum DUO-VIS was added smoothly and strictly on time.

Comparative analysis of dynamic shear stress of xanthan gum solution, with and without glyoxal bactericides additives (figure 3) and MI CIDE (figure 4), showed that glyoxal as well as starch, demonstrate good stabilizing properties. Stability of polymeric solutions increases with growth of bactericide concentration. However, xanthan gum DUO-VIS has sufficient resistance to bacteria (gum solution is resistant to 15 days), so it is possible to assume, without bactericides addition, that present reagent has been already modified by the producer. Static gel stress of researched solutions varies slightly within a month.

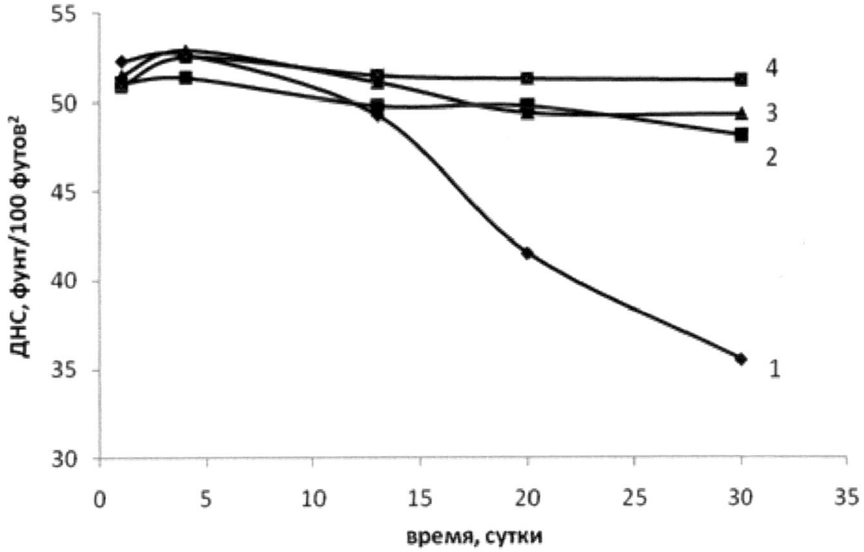


**Figure 3.** Time variation of the dynamic shear stress limit of 1% water xanthan solution DUO-VIS, processed by glyoxal bactericide (1 - xanthan solution without bactericide, 2 - 0.25% glyoxal, 3 - 0.5% glyoxal, 4 - 1 0% glyoxal).

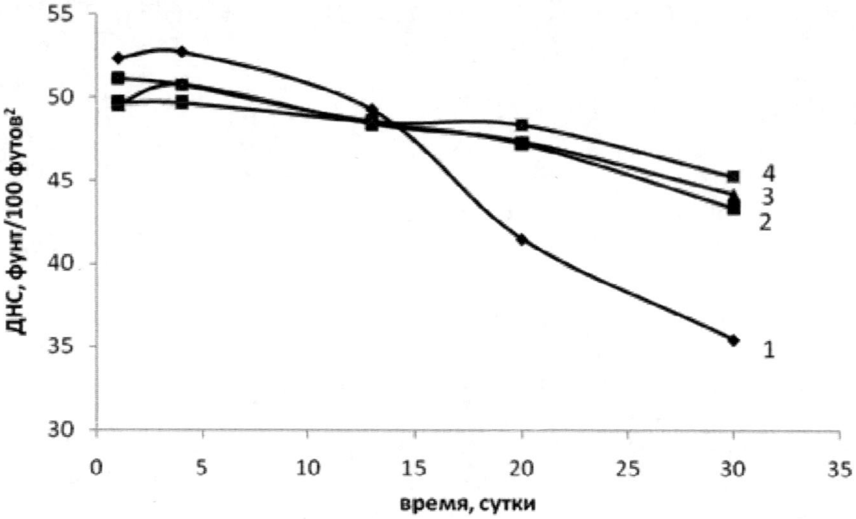


**Figure 4.** Time variation of the dynamic shear stress limit of 1% aqueous xanthan DUO-VIS, processed by bactericide MI CIDE (1 - xanthan solution without bactericide, 2 - 0,25% MI CIDE, 3 - 0,5% MI CIDE, 4 - 1,0% MI CIDE).

An important aspect of polysaccharides polymeric reagents application is temperature resistance. Study of xanthan reagent solution thermal stability, before and after bactericide addition, was carried out by aging test and cylinder stove, simulating the circulation of drilling mud in the borehole during the process of drilling. Temperature was maintained at 125 ° C and pressure at 2 atm. during 24 hours, after which flow properties of polymer solutions were analyzed. Studies have shown that initial solution, not containing the studied bactericides, retains rheological characteristics to a greater degree than the solution with the addition of glyoxal and MI CIDE, therefore it can be concluded that the addition of bactericide solution of xanthan gum results in the decrease of solution thermal stability.

## 3. Conclusion

1. Liquefying effect of glyoxal on colloidal clay solutions was determined.
2. In the presence of starch, along with clay mud liquefying, glyoxal textural properties are observed.
3. Glyoxal adding into the drilling mud results in a significant inhibition of shale swelling.
4. Glyoxal modified starch demonstrates biocidal properties and saves the original rheological parameters for a long period of time.
5. Glyoxal at equal concentrations shows better bactericidal properties for polysaccharides under study as compared to MI CIDE bactericide.
6. Processing of xanthan gum with bactericides results in a reduction of polymer solution thermal stability.

## References

[1] Morozov YU D and Molodkin S V 2009 Applying of Bactericides and Corrosion Inhibitors During Oil Production *Journal Ekspozitsiya Neft' Gaz* 2 23–25 [in Russian]
[2] Kletter V YU 2010 Improving of Drilling Mud for Well Construction on Arctic Shelf: Author's thesis….PhD p 24
[3] Pat. 2272900 Russian Federation, MPK E21 B43/22 3. Components for Preparing Water-soluble Inhibitor of Microbiological and Hydrogen Corrosion / Minenkov V.M.; patentee OOO NPK "EKSBUR Ko" published 27th March 2006 Bulletin № 23 (part II) p 3
[4] Vodyankina O V et al 2007 *Glioxal'* (Moscow: Akademiya) p 248
[5] Osipov V I 1989 *Microstructure of Clay Rocks* (Moscow: Nedra) p 211

# A Study of Specific Fracture Energy at Percussion Drilling

**Shadrina A[1, a], Kabanova T[2, b], Krets V[1, c] and Saruev L[1, d]**
[1]Tomsk Polytechnic University, Tomsk, Russia
[2]Tomsk State University, Tomsk, Russia

E-mail: [a]avshadrina@rambler.ru, [b]tvk@bk.ru, [c]Krets@tpu.ru, [d]saruevla@tpu.ru

**Abstract.** The paper presents experimental studies of rock failure provided by percussion drilling. Quantification and qualitative analysis were carried out to estimate critical values of rock failure depending on the hammer pre-impact velocity, types of drill bits and cylindrical hammer parameters (weight, length, diameter), and turn angle of a drill bit. Obtained data in this work were compared with obtained results by other researchers. The particle-size distribution in granite-cutting sludge was analyzed in this paper. Statistical approach (Spearmen's rank-order correlation, multiple regression analysis with dummy variables, Kruskal-Wallis nonparametric test) was used to analyze the drilling process. Experimental data will be useful for specialists engaged in simulation and illustration of rock failure.

## Notation accepted

***A*** is the type of drill bits: *A*1 is the insert bit; *A*2 is the crosscut bit; *A*3 is the spade bit.
***B*** is the hammer pre-impact velocity: *B*1 is 5 m/s; *B*2 is 7 m/s.
***C*** is the type of hammers: *C*1 is the hammer with 0,350 m length, 0,035 mm diameter, and 2.5 kg weight; *C*2 is the hammer with 0.45 m length, 0.045 mm diameter, and 5,48 kg weight; *C*3 is the hammer with 0.70 m length, 0.035 mm diameter, and 5.48 kg weight; *C*4 is the hammer with 0.253 m length, 0.075 mm diameter, and 8.32 kg weight.
***D*** is the turn angle angular rotation of a drill bit: *D*1 is 20°; *D*2 is 30°; *D*3 is 40°.

## 1. Introduction

All previous investigations conducted by the authors into wave actions associated with percussion drilling related to the drill string rather than the rock. Investigations included the principles of formation and propagation of power pulses via the drill string towards the drill bit [1].

This paper mainly focuses on estimation of the interaction between power pulses and rock, and the specific fracture energy produced by different types of drill bits.

The principles of percussion drilling were determined for coarse-grained granitic rock which is widely used in this type of research allowing to compare experimental results with obtained results by other researchers.

The efficient rock failure was studied in many works [3, 4, 5, 7, 10]. It is worth noting that some of them present conflicting data on results of rock failure affected by the above stated parameters. Thus, Ivanov et al. [2] suggest increasing the efficiency of rock disintegration up to 50% via changing the pulse configuration produced by the hammer. At the same time, Baron et al. [2] notes that the change of the pulse configuration cannot improve the efficiency of the dynamic rock failure. In works [2, 3] it was shown that at the critical percussive energy affecting the granitic rock (40 *J/cm* and higher), the hammer impact rate in the range of 5-9 m/s did not practically change the specific fracture energy.

Mavljutov [8] proved that the dependence between the specific fracture energy and the percussive energy was characterized by the existence of several extreme values. This, in turn, contradicts the results obtained by other researchers concerning the existence of critical percussive energy characterized by the optimum value of specific fracture energy.

Based on the above statements, the rock failure tests should be conducted to sort out the conflicting cases. The granitic rock workpiece was selected to define the degree of impact of a power pulse propagating along the drill string and that of a drill bit on the efficiency of rock failure.

## 2. Research Methodology

The impact testing machine was designed to study the specific fracture energy of coarse-grained granitic rock failure, permitting to transmit the impact and static loads to the rock. The cylindrical hammers with various parameters were used to apply impact loads to the drill string equipped with a drill bit. A static load was being constant during all tests that provided a close contact between the drill bit and the rock.

The percussive energy ranged between 31.25 – 207.35 *J*. The hammer pre-impact velocity was provided by the change of its drop height.

The dependence between the number of independent and dependent variables was established during this experiment. Herewith, independent variables were presented by the drill bit type (*A*), the hammer pre-impact velocity (*B*), the hammer type (*C*), and turn angle of the drill bit (*D*), while dependent variables included the amount of disintegrated granitic rock (weight ratio between average (per impact) and specific values), specific fracture energy, and drilling deeper per single impact (figure 1). The depth of the well was measured with a beam compass for ten times to find an average value.

The statistical approach was used to analyze the specific fracture energy. Reproducibility of test results was considered to be sufficient at values rather close to each other (±10%) obtained from 15 or 20 tests (each test included ten impacts). In some cases, the number of tests achieved 30 or 40.

Arithmetical averages were determined for the following parameters: the amount of disintegrated granitic rock, drilling deeper per single impact, and specific fracture energy. Coefficients of variations made up 6 – 9 %.

Obtained results during the experiment were analyzed using methods of mathematical statistics with probability belief of 0.95. This analysis showed that the distribution of obtained random values did not exactly conform to the normal law of distribution that required conducting nonparametric tests, namely: Spearmen's rank-order correlation, multiple regression analysis with dummy variables, and Kruskal-Wallis nonparametric test.

## 3. Results and Discussion

Table 1 presents the summary of experimental results that have proved the insignificant difference between specific fracture energy values at impact rates of 5 and 7 m/s, and types of dependencies were also identical. However, the hammer weight had a substantial effect on the specific fracture energy and drilling deeper per single impact.

At the same time, a substantial impact of load application rate on the specific fracture energy and drilling deeper per single impact is obvious.

Hammers *C*2 and *C*3 both having 5.48 kg weight but different length, showed unexpected results. Thus, a widely spread opinion that longer hammers provide a longer power pulse duration and are 1,4 – 1,9 times energy efficient than that shorter and thicker ones [2] was proved only for the insert bit (figure 3 *a*). The increase of the hammer length from 450 to 700 mm resulted in 6–10% decrease of the specific fracture energy.

Another situation was observed, however, for crosscut and spade bits. For crosscut and spade bits the increase of the hammer length from 450 to 700 mm resulted in 9–16% and 5% increase of the specific fracture energy, respectively. It was assumed that the hammer length of 700 mm leads to the increase of the reflected power pulse energy.

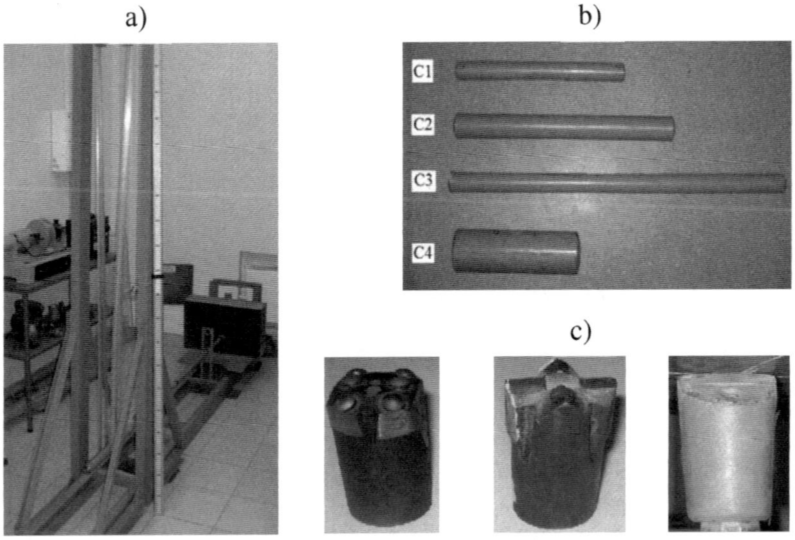


**Figure 1.** Test bench: *a)* – impact testing machine; *b)* – cylindrical hammers; *c)* – drill bits (43 mm diameter).

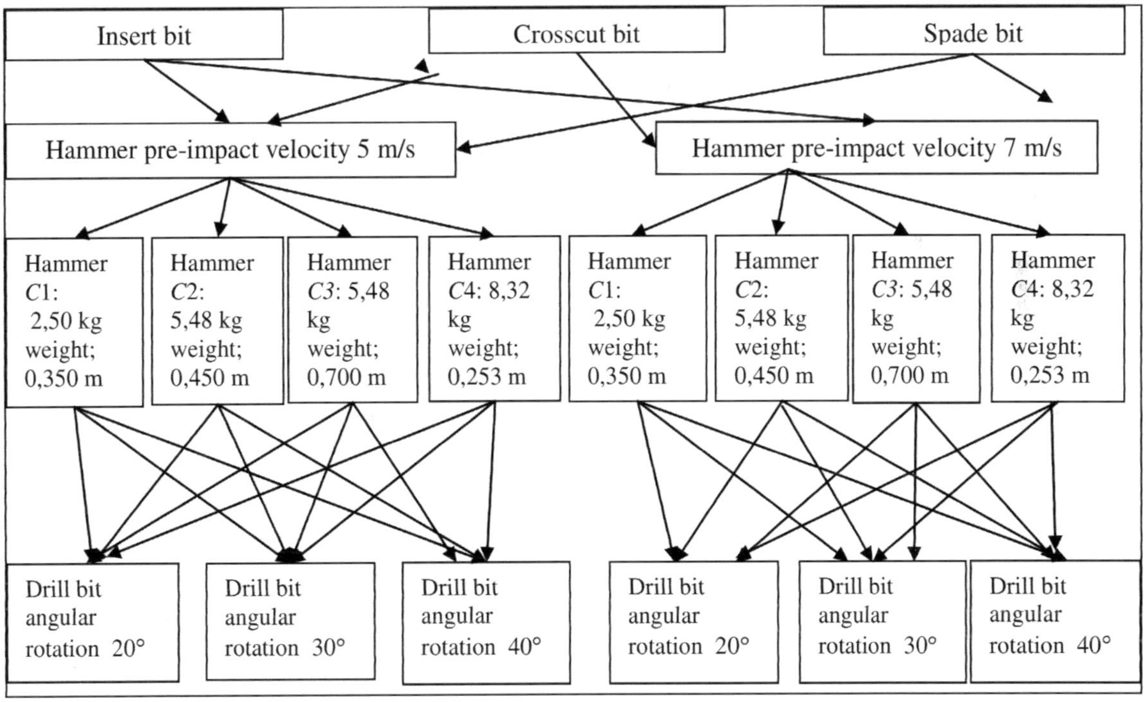


**Figure 2.** Range of independent variables.

**Table 1.** Summary of experimental results for drill bits (insert A1, crosscut A2, spade A3) [9]

| Set of tests | | | | Average amount of disintegrated granitic rock per impact, $cm^3$ | Specific fracture energy, $J/cm^3$ | Drilling deeper per single impact, mm | | Average amount of disintegrated granitic rock per impact, $cm^3$ | Specific fracture energy, $J/cm^3$ | Drilling deeper per single impact, mm | | Average amount of disintegrated granitic rock per impact, $cm^3$ | Specific fracture energy, $J/cm^3$ | Drilling deeper per single impact, mm |
|---|---|---|---|---|---|---|---|---|---|---|---|---|---|---|
| **B1** | **C1** | **D1** | **A1** | 0,130 | 242,978 | 0,09 | **A2** | 0,115 | 274,022 | 0,08 | **A3** | 0,139 | 225,671 | 0,08 |
| **B1** | **C1** | **D2** | **A1** | 0,130 | 241,107 | 0,09 | **A2** | 0,115 | 273,821 | 0,08 | **A3** | 0,132 | 239,162 | 0,08 |
| **B1** | **C1** | **D3** | **A1** | 0,126 | 249,552 | 0,09 | **A2** | 0,125 | 252,207 | 0,08 | **A3** | 0,127 | 236,146 | 0,08 |
| **B2** | **C1** | **D1** | **A1** | 0,255 | 242,029 | 0,15 | **A2** | 0,241 | 255,189 | 0,15 | **A3** | 0,253 | 242,306 | 0,11 |
| **B2** | **C1** | **D2** | **A1** | 0,225 | 273,948 | 0,15 | **A2** | 0,236 | 260,109 | 0,14 | **A3** | 0,261 | 235,108 | 0,15 |
| **B2** | **C1** | **D3** | **A1** | 0,237 | 259,258 | 0,15 | **A2** | 0,258 | 238,406 | 0,16 | **A3** | 0,257 | 239,588 | 0,16 |
| **B1** | **C2** | **D1** | **A1** | 0,212 | 328,403 | 0,14 | **A2** | 0,178 | 391,335 | 0,10 | **A3** | 0,238 | 290,467 | 0,14 |
| **B1** | **C2** | **D2** | **A1** | 0,212 | 325,949 | 0,14 | **A2** | 0,191 | 361,556 | 0,11 | **A3** | 0,237 | 289,361 | 0,13 |
| **B1** | **C2** | **D3** | **A1** | 0,237 | 290,315 | 0,16 | **A2** | 0,219 | 314,817 | 0,13 | **A3** | 0,236 | 290,034 | 0,15 |
| **B2** | **C2** | **D1** | **A1** | 0,432 | 312,814 | 0,30 | **A2** | 0,401 | 336,304 | 0,21 | **A3** | 0,401 | 336,187 | 0,23 |
| **B2** | **C2** | **D2** | **A1** | 0,409 | 328,942 | 0,29 | **A2** | 0,430 | 312,739 | 0,25 | **A3** | 0,483 | 276,518 | 0,27 |
| **B2** | **C2** | **D3** | **A1** | 0,455 | 295,113 | 0,25 | **A2** | 0,437 | 309,166 | 0,26 | **A3** | 0,488 | 311,798 | 0,25 |
| **B1** | **C3** | **D1** | **A1** | 0,231 | 297,082 | 0,16 | **A2** | 0,183 | 378,152 | 0,11 | **A3** | 0,218 | 313,909 | 0,13 |
| **B1** | **C3** | **D2** | **A1** | 0,240 | 287,519 | 0,17 | **A2** | 0,167 | 414,392 | 0,10 | **A3** | 0,227 | 302,077 | 0,14 |
| **B1** | **C3** | **D3** | **A1** | 0,259 | 264,772 | 0,16 | **A2** | 0,187 | 370,255 | 0,11 | **A3** | 0,229 | 300,487 | 0,13 |
| **B2** | **C3** | **D1** | **A1** | 0,460 | 292,465 | 0,29 | **A2** | 0,370 | 370,372 | 0,25 | **A3** | 0,376 | 357,278 | 0,20 |
| **B2** | **C3** | **D2** | **A1** | 0,456 | 295,051 | 0,29 | **A2** | 0,366 | 367,909 | 0,23 | **A3** | 0,453 | 296,905 | 0,25 |
| **B2** | **C3** | **D3** | **A1** | 0,455 | 295,659 | 0,3 | **A2** | 0,366 | 370,510 | 0,19 | **A3** | 0,487 | 277,132 | 0,25 |
| **B1** | **C4** | **D1** | **A1** | 0,289 | 361,199 | 0,19 | **A2** | 0,251 | 415,054 | 0,15 | **A3** | 0,284 | 365,519 | 0,17 |
| **B1** | **C4** | **D2** | **A1** | 0,278 | 374,564 | 0,19 | **A2** | 0,224 | 464,383 | 0,13 | **A3** | 0,299 | 347,567 | 0,17 |
| **B1** | **C4** | **D3** | **A1** | 0,282 | 373,081 | 0,16 | **A2** | 0,241 | 431,340 | 0,16 | **A3** | 0,302 | 343,850 | 0,18 |
| **B2** | **C4** | **D1** | **A1** | 0,479 | 427,043 | 0,31 | **A2** | 0,471 | 432,805 | 0,27 | **A3** | 0,513 | 397,168 | 0,29 |
| **B2** | **C4** | **D2** | **A1** | 0,519 | 392,444 | 0,34 | **A2** | 0,478 | 428,141 | 0,27 | **A3** | 0,523 | 390,644 | 0,30 |
| **B2** | **C4** | **D3** | **A1** | 0,499 | 408,313 | 0,37 | **A2** | 0,467 | 438,524 | 0,28 | **A3** | 0,517 | 394,356 | 0,30 |

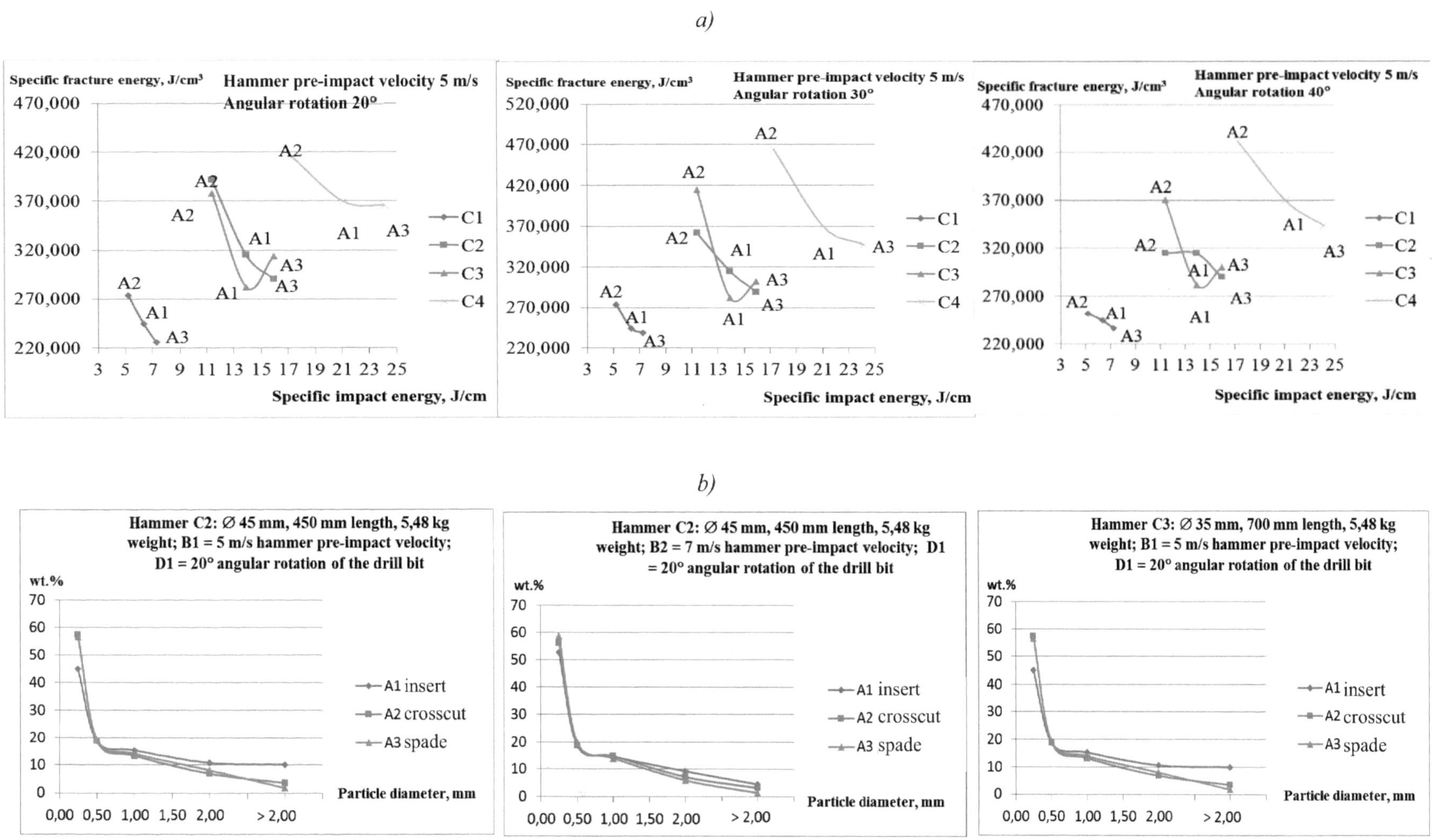


**Figure 3.** Empirical relations between the specific fracture energy and specific impact energy and the particle-size distribution in granite-cutting sludge:
*a)* – relation between the specific fracture energy and the specific impact energy. Hammer C2. Impact rate 5 m/s. Angular rotation 20°, 30°, 40°
*b)* – particle-size distribution in granite-cutting sludge (hammers C2 and C3)

As shown in figure 3 *b*, the percentage of particles having 0.25 mm diameter is maximum (40–60 wt.%). Particles having 0.5 and 1,0 mm diameter make up 10–20 wt.%; and particles with 2.0 mm diameter and larger are 5-10 wt.% and 2–14 wt.%, respectively.

With increase of the percussive energy the insert bit (see table 1 and figure 3 *b*) has the advantage over other drill bit types. The insert bit produced granite-cutting sludge of larger size in all cases and showed perfect results at different impact rates. Thus, at 5 m/s impact rate of the hammer the percentage of particles having the diameter over 2 mm is about 10 wt.%, while that of particles with diameter less than 0.25 mm does not exceed 49 wt.%. Results obtained at 7 m/s impact rate of the hammer are comparable with those of the crosscut and spade drill bits.

The lowest specific fracture energy is observed at 20° turn angle angular rotation of the spade drill bit and the load applied by hammer *C*1. With the increase of the percussive energy from 68.5 to 103.7 *J*, 30° and 40° angular rotations show practically identical results in relation to the specific fracture energy.

The crosscut bit is characterized by the best specific fracture energy value at 40° turn angle at the percussive energy of 30-70 *J*. At higher values of percussive energy the difference between the specific fracture energy values is substantially reduced at 30° and 40° turn angle angular rotations.

**References**


[1] Shadrina A and Saruev L 2014 *International Journal of Mining Science and Technology* **24** (2) 245–249

[2] Ivanov K I, Latyshev V A and Andreev V D 1987 Drilling *technique for exploitation of mineral deposits* (Moscow: Nedra) p 272

[3] Kulikov I V, Voronov V N and Nikolaev I I 1989 *Pneumatic impact drilling of prospect holes* (Moscow: Nedra) p 235

[4] Li X B, Summers D A, Rupert G and Santi P 2001*Tunnelling and Undegraund Space Technology* **16** 107-114

[5] Luiz F.P. Franca 2011 *International Journal of Rock Mechanics and Mining Sciences* **48** (5) pp. 827-835

[6] Lundberg B and Collet P 2010 *International Journal of Impact Engineering* **37** (8) 901-906

[7] Lundberg B and Okrouhlik M 2006 *International Journal of Impact Engineering* **32** (10) 1573-1583

[8] Mavljutov M R 1978 Drilling rocks failure (Moscow: Nedra) p 215

[9] Shadrina A and Kabanova T 2014 *Mining Informational and Analytical Bulletin (scientific and technical journal)* **6** 170–175

[10] Thuro K 1997 *International Journal of Earth Sciences* **86** (2) 426-438

IOP Conf. Series: Earth and Environmental Science **21** (2014) 012037 doi:10.1088/1755-1315/21/1/012037

# Simulation of pipeline in the area of the underwater crossing

**P Burkov[a], D Chernyavskiy, S Burkova [b] and E C Konan**
Tomsk Polytechnic University, Tomsk, Russia

E-mail: [a] burkovpv@mail.ru, [b] burkovasp@tpu.ru

**Abstract.** The article studies stress-strain behavior of the main oil-pipeline section Alexandrovskoye-Anzhero-Sudzhensk using software system Ansys. This method of examination and assessment of technical conditions of objects of pipeline transport studies the objects and the processes that affect the technical condition of these facilities, including the research on the basis of computer simulation. Such approach allows to develop the theory, methods of calculations and designing of objects of pipeline transport, units and parts of machines, regardless of their industry and destination with a view to improve the existing constructions and create new structures, machines of high performance, durability and reliability, maintainability, low material capacity and cost, which have competitiveness on the world market.

Development of oil and gas industry on the basis of heavy-duty pipelines and their operation in the areas with difficult natural conditions created the agenda of control and evaluation of strength and performance capability of these structures. In this case the theoretical strength calculations for main pipelines are tentative as they essentially cannot take into account all the operation factors.

In Russia the major amount of oil is produced in northern deposits of Western Siberia, Yamalo-Nenets Autonomous Okrug, the Republic of Komi, and their vast territories are swampy. Main pipelines delivering oil to the country's central regions and for export, cross water courses of various length. Water courses crossing is mainly solved by building underwater crossings.

Underwater crossings of main oil-pipelines, although making up a relatively small part of the total construction volume, belong to the most critical sections of these structures. So there are high requirements to reliability of underwater crossings, as even minor damages cause heavy environmental aftermaths.

Increasing reliability of pipelines becomes a pressing issue at all stages: design, construction and operation of pipeline systems. It is very important to determine adequacy of behavior for the constructed pipeline under operational and environmental loads to the model calculation adopted by the rules and regulations.

The main oil-pipeline Alexandrovskoye-Anzhero-Sudzhensk crosses in general 98 water courses. The section is undeveloped, covered by mixed forest along the pipeline route situated at the 208$^{th}$ km of the oil-pipeline Alexandrovskoye-Anzhero-Sudzhensk. The flood plain is wide, even and swampy. It consists of stretched interchanging low ridges and swaps covered by ledum and osier shrubs and birch trees of medium density, with shrub and sedge complex along the river bed. The river bed bogginess makes up 20%; its forest coverage is 90%. The river bed at the underwater crossing area is meandering, one-armed and sandy. The river bed width in low water period is 10-15 m the depth is 1.5-2 m.

The aim of this work is to research the stress-strain behavior (SSB) of the underwater crossing at the swampy pipeline area.

The calculation of the SSB pipeline structures based on material resistance factors and structural mechanics does not allow making an adequate analysis of fuel and energy sector pipelines strength with the required accuracy and in some cases can give a wrong picture of SSB structures. Nowadays numerical methods are intensively developed, which allow extending considerably the class and statement of solved problems due to a more complete consideration of real loading conditions and the properties of used materials. The finite element analysis (FEA) method is the most widely-spread among these ones. The advantages of FEA are minimal requirements to the source data and optimal presentation of the results. Considering the temperature influence and structure operation does not create any major difficulties for implementing the method.

Using FEA in structural analysis is actually the world's standard now for structure strength and other related types of calculations. This is based on FEA universality, which allows using a unified method for calculating different structures with different materials properties. The information obtained as a result of pipelines SSB evaluation allows determining sections in pre-emergency state (even before damages appear) and take all the necessary measures for eliminating them, thus increasing the pipeline system reliability. [2-3]

The diameter of the studied pipeline is 1,220 mm, the operating pressure is 3.7 MPa. The pipeline crosses a swampy area. The pipe material characteristics: 13GS steel with the following mechanical properties $\sigma_{вр} = 510 МПа$, $\sigma_{т} = 335 МПа$. *[4-6]* The pipeline is affected by the following loads except for the operating pressure: distributed load of the pipe with isolation and transportation oil – q, buoyancy in water – $q_w$, hydrostatic pressure – $q_{hyd}$. It is assumed in the calculations that there are no loads acting in winter period and temperature influence on the pipeline.

The calculation model is shown in figure 1 and represents the transition place of the pipeline from ground to water.

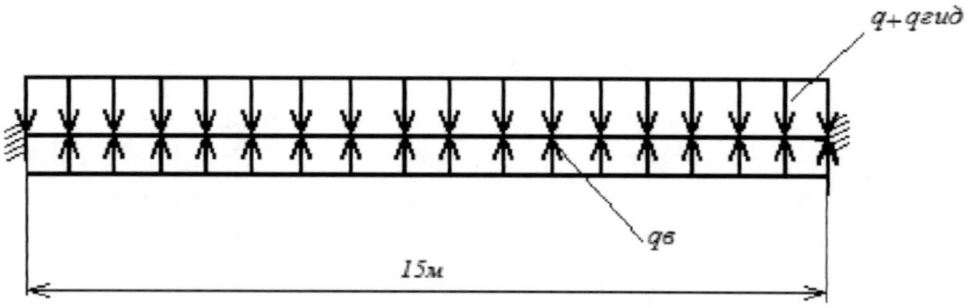


**Figure 1.** Pipeline calculation model.

The calculation results are shown in figures 2 and 3.

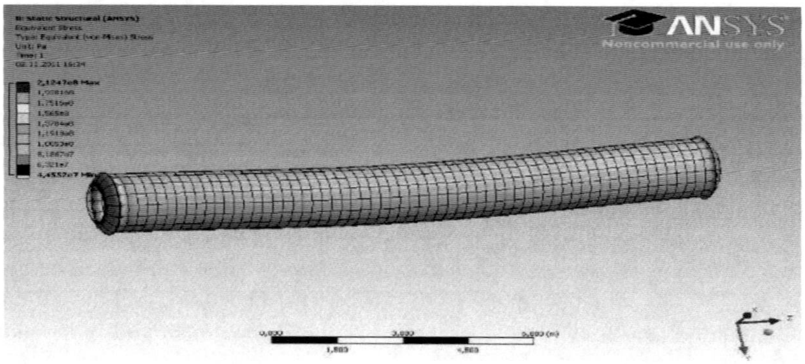


**Figure 2.** Von Mises design strains of the studied section.

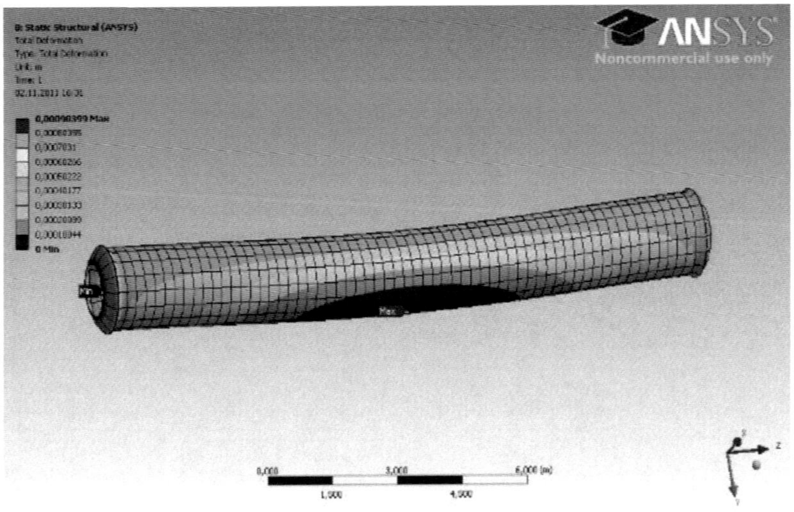


**Figure 3.** Total shifts.

Having calculated the studied pipeline section, we'll check if the Ansys model corresponds to the conventional calculation method results. So we take the created pipeline section model, apply only the design pressure and make a calculation of circular strains.

The calculation results are shown in figure 4.

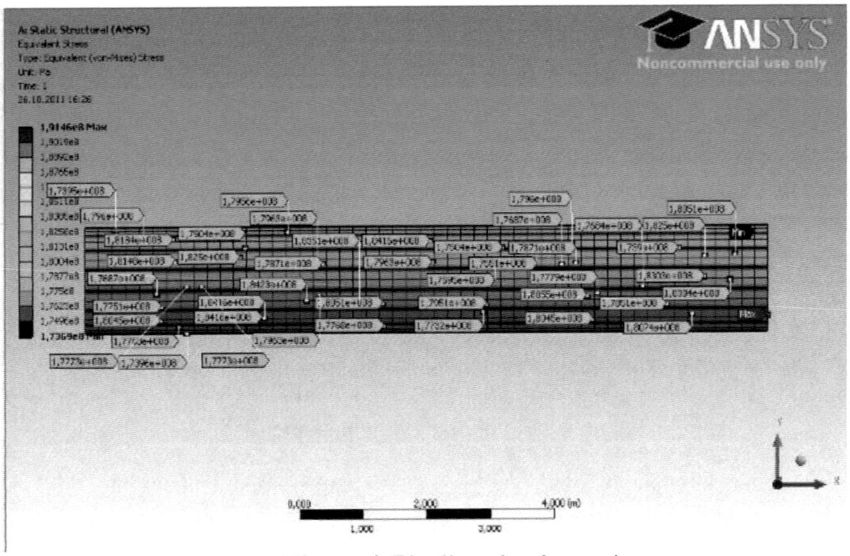


**Figure 4.** Pipeline circular strains.

Maximal circular strains of the pipeline section under design pressure by FEA method is $\sigma_{\text{эксп}} = 191\text{МПа}$.

Calculated circular strains are $\sigma_{\text{рас}} = 180\text{МПа}$.

The calculation error:

$$\Delta = \frac{\sigma_{\text{эксп}} - \sigma_{\text{расч}}}{\sigma_{\text{эксп}}} \cdot 100\% = \frac{191 - 180}{191} \cdot 100\% = 5{,}7\% \qquad (1)$$

This means the adopted model gives the adequate result.

IOP Conf. Series: Earth and Environmental Science **21** (2014) 012037 doi:10.1088/1755-1315/21/1/012037

Distribution of the total shifts and strains obtained by the calculation is shown in figures 5 and 6.

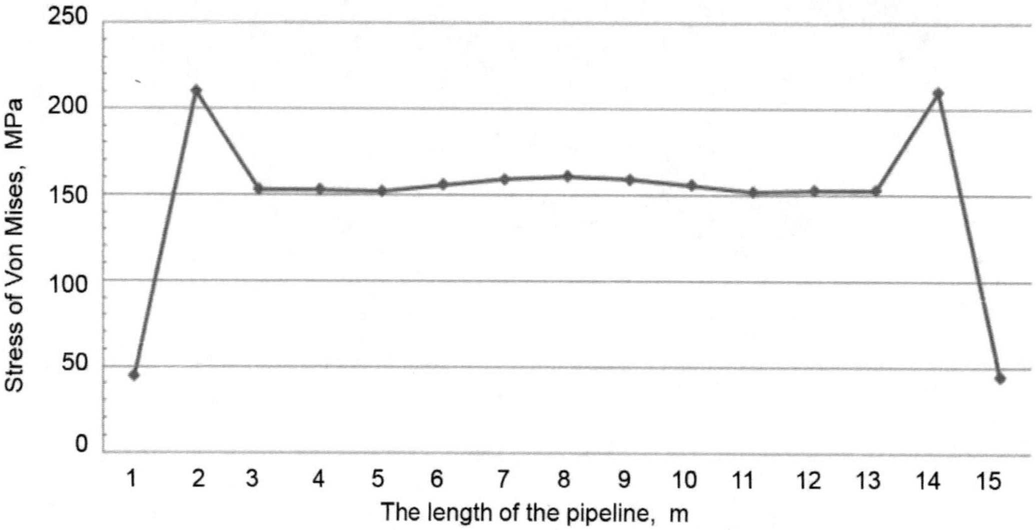


**Figure 5.** Distribution of strains along the pipeline.

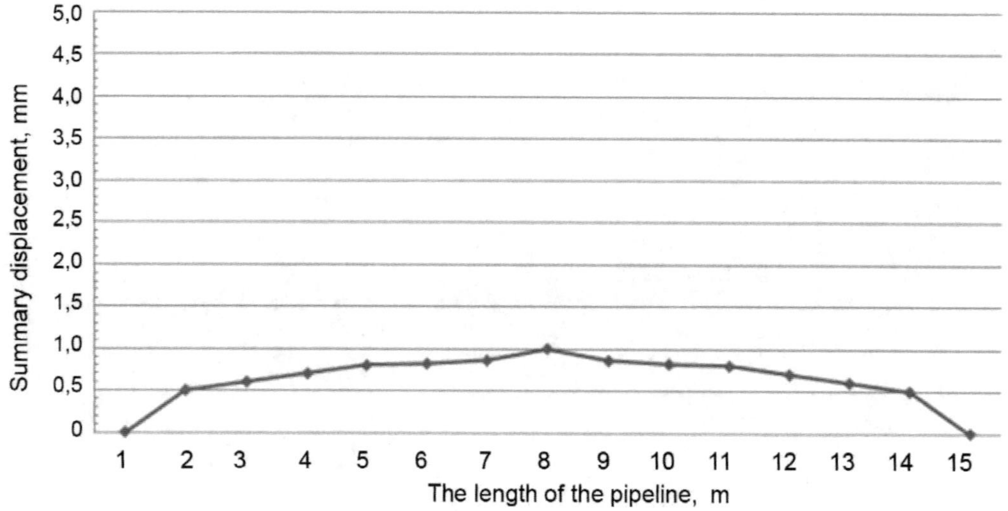


**Figure 6.** Distribution of the total shifts along the pipeline.

The following conclusions can be made from the obtained and presented results:

- the most critical pipeline cross-section is situated at the place of transition from one environment to another; here we observe maximal strains decreasing the pipeline reliability level;
- total shifts are variable along the tube length, and they considerably depend on the effect of distributed loads from the weight of the tube and the transported oil, as well as hydrostatic pressure and water buoyancy . The most critical cross-section is situated in the middle of the pipeline.

## References

[1] Shammazov A M, Mugallimov F M and Nefedova N F 2000 *Main Oil-Pipelines Underwater Crossings* (Moscow: Nedra-Business-Center LLC) p 237: il. – ISBN 5-8365-0049-5

[2] Basov K A 2002 *ANSYS in Examples and Problems* (Moscow: Computer Press) p 224

[3] Chigarev A V, Kravchuk A S and Smalyuk A F 2004 *Ansys for Engineers* (Moscow: Mashinostroyeniye) p 512

[4] Methodological Instructions on Examination and identification of Steel Tubes for Oil and Gas Pipelines. Industrial Guideline 39-1.11-014-2000.

[5] Burkov P V and Yepifantsev K V 2009 Study of Stress-Strain Behavior of the Powered Support Canopy MKYu.2Sh-26/53 Produced by Yurginskiy Mashzavod LLC *Innovative Technologies and Economics in Engineering: Works of VII All-Russia Applied Science Conference with International Participants* – Yurga, UTI of TPU, 21-22 May. (Tomsk: Publ. TPU) 637-640.

[6] Burkov P V, Burkova S P, Timofeev Yu, Ashcheulova A A and Klyus O V 2013 Analysis of the stress-strain state of the pipeline in permafrost *Herald Kuzbass State Technical University* **6** 77-79

IOP Conf. Series: Earth and Environmental Science **21** (2014) 012038 doi:10.1088/1755-1315/21/1/012038

# Specific Aspects of Turbulent Flow Separation in Nods and Junctions of Pipelines

**S N Kharlamov** [1,2,a] **and R A Alginov**[1,3,b]
[1] Tomsk Polytechnic University, Tomsk, Russia
[2] Tomsk State University, Tomsk, Russia
[3] LLC "Gazprom transgaz Tomsk", Tomsk, Russia

E-mail: [a]felic@tpu.ru, [b]alginov@mail.ru

**Abstract.** The paper is devoted to numerical simulation of space effects appearing with flow separation in pipeline junctions. Gas and oil flows through T-bends are considered. Two-parametric turbulence models are analyzed, among them k-ω is chosen. Results of numerical simulation are analyzed and explained, practical recommendations on appropriate t-bend geometry are given.

## 1. Introduction

Pipeline transportation anywhere in the world demands significant expenditures, both capital in construction / reconstruction and operational for equipment maintenance. The age of pipeline networks grows and risks of environmental hazards grow accordingly. That's why one can observe increasing interest in technologies that can make possible complete utilization of pipeline network capacities and extend its' accident-free operation. Under these conditions problems of complex physico-mathematical and numerical simulation of three-dimensional liquid and gaseous viscous flows in pipelines with complex boundaries of current represent broad interest.

Complete and particular enough analysis of flow structure one can obtain in terms of second order statistical models for Reynolds shear stresses (RSS), K-theory of turbulence, including two-parametric dynamic models of turbulence of k-ε, k-τ, k-ω, k-L types. This approach makes possible an insight into the momentum transfer process under conditions of low and moderate intensity of velocity pulsations and allows to predict shear flows with adequate success. Understanding of redistribution mechanisms in case of space deformation of viscous flow will develop effective solutions for flow control, decrease of accident rate, effective maintenance of pipeline in many branches of manufacturing – oil –and-gas industry, heat and water supply [1-6].

Investigation of non-linear three-dimensional hydrodynamic processes is commonly carried out in terms of semiempirical theory of turbulence with attraction of Reynolds averaged Navier-Stokes equations (RANS). The method has strong problem of appropriate turbulence model selection. Moreover in calculations of inner low-Reynolds flows it is significant for turbulence model to exert "computational flexibility" and stability against perturbations caused by near-transient regimes of low turbulence [5].

## 2. Aim of investigation

The aim of the investigation is a detailed mathematical simulation and investigation of separating shear flows appearing in pipeline T-junctions under conditions of low and moderate turbulent vortex intensity in terms of low-Reynolds second order closure schemes. Investigation aims also include prediction of deflected zones in pipelines, estimation of turbulent models' efficiency in prediction of transient turbulence structure in T-junctions.

## 3. Statement of problem

*Theoretical base*

Simulation of flow separation was performed with approximation of stationary turbulent and near-transient flows of incompressible and weakly compressible dropping Newtonian viscous media. Geometry was considered three-dimensional. Heat exchange in system was defined with moderate heat fluxes localized at pipeline wall.

Common mathematical statement of problem includes system of full dynamic Reynolds equations (RANS), equations of state, continuity, energy in Cartesian coordinates [4,6], which is incomplete with respect to Reynolds shear stress tensor $\overline{u_i'u_j'}$ and specific heat turbulent flows $\overline{u_i't'}$ [4-6]. The closure of governing equations was performed in terms of Boussinesq hypothesis and two-parametric dynamic turbulence models: standard two-parametric k-ε model of Johns-Launder [4], k-ω models of Wilcox and Menter [5] (where k – turbulence kinetic energy, ε – dissipation rate of k, ω – turbulence frequency, also called pseudovorticity).

Numerical solution was built in terms of control volume approaches (SIMPLE and its' modifications [7]).

*T-bend geometry and oil properties*

Computational model was represented by flush joint T-junctions (T-bends) according to GOST 17375-2001 [8]. The "neck" of the T-bend – the place of lateral branch connection into the main line – was expressed by rounding with radius over 5 mm (agreeably to GOST). T-bend models were appended with direct regions of 160 inner diameter length to provide "thin" turbulence parameter stabilization at inlet into T-bend.

Calculations were performed at pipe outer diameters up to 820 mm. Reynolds criterion Re was taken from $3 \cdot 10^3$ up to $10^6$.

Working medium was represented with low- and middle-viscous oils, the Newtonian assumption for which is valid with certain permissible variations excluding low-temperature conditions near the chilling (congelation) or wax deposition temperature.

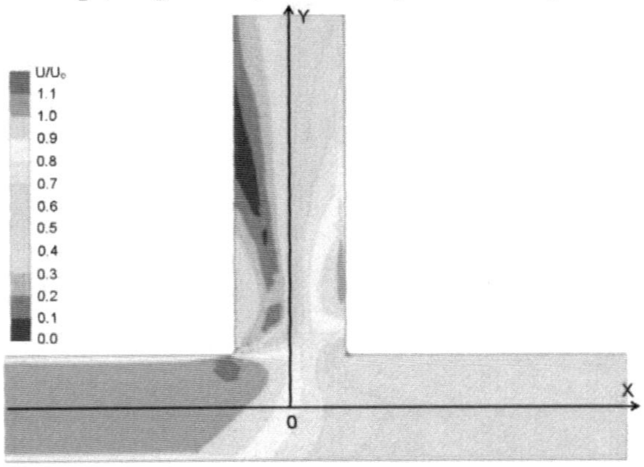


**Figure 1.** Flow velocity distribution at symmetry plane for turbulent mode.

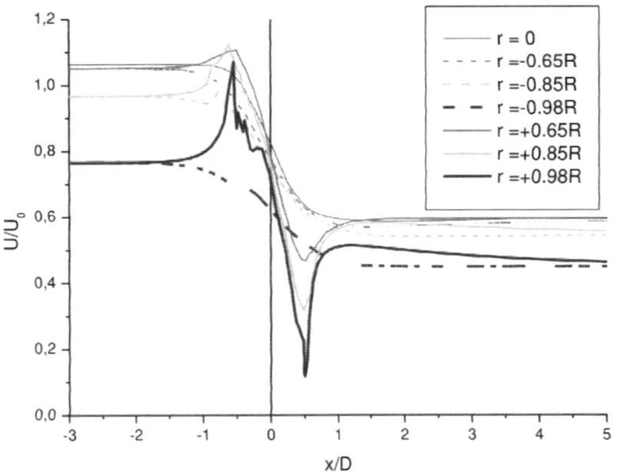


**Figure 2.** Axial velocity distribution along main line axis versus distance to side branch, "+" and "-" correspond to bringing closer and move away from branch.

## 4. Results and discussion

Results of T-bends' numerical simulation are represented at figure 1-4. Working medium at given maps flows along OX axis, in T-junction flow separates and a part of it enters lateral branch (OY direction). According to figure 1 zones of maximal motion intensification appear in points of flow separation from the main line at the T-bend's neck and on inner surface of side branch in place of separated flow strike. At the opposite wall of branch there appear decelerated flow zone. I.e. at the T-bend neck flow sufficiently accelerates, separates from the main line and strikes the lateral branch.

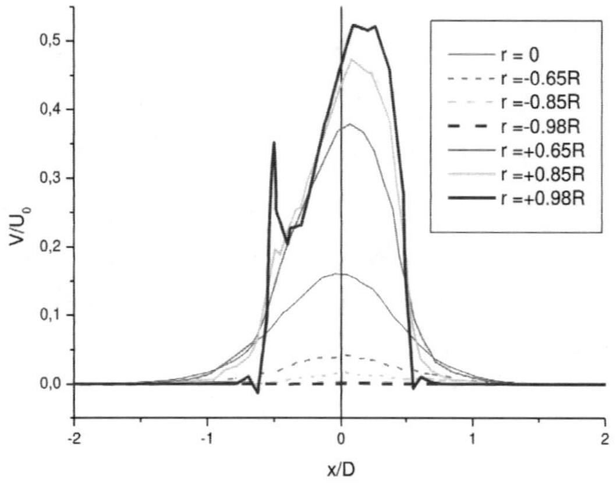


**Figure 3**. Radial velocity distribution along main line axis versus distance to side branch (notations according to figure 2).

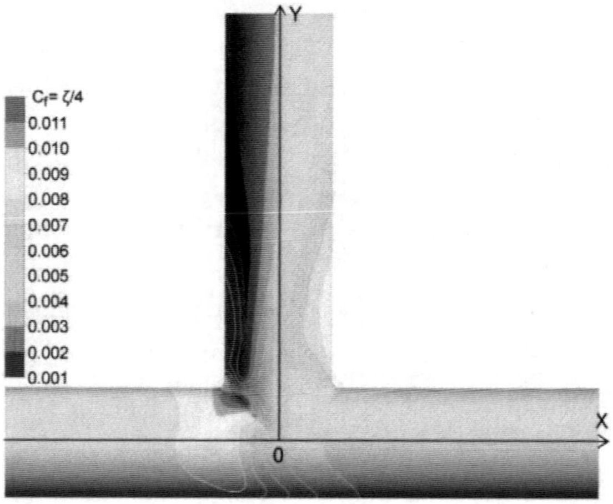


**Figure 4.** Dimensionless friction factor on T-bend wall at Re=$10^4$.

Analyzing velocity profile on lines, parallel to OX, one can observe disruption of flow axial symmetry near the flow separation zone, with further reconstruction to axial symmetry at the T-bend outlet (figure 1,2). In flow redistribution area axial symmetry disruption appears in the following way: at fluid flow lines situated under the main line axis one can observe smooth flow rate change (figure 2, solid lines). At the same time at lines close to lateral branch (dashed lines) axial velocity growth is observed, with following sharp decrease to extreme values when crossing the lateral branch axis and further restoration to values corresponding to axisymmetric flow. Radial velocity redistribution in T-bend (figure 3) is also of serious interest: it is sufficiently asymmetric across lateral branch axis (OY) and reaches maximal values in the zone of separated flow strike on branch wall.

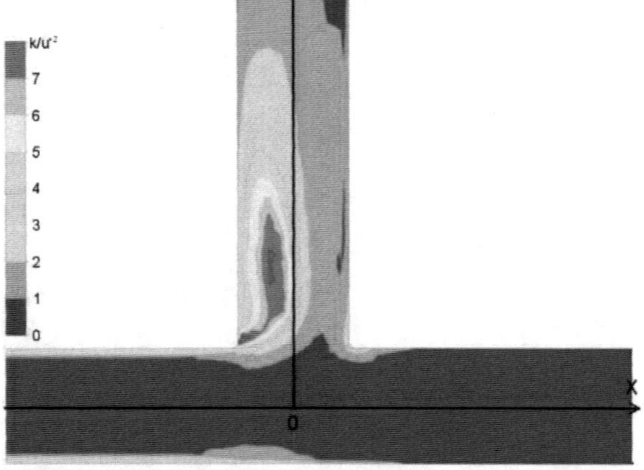


**Figure 5**. Turbulence kinetic energy on symmetry plane.

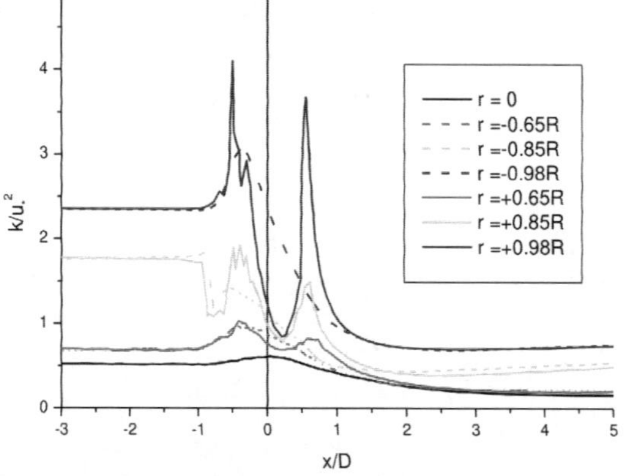


**Figure 6.** Turbulence kinetic energy distribution along main line axis versus distance to side branch (notations according to figure 2).

IOP Conf. Series: Earth and Environmental Science **21** (2014) 012038 doi:10.1088/1755-1315/21/1/012038

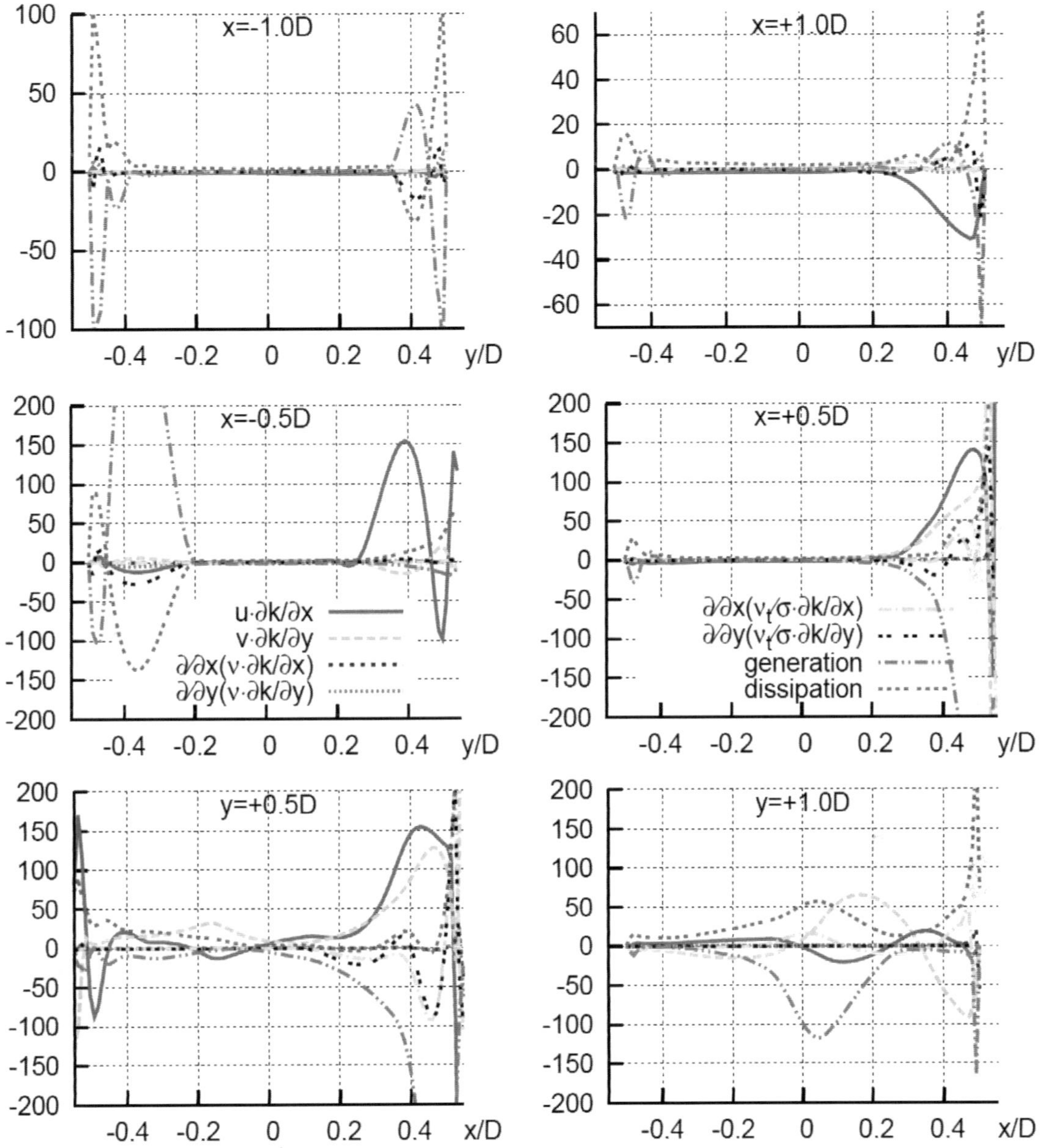


**Figure 7.** Turbulence kinetic energy balance in cross-sections: x=-1.0D,-0.5D correspond to entrance region of T-bend, x=+0.5D,+1.0D – output from T-bend on the main line, y=+0.5D,+1.0D – side branch.

Mentioned above assumption about the separation character of flow in T-bend is approved with shear stress' distribution on T-bend inner surface (figure 4). One can see that zones of potential risk are T-bend's neck and initial area of lateral branch in which girth weld, connecting pipe to T-bend, is usually located. However with neck rounding growth separated flow intensity decreases that let to reduction of dynamic load on lateral branch.

Analysis of steady flow at the outlets of T-bend shows that flow separation appears significantly not uniform – flow rate in main line after T-bend can hold up to 65% of all flow – thus exceeding flow rate in junction in two times (of course, characteristics of side branch and main line after the T-bend are taken to be hydraulically equal). Obtained result can be explained with distribution of turbulence parameters. Thus, according to turbulence kinetic energy figure 5, turbulent vortex occur in lateral branch, creating additional hydraulic resistance to liquid flow into branch.

Turbulence kinetic energy profile analysis on lines, parallel to OX (figure 6), correspond to that of velocity – while along distant from side branch dashed lines k smoothly decreases, near the lateral branch k appears two maximums. This maximums correspond to extreme values of velocity and occur at $x = \pm 0.5$ R – i.e. in the beginning and at the end of lateral branch.

Balance of turbulence kinetic energy is represented at figure 7. In the area before the inlet (X=-1.0D), as it takes place for steady flows [9], in near wall region generation of turbulence kinetic energy is balanced with its dissipation. But in the inlet of T-bend (X=-0.5D), at T-bend's neck $Y \approx 0.4 \div 0.5$D convective mechanisms begin to play governing role, improving flow separation from the wall.

Outlet from the T-bend on the main line (X=+0.5D) is also significantly not symmetric. While at the wall opposite the branch balance is quiet to that of steady flow, area of flow attachment characterizes with large scale perturbations of convective and turbulent diffusive components.

In the side branch there observed a very complicated structure of turbulence kinetic energy balance. In both sections (Y=+0.5D,+1.0D) convective components play significant role, being balanced with generation and turbulent diffusion at Y=+0.5D and generation and dissipation at Y=+1.0D.

Obtained results lead to following practical recommendations: elongation of lateral branch away from recirculation zone and maximization of neck curvature radius. It let not only to increase durability of T-bends but also reduce its hydraulic resistance. Obtained results also illustrate practically important fact of less reliability of welded T-joints [10] in comparison with T-bends. Its explanation partially consists in flow separation (because T-joint has now neck curvature) and intensive dynamic strike in lateral branch wall. In this case preferred utilization of prefabricated T-bends is recommended.

## References

[1] Kharlamov S N 2010 *Mathematical Modelling of Thermo- and Hydrodynamical Processes in Pipelines* (Rome: "Ionta")

[2] Hadzic I and Hanjalic K 1999 Separation-induced transition to turbulence: second-moment closure modeling *Flow, Turbulence and Combustion* 153-173

[3] Koizumi H 2002 Laminar-turbulent transition behavior of fully developed air flow in a heated horizontal tube *International Journal of Heat and Mass Transfer* 937-949

[4] Bubenchikov A M and Kharlamov S N 2001 Matematicheskiye modeli neodnorodnoy anizotropnoy turbulentnosti vo vnutrennikh techeniyakh (Tomsk: Tomsk State University) [in Russian]

[5] Kharlamov S N 2012 Actual Problems of Hydrodynamics at Internal not Isothermal Flows in Fields of Mass Forces, Intech-Open, Rijeka, Croatia

[6] Kharlamov S N and Silvestrov S I 2010 *Hydrodynamics and Heat Transfer of Complex Internal Streams With Recirculated Fields* (Rome: "Ionta")

[7] Patankar S V 2003 *Chislennoye resheniye zadach teploprovodnosti i konvektivnogo teploobmena pri techenii v kanalakh* (Moskva: Izdatelstvo MEI) s 312 [in Russian]

[8] GOST 17376-2001 Detali truboprovodov besshovnyye privarnyye iz uglerodistoy i nizkolegirovannoy stali Troyniki Konstruktsiya [in Russian]

[9] Laufer J 1953 The structure of turbulence in fully developed pipe flow // NACA Report no. TN 2954

[10] Vedomstvenniye Stroitelniye Normi VSN 1-84. Troyniki i troynikovyye soyedineniya svarnyye iz stalnykh trub na Pu 5,5 i 7,5 MPa (55 i 75 kgs/sm2) [in Russian]

IOP Conf. Series: Earth and Environmental Science **21** (2014) 012039 doi:10.1088/1755-1315/21/1/012039

# Research of stress-deformed state of main gas-pipeline section in loose soil settlement

**P Burkov [a], K Kalmykova, S Burkova [b] and Do Thi Thu Tien**
Tomsk Polytechnic University, Tomsk, Russia

E-mail: [a] burkovpv@mail.ru, [b] burkovasp@tpu.ru

**Abstract.** The article represents the analysis of a main gas-pipeline section in loose soil settlement using ANSYS software.

The objective characteristics of the Russian gas network are difficult climatic conditions that have negative impact on gas pipe lines operation, which may increase risk of ecological or technical disasters. The forces that are not provided by pipeline specifications can cause significant deviations of stress-deformed state (SDS) of gas pipelines. These deviations can result in changes of gas pipeline spatial positions [1]. It can be attributed to ground settlement which appears as a result of soil consolidation due to moisture draining; weight of soil and the vibrations of gas pipe line. These forces lead to greater pipe curvature, its excessive strain and, as a result, to pipe line damage (weld joint ruptures and wormholes in the pipe walls). The repair of such damage requires repair works with temporary gas transmission shutdown. Besides, each damage, depending on its character, can lead to great gas losses [2]. Such failures more often occur in the so called loose grounds [3]. The SDS calculation of an underground main gas pipeline should be carried out taking into account both pipeline specifications and foundation soil properties as well as their interaction [4].

The purpose of this work is to study main gas pipeline behavior in loose ground settlement and its stress-deformed state.

Let us consider the section of an underground straight gas pipeline. Suppose that it is buried in dry ground which won't be encroached during the operation period. In this case vertical displacement (ground settlement) will appear as a result of soil packing under the pipe. This packing, according to the design parameters, is insignificant because the pressure of the pipe on the ground is no greater than 0.5 N/cm2 and there is no need to take it into account.

Let us suppose that the pipeline is buried in the loose water-saturated ground or the ground is periodically flooded. The water-saturated ground settlement is supposed to be soil packing under the pipe. It is known from the mechanics of soil that fully water-saturated ground can be considered as a two-phase system, and its compaction is conditioned by filtering out water from the cavities of the soil skeleton under the pressure of compaction load [3].

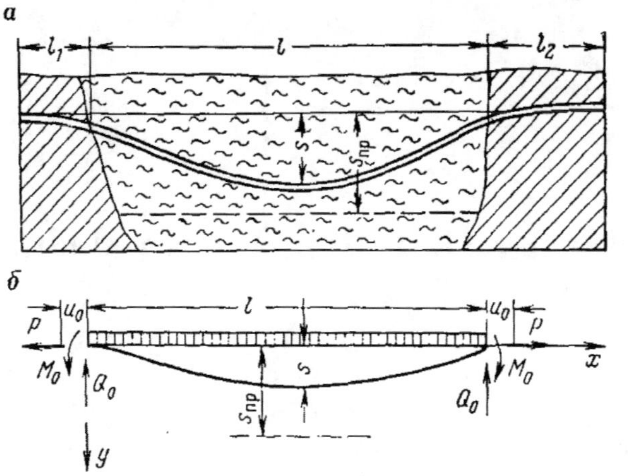

**Figure 1.** Motion of settlement type: a – general view; b – scheme.

Let the ultimate settlement be $S_u$. This settlement is called stabilized. $S_u$ along the pipe line is drawn with the dashed line in figure 1. If the loose ground would spread proportionally throughout the length of pipeline, the settlement would be equal throughout the length. However, in practice we can see alteration of the loose grounds and grounds with greater bearing capacity. The last ones can have no settlement. In the middle of the loose ground section settlement could reach the ultimate value $S_u$. The gas pipeline on section 1 curves, as shown in figure 1. Since the pipe elongation can occur only at the expense of its stretching at section 1 and neighboring sections $l_1$ and $l_2$, there appears a stretching lateral force P, and section l begins to work as a hard thread. The real settlement S appears to be substantially less then $S_u$ [3]. There emerge lateral force stresses, bending stresses from the overlaying ground, and stresses resulted from the weight of the pipe itself, the inner pressure and pipe walls temperature drop [5].

The software ANSYS is used for analysis of stresses caused by ground settlement to determine the stresses inside the pipe that exceed acceptable values and the variation range of numerical characteristics of the processes.

The scheme of lowering of a pipeline section with the following parameters is studied: outer diameter 530 mm, wall thickness 11 mm, length 12 m, operating pressure 9.0 MPa. Pipe line is made of steel 17G1S with following mechanical characteristics: short-term strength limit $\sigma_s$ – 490 MPa, residual deformation fluctuation limit $\sigma_r$ – 350 МПа. Apart from the operating pressure there are following loads on the pipeline: distributed load caused by the weight of pipe with insulation, the gas and the pressure of overlaying ground – $q_{gen}$, ground pressure on the pipe from underneath– $q_r$.

The loads, affecting the pipe line, are estimated due to Construction Norm and Rules 2.05.06-85* «Main pipelines» [6].

There are assumptions that correspond to the easiest option of pipe-ground interactions: in the beginning and in the end of the pipeline there are no X-movements, the ground pressure is 20% from the overlaying load, the pipe wall temperature rate is not considered.

IOP Conf. Series: Earth and Environmental Science **21** (2014) 012039 doi:10.1088/1755-1315/21/1/012039

Considering this hypothesis and the scheme in figure 1, the following model is shown in

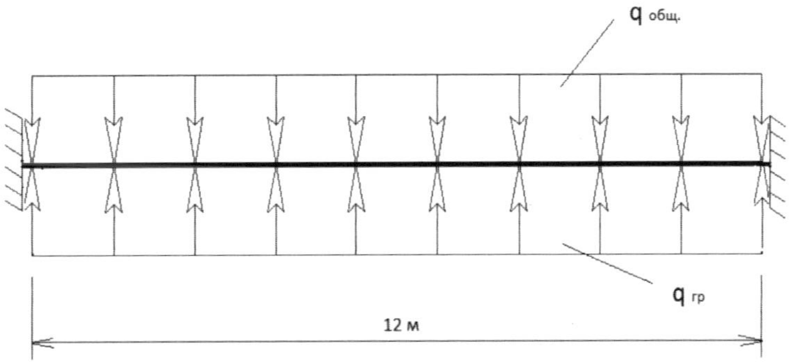


**Figure 2.** Pipeline model of structure.

Figures 3 and 5 show the pipe line section calculation made with the ANSYS, and figures 4 and 6 show the graphic data interpretation.

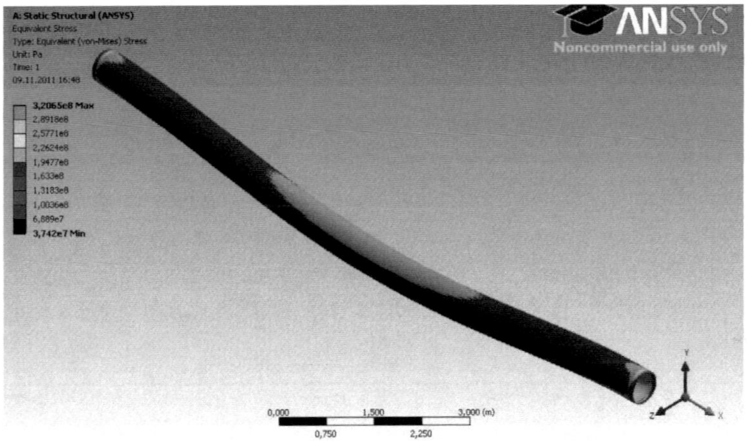


**Figure 3.** Stresses caused by ground settlement.

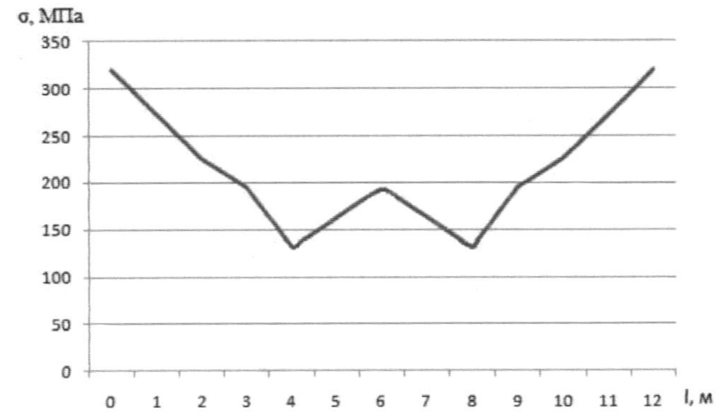


**Figure 4.** Distribution of stresses (σ) lengthwise (l) of the pipeline section.

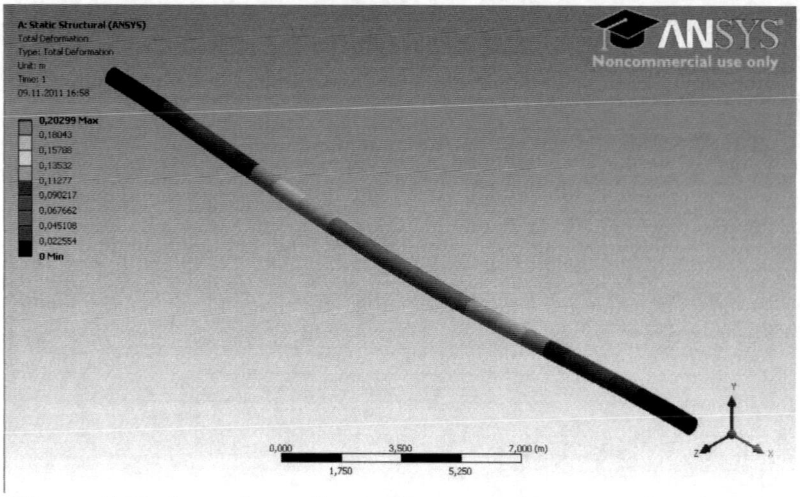


**Figure 5.** Deformations of a pipeline caused by ground settlement.

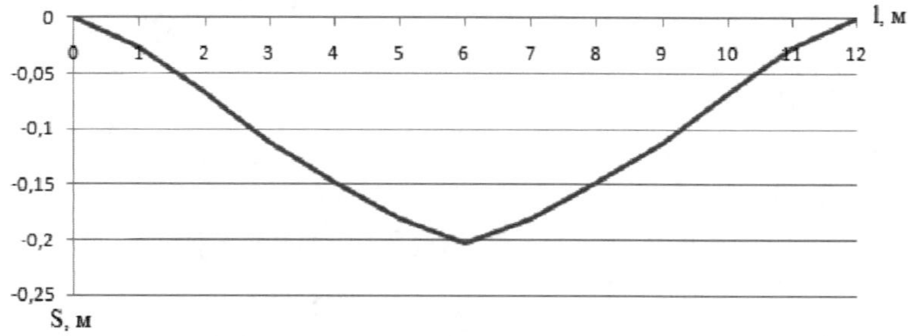


**Figure 6.** Stress distribution (S) lengthwise (l) in the pipeline section.

## Conclusion

The obtained results allow us to make the following conclusions:

- changing of the stress values, which occur during the pipeline settlement, can reach the values which are close to steel flow limit. It decreases the pipe line safety level;
- intensity of stress is changeable lengthwise the pipeline. The sections of pipeline that are situated in the zone of transition from loose grounds to the grounds with greater load bearing capacity, is characterized with great stress level;
- these results are not sufficient for making the design decision. It is necessary to conduct more detailed study of stress deformed state of pipeline, taking into account the physical and mechanical soil properties.

## References

[1] Zakharkin F I and Fomin V A 2008 The definition of stress deformed state of the main pipeline sections, which are in non-design state *Gas industry* **12** 40-43

[2] Balson F S 1991 *Embedded structure* (Moscow: Stroyizdat) p 239

[3] Borodavkin P P 1982 *Underground gas pipelines (design and building)* (Moscow: Nedra) p 384
[4] Dimov D A 2008 Ground deformation capacity and the calculation of underground main gas pipe lines *Gas industry* **2** 82-85
[5] Burkov P V, Burkova S P, Timofeev Yu, Ashcheulova A A and Klyus O V 2013 Analysis of stress-strain state of the pipeline in permafrost *Herald Kuzbass State Technical University* **6** 77-79
[6] Construction Norm and Rules 2.05.06-85* «Main pipelines» 1997 (Moscow: GUP CPP)

IOP Conf. Series: Earth and Environmental Science **21** (2014) 012040 doi:10.1088/1755-1315/21/1/012040

# Hydrodynamic, Heat and Acoustic Processes Modelling in Tranport of Rheologically Complex Viscous Media Technology in Pipelines

**Sergey N. Kharlamov**[1], **Nikita S. Kudelin**[2], **Pavel O. Dedeyev**[2]

[1] DSc. (Phys.-Math), professor, National Research Tomsk Polytechnic University, Tomsk, Russian Federation

[2] Student, National Research Tomsk Polytechnic University, Tomsk, Russian Federation

E-mail: kharsn@mail.ru, kudelin@tpu.ru

**Abstract.** The paper describes the results of mathematical modelling of acoustic processes, hydrodynamics and heat exchange in case of oil products transportation in pipelines with constant and variable cross-section. The turbulence model features of RANS approach and intensification of heat exchange in substances with anomalous rheology are reviewed. It is shown that statistic second order models are appropriate to use for forecasting details of the pulsating flows. The paper states the numerical integration features of determining equations. The properties of vibratory effect influence are determined. Vortex and heat perturbations, rheological changes impact on resistance regularities and intensity of heat exchange are analyzed.

## 1. Introduction

Nowadays oil and gas, chemical, technological, heat-and-power engineering and space industries are facing problems of modern, efficient and safe equipment inculcation. State-of-the-art equipment is expected to function without failures in the intensive thermal and hydrodynamic effects regime, accompanied by the transitional phase, vortex and acoustic wave processes, heat and mass exchange and chemical reactions and etc. Due to that the problem of adequate description of momentum, heat and mass transfer in difficult shear flows of viscous media (oil products, drop and gaseous homogeneous and heterogeneous mixes) in channels and the pipelines with constant and variable cross-section is the most difficult one. The details of noted flows are demanded for making metal-consuming equipment that is conservative to high stresses. The optimization of its work during crude oil with abnormal rheological properties transportation is quite vital as well.

The references analysis of the abovementioned issues shows (for example, [1–4]), that, despite Russian and foreign researchers' wide experience of studying various aspects of flows with heat and mass exchange in internal systems, there is no agreement concerning a complex assessment of attendant factors defining complex substances flows such as influence of non-stationarity, acoustics, phase transitions, flow contamination, turbulization, relaminarity of an entering flow, heat exchange with environment. Usually researches of regularities in changes of dynamic flow structure, resistance and heat exchange are carried out simplifying real process assumptions. In addition, it should be mentioned that almost all real non-isothermal flows of viscous substances are turbulent in the pipelines functioning under high stresses. Their full direct numerical modelling forecast is linked with well-known problems of numerical character, restrictions of modern computers [4]. In these conditions RANS method looks more preferable, especially considering positive experience of statistical multiple parameter models of the second rank in tension and flows computing. Despite the great potential opportunities, adaptation of the mentioned models to calculation of technological processes in the oil and gas industry faces serious difficulties.

In this regard the objective of this paper is the creation of complex mathematical model of the physical processes accompanying flows and heat and mass transfer in pipelines; explanation of acoustic, vortex and thermal processes influence on resistance and heat return, generation of large-scale formations in carrying agent; establishment of energy intense zones' sizes and rate of their intensity.

### *1.1. Acoustic-hydrodynamic interactions in the pipeline*

The solution of heat and mass transfer intensification becomes seriously complicated in case of the attempt to account effects of interaction of acoustic, hydrodynamic and thermal processes in the carrying agent and a pipeline wall. Particularly, problems of acoustic processes in pipeline system are fully viewed in papers [5, 6] where analytical model expressions of pulsations in carrying fluid have been received for the system with established fluctuations [5], also the mathematical model of vibroacoustic processes in pipeline systems of difficult spatial configuration, considering simultaneous impact on system of power irritation of carrying liquid and kinematic energizing of the attached objects.

According to the idea of [5], it is supposed in the analysis of pipeline wall reaction on dynamics of the internal processes to investigate effect of axis deflections of the pipeline with cross-section vibrations on the differential equation (1):

$$EI\frac{\partial^4 L_y}{\partial x^4} = F_P - F_N\frac{\partial^2 L_y}{\partial x^2} - 2\frac{\partial F_N}{\partial x}\frac{\partial L_y}{\partial x} - L_y\frac{\partial^2 F_N}{\partial x^2}, \tag{1}$$

where $E$ - the Young's modulus, $I$ – the inertia moment of section relative to neutral axis, $L_y$ – a deflection of the pipeline, $F_N$ - longitudinal compressing force, $F_P$ – load of unit of length.

Within assumptions of one-dimensional character of a viscous fluid flow, this equation is simplified to the kind, that is ready for parametrical analysis:

$$\begin{aligned}\frac{\partial^4 L_y}{\partial x^4} + a_{10}(x,t)\frac{\partial^2 L_y}{\partial x^2} + a_{20}(x,t)\frac{\partial^2 L_y}{\partial x\partial t} + a_{30}(x,t)\frac{\partial^2 L_y}{\partial t^2} + \\ + a_{40}(x,t)\frac{\partial L_y}{\partial x} + a_{50}(x,t)L_y + a_{30}(x,t) = 0,\end{aligned} \tag{2}$$

where $a_{i0}$ – known parametrical functions [5].

With the help of (2) the Bubnov-Galyorkin method [7], we will receive the system consisting of $n$ the differential equations:

$$\sum_{i=1}^{n} A_{ik}f''_k + B_{ik}f'_i + C_{ik}f_k - D_{ik} = 0; n = 1, .2, ...n, \tag{3}$$

where factors $A_{ik}$, $B_{ik}$, $C_{ik}$, $D_{ik}$ have a certain expression, and $f_k$ are time functions which are received with approach of $L_{byy}(x,t) = \sum_{i=1}^{n} f_i(t)\varphi_i(x)$. It is necessary to notice that the analytical calculation of such system is rather difficult.

In this paper the generalization (2) of the spatial character of velocity field change is described. It should be noticed that the change of the velocity field is received from the solution of a multidimensional hydrodynamic task, intensified by the difficult shift flow of the hydrocarbonic fluids in the pipeline. In turbulent mode the flow structure was predicted on the basis of two-parametrical models by G. S. Glushko and D. Wilkox. It should be noted that the defining equations system of acoustic and hydrodynamic processes interaction is based on the method of final differences with the approximation of derivatives on spatial co-ordinates with the second order of accuracy. In the definition of the pressure field L.M. Simuni's generalised algorithm is used [1], because that algorithm is well-proved in forecasts of internal flows. The results of calculation show that in case of the developing flows convective and diffusive mechanisms of velocity field change cause an intensification of acoustic processes in the pipeline. Some ideas of advantages of model and complex flow calculation method can be received in extract 2 of the paper.

## 2. Hydrodynamics and heat exchange in constant non-Newtonian oil products flow conditions

As it was mentioned above, most of hydrocarbon media flows are related to the complex rheology flows. Due to modern computer technology development an interest in its mathematical modelling is growing [2, 4, 8].

This paper discusses both laminar and turbulent isothermal flow in pipelines, to illustrate the mathematical model of the flow and heat and mass transfer equations are given below describing the axisymmetric case flow.

The mathematical model of nonisothermal viscous-inertial vortex flow of hydrocarbon fluids is described by equations representing conservation laws: mass (continuity equation), momentum (Navier-Stokes equation), energy (for the carrier media) and thermal conductivity for the walls.

Mathematical formulation meets the conditions of the flow with the axial symmetry. It is assumed that the nature of the movement throughout the range of variation of perturbation corresponds to the stable laminar and turbulent transport processes, complicated conjugate heat transfer. Below, for simplicity, the vector and index form in generally accepted notation is used in constitutive equations [1]:

$$div\overline{v} = 0; \tag{4}$$

$$\frac{D\overline{v}}{Dt} = Div\overrightarrow{\overrightarrow{\tau}}; \tag{5}$$

$$\frac{DT}{Dt} = div(agradT) + S_T; \tag{6}$$

$$\frac{\partial T_w}{\partial t} = div(a_w gradT_w); \tag{7}$$

In this case the velocity vector $\overline{v}$ and its components are as following: $u_1$ - axial velocity, $u_2$ - radial velocity; $\frac{D}{Dt} = \frac{\partial}{\partial t} + (\overline{v} \cdot \nabla) \equiv \frac{\partial}{\partial t} + u_j \cdot \frac{\partial}{\partial x_j}$ - total derivative; $T$, $T_w$ - temperature of transported media and pipeline wall as accordingly; $a$, $a_w$ - media and wall thermal diffusivity coefficients; $S_T$ - volumetric heat source; $\overrightarrow{\overrightarrow{\tau}}$ - stresses tensor, which is described by three-constant equation of nonlinear-viscoplastic liquids in the terms of abnormal rheology model:

$$\begin{aligned} \tau_{ik} &= -p\delta_{ik} + 2\left[\tfrac{\tau_0(T)}{h} + \kappa(T)h^{\frac{1}{m-1}}\right] \cdot \dot{S_{ik}}, \\ \dot{S_{ik}} &= \tfrac{1}{2}\left(\tfrac{\partial u_i}{\partial x_k} + \tfrac{\partial u_k}{\partial x_i}\right). \end{aligned} \tag{8}$$

In the terms of formulation and independent variables $\{t, x, r\}$ described in the paper, components of stress tensor deviator $\left(\overrightarrow{\overrightarrow{\tau}}\right)$ quasi-Newtonian viscosity $(\mu^*)$ in summarized rheological equation (8) are formulated as:

$$\begin{gathered} \tau_{xx}^{(d)} = 2\mu^* \cdot \tfrac{\partial u}{\partial x},\ \tau_{rr}^{(d)} = 2\mu^* \cdot \tfrac{\partial v}{\partial r},\ \tau_{xr}^{(d)} = \mu^*\left(\tfrac{\partial u}{\partial r} + \tfrac{\partial v}{\partial x}\right), \\ \mu^* = \tfrac{\tau_0(T)}{h} + \kappa(T)h^{\frac{1}{m-1}}. \end{gathered} \tag{9}$$

The rheology of Newtonial liquid is determined by a known law:

$$\tau_{ik} = -p\delta_{ik} + 2\mu\dot{S_{ik}} + (\zeta + \mu_S)\frac{\partial u_j}{\partial x_j}\delta_{ik}, \tag{10}$$

where $\mu, \zeta, \mu_S$ - dynamic, volume and second viscosity coefficients accordingly, $p$ - hydrodynamic pressure.

It is also expected that $S_T$ defines the value of inner heat sources taking dissipation processes into account:

$$S_T = \mu^* \cdot h,\ h = 2\left[\left(\frac{\partial u}{\partial x}\right)^2 + \left(\frac{\partial v}{\partial r}\right)^2 + \left(\frac{v}{r}\right)^2\right] + \left(\frac{\partial u}{\partial r} + \frac{\partial v}{\partial x}\right)^2. \tag{11}$$

In the terms of forecasting of the turbulent flow and heat exchange with the known values ($u,\ v,\ p,\ T$) the system (4) - (11) uses its average analogue (time Reynolds averaging). To define additional unknown values [in this case - Reynolds turbulent stresses $\left(\overline{u_i' u_j'}\right)$, specific turbulent heat fluxes $\left(\overline{u_i' t'}\right)$] modern second order statistical models that use RANS method are engaged. The models use original and rarely used in practice base of applied caculations of turbulence kinetic energy ($k$) and pulsation time ($\tau$) equations, originally developed by C. Speziale and A.F. Kurbatskiy [9, 10] and common to the class of internal flows.

General view of defining equations of RSS-TF-$k\tau$ turbulent model in index notation is presented below [1]:

$$\begin{aligned}\frac{D\left(\overline{u_i' u_j'}\right)}{Dt} &= \frac{\partial}{\partial x_\alpha}\left[\left(\nu + c_{\mu 1} f_\mu \overline{u_i' u_j'}\tau\right)\frac{\partial \overline{u_i' u_j'}}{\partial x_\alpha}\right] - \frac{d_2}{\tau}\left(\overline{u_i' u_j'} - \frac{2}{3}k\delta_{ij}\right) - \\ &-d_3\nu\frac{\overline{u_i' u_j'}}{\tau^2 k} - \frac{2}{3}d_4\frac{k}{\tau}\delta_{ij} + P_{ij} - \frac{\partial \nu}{\partial x_\alpha}\frac{\partial \overline{u_i' u_j'}}{\partial x_\alpha};\end{aligned} \tag{12}$$

$$\begin{aligned}\frac{D\left(\overline{u_i' t'}\right)}{Dt} &= \frac{\partial}{\partial x_\alpha}\left[\left(\nu + \frac{a-\nu}{n_1+2} + c_{\mu\theta} f_{\mu\theta}\overline{u_i' u_j'}\tau\right)\frac{\partial \overline{u_i' t'}}{\partial x_\alpha}\right] - \frac{c_2}{\tau}\overline{u_i' t'} - \\ &-c_3(\nu + a)\frac{\overline{u_i' t'}}{l^2_{u_i' t'}} - \overline{u_i' u_\alpha'}\frac{\partial T}{\partial x_\alpha} - \overline{u_i' u_\alpha'}\frac{\partial U_i}{\partial x_\alpha} - \frac{\partial \nu}{\partial x_\alpha}\frac{\partial \overline{u_i' t'}}{\partial x_\alpha};\end{aligned} \tag{13}$$

$$\frac{Dk}{Dt} = \frac{\partial}{\partial x_\alpha}\left[\left(\nu + c_{\mu 2} f_\mu \overline{u_i' u_j'}\tau\right)\frac{\partial k}{\partial x_i}\right] - c_{kl}\overline{u_i' u_j'}\frac{\partial U_i}{\partial x_j} - \frac{k}{\tau}; \tag{14}$$

$$\begin{aligned}\frac{D\tau}{Dt} &= \frac{\partial}{\partial x_i}\left[\left(\nu + c_{\mu 2} f_\mu \overline{u_i' u_j'}\tau\right)\frac{\partial \tau}{\partial x_i}\right] - \frac{2}{\tau}\left(\nu + c_{\mu 2} f_\mu \overline{u_i' u_j'}\tau\right)\frac{\partial \tau}{\partial x_i}\frac{\partial \tau}{\partial x_i} + \\ &+(c_{\varepsilon 2} f_2 - 1) - (1 - c_{\varepsilon 1})\frac{\tau}{k}\overline{u_i' u_j'}\frac{\partial U_i}{\partial x_i} + \frac{2}{k}\left(\nu + c_{\mu 2} f_\mu \overline{u_i' u_j'}\tau\right)\frac{\partial k}{\partial x_i}\frac{\partial \tau}{\partial x_i}.\end{aligned} \tag{15}$$

Boundary conditions for the integration of the complete system of constitutive equations describing the flow and heat transfer in a carrier media and the wall are: inlet - homogeneous distribution of desired characteristics (or their known experimental field); output - "firm" conditions; on the inner wall surface of the pipe - for the dynamic slip condition parameters conjugation - a field temperature; on the axis of symmetry - symmetry conditions required parameters; on the outer wall of the pipe (for the heat equation) - the boundary conditions of the 1st or 2nd kind, determining the environmental impact.

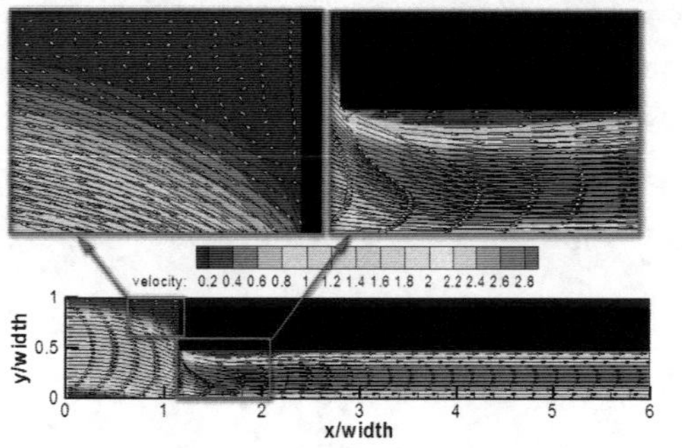


**Figure 1.** Calculated velocity field of liquid flow in a flat channel with step for $Re$ = 1800. The background depicts dimensionless velocity $u/u_0$, where $u_0$ is the average velocity at the channel entrance

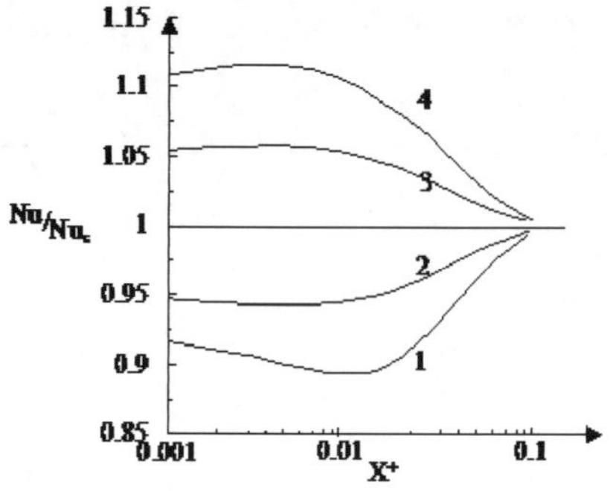


**Figure 2.** Change of reduced heat exchange factor along the channel length depending on the dimensionless reduced length $X^+ = x/DRe_m$ in variable heat conditions (of flow heating and cooling)

## 3. Results and discussion

The research of the flows and the heat exchange in pipelines was conducted in wide range of defining parameters changes, particularly: $L/D \sim 10 \div 1000$, $D/d \sim 1 \div 2$, $T_c/T_w \sim 1 \div 2$, $\mathrm{Pr}_c/\mathrm{Pr}_w \sim 0.25 \div 6$, $\mathrm{Re} = 10 \div 10^4$; heat transfer agents were oil and water; the wall was manufactured from Steel 14XTC.

The conclusion about steadiness, accuracy and reliability of the algorithm and features of closure correlations for RSS-TF-$k\tau$ model were acquired using data of numerous comparisons of the process local and integral parameters calculations with availiable experimental data in wide range of input characteristics on laminar and turbulent flows class [12].

On figures 1, 2 below individual results of dynamic heat disturbance from the entrance and the wall on the structure of transported Newtonian media with the variable heat and physical properties are depicted [$a = a(T)$ according to Reynolds-Philonov law].

Thus, figure 1 represents features of evolution of recirculating zones of vortex flow and its intensity in the pipeline including sudden enlargement places/sections enlargement. Tendencies typical for the developed flow are observed in the near-wall zones ($X^+ \leq 0.1$). The experience shows that in such conditions the structure of the turbulent eddy becomes significantly heterogeneous, tendencies to anisotropic mechanisms in processes of heat and momentum transfer are growing. In these situations it is unlikely that the forecasting of resistance and heat exchange patterns would be satisfactory with inclusion of simple turbulence models that use scalar formulation of molar viscosity and thermal diffusivity coefficients. That is why it is quite valid that the RSS-TF models of turbulence and evolution data of "firm" flow parameters in energy-strained zones may be useful.

It is necessary to point out that regarding heat processes, as it seen from figure 2, the impact from heat and physical properties' variability in the working media and pipeline wall on heat exchange intensity ($Nu\ /Nu_c$) is to be expected in X<0.1 zones [both during heating and cooling of the flow by the wall: $\mathrm{Pr}_c/\mathrm{Pr}_w = 0.25$ (line 1); 0,5(2); 2(3); 4(4)]. $Nu_c$ is a characteristic value of Nusselt number that corresponds to the flow conditions in "working media - wall" under variable heat and physical properties. Note that the comparative analysis of transfer process intensity in high-viscous systems have shown a rapid attenuation of dynamic disturbances. In case of oil or water transportation the impact of properties variability on friction and heat exchange is expressed in pipeline proximal zone where the disturbances are form the dynamic and heat pattern ($X \leq 0.01$). The calculations show that one variability of heat and physical properties in the media may intensify up to 10-15% the exchange processes in heat and momentum transfer. However, models with constant properties are quite successful to use in zones with $X > 0.1$ [12].

The evaluation of perspectives of multiparametric turbulence models usage has stated that the model is very sensitive to disturbance character coming from the walls and entrance and requires a precise definition of full spectrum of pulsar values in working zone entrance. In addition, these models may be a rational compromise from the point of expenditures calculation and its precision in comparison with less complex (but quite reduced in the forecasting of flow pattern and emergency regimes of pipeline functioning) algebraic models. This fact approves the advantages of RANS approach and the second order statistical models that include transport equations for stresses and flows of velocity and the temperature pulsar fields in relation to the forecasting of the complex processes of crude oil transport.

## References

[1] Kharlamov S N 2012 *Hydrodynamics - Theory and Model* (Croatia: Intech) 183-232

[2] Marusic I, McKeon B J, Monkewitz P A, Nagib H M, Smiths A J 2010 Wall-bounded turbulent flows at high Reynolds numbers: Recent advances and key issues *Physics of Fluids* **22** 065103

[3] Kharlamov S N, Dedeyev P O 2013 *Energosberezhenie i innovacionnye texnologii v toplivno-energeticheskom komplekse: materialy Vserossijskoj nauchno-prakticheskoj konferencii studentov, aspirantov, molodyx uchenyx i specialistov, posvyashhennoj 50-letiyu sozdaniya Tyumenskogo industrialnogo istituta* (Tyumen: TyumGNGU) pp 241-4

[4] Kharlamov S N, Tereschenko R E, Kudelin N S 2013 *Energosberezhenie i innovacionnye texnologii v toplivno-energeticheskom komplekse: materialy Vserossijskoj nauchno-prakticheskoj konferencii studentov, aspirantov, molodyx uchenyx i specialistov, posvyashhennoj 50-letiyu sozdaniya Tyumenskogo industrialnogo instituta* (Tyumen: TyumGNGU) pp 244-8

[5] Prokofiev A B 2001 *ssledovanie rocessov vibroakusticheskogo vzaimodejstviya v elementax gidromexanicheskix sistem dvigatelej letatelnyx apparatov: dissertaciya na soiskanie uchenoj stepeni kandidata texnicheskix nauk* (Samara)

[6] Shakhmatov E V, Prokofiev A B, Mironova T B 2006 Vozbuzhdenie pulsacii davleniya v rabochej zhidkosti i vibracii truboprovoda *Vestnik Samarskogo gosudarstvennogo aerokosmicheskogo universiteta imeni akademika S.P. Koroleva* **2-2** pp 161-4

[7] Kantorovich L V, Krilov V I 1962 *Priblizhennye metody vysshego analiza* (L.-M.: Fizmatlit)

[8] Kharlamov S N, Alginov R A, Tereschenko R E and Pavlov S A 2013 *Proceedings of IFOST 2013: The 8th International Forum on Strategic Technology* pp 714-9

[9] Kurbatskiy A F 1999 Uravnenie perenosa dlya masshtaba vremeni turbulentnogo skalyarnogo polya *TVT* vol 77 **4** 589-94

[10] Speciale C G 1998 Turbulence modeling for time-dependent RANS and VLES: A review *AIAA Journal* vol 36 **2** 173-84

[11] Patancar S 1984 *Chislennye metody resheniya zadach teploobmena i dinamiki zhidkosti* (M.: ENERGOATOMIZDAT)

[12] Kharlamov S N, Silvestrov S I 2007 *Trudy XVI shkoly-seminara molodyx uchenyx i specialistov pod rukovodstvom akademika RAN A.I. Leonteva «Problemy gazodinamiki i teploobmena v energeticheskix ustanovkax»* (Sankt-Peterburg)

IOP Conf. Series: Earth and Environmental Science **21** (2014) 012041 doi:10.1088/1755-1315/21/1/012041

# Prospects of Frequency-Time Correlation Analysis for Detecting Pipeline Leaks by Acoustic Emission Method

**V A Faerman[1], A G Cheremnov[2], V V Avramchuk[3] and E E Luneva[4]**
Tomsk Polytechnic University, Tomsk, Russia

E-mail: [1]csevladfaerman@gmail.com, [2]8xandr@gmail.com, [3]avs@tpu.ru, [4]lee@tpu.ru

**Abstract.** In the current work the relevance of nondestructive test method development applied for pipeline leak detection is considered. It was shown that acoustic emission testing is currently one of the most widely spread leak detection methods. The main disadvantage of this method is that it cannot be applied in monitoring long pipeline sections, which in its turn complicates and slows down the inspection of the line pipe sections of main pipelines. The prospects of developing alternative techniques and methods based on the use of the spectral analysis of signals were considered and their possible application in leak detection on the basis of the correlation method was outlined. As an alternative, the time-frequency correlation function calculation is proposed. This function represents the correlation between the spectral components of the analyzed signals. In this work, the technique of time-frequency correlation function calculation is described. The experimental data that demonstrate obvious advantage of the time-frequency correlation function compared to the simple correlation function are presented. The application of the time-frequency correlation function is more effective in suppressing the noise components in the frequency range of the useful signal, which makes maximum of the function more pronounced. The main drawback of application of the time-frequency correlation function analysis in solving leak detection problems is a great number of calculations that may result in a further increase in pipeline time inspection. However, this drawback can be partially reduced by the development and implementation of efficient algorithms (including parallel) of computing the fast Fourier transform using computer central processing unit and graphic processing unit.

## Introduction

Pipeline networks constructed for different purposes are an integral part of oil and gas production infrastructure. The exceptional value of the pipeline transportation for oil industry is explained by relatively transport costs, high reliability and the ability to maneuver flows [1]. Currently, crude oil and almost all petroleum products are transported through pipelines [2]. The environmental safety requirements relating to oil pipeline transport, especially to the places where pipelines cross rivers, railways, highways and ecologically sensitive areas, have been recently become much more stringent [3]. In connection with the above, failure-free operation of pipeline networks is a priority goal and rapidly developing field.

Failure-free operation of pipelines is achieved by the activities aimed at early detection and elimination of various pipe defects, such as destruction of the based metal, weld damage, defects in pipe insulation coating, etc. [1]. For these purposes, a wide variety of various domestic and foreign inspections techniques based on different operation principles is applied [2]. The most common inspection techniques are electromagnetic (eddy current) and acoustic emission testing which are

IOP Conf. Series: Earth and Environmental Science **21** (2014) 012041 doi:10.1088/1755-1315/21/1/012041

usually used together. For example, the pipeline inspection technique presented in [4] allows accurate estimation of residual wall thickness in corroded pipeline areas. In addition, such inspection is carried out in three stages:

- the damaged area of pipeline is detected by the long-wave method, with accuracy being from 3 to 4 meters;
- the corroded area of pipeline is defined by the eddy current method, with accuracy being from 2 to 4 square centimeters square centimeters;
- the residual wall thickness is estimated by the ultrasonic inspection.

Along with periodic pipeline inspection, the most important sections of pipelines can be continuously monitored by the stationary acoustic emission units [2].

However, despite pipeline requirements strengthening (particularly corrosion resistance specifications) and enhancement of early detection tools and control methods, accidents could not be completely avoided. This can be explained by the fact that the causes of pipeline accidents are not always related to the operation itself. Particularly, according to [5], 83 accidents were registered by ROSTECHNADZOR (Federal Environmental, Industrial and Nuclear Supervision Service) in the period from 2004 to 2012. The cause of 52 accidents was illegal tapping. Therefore, it is necessary to develop techniques for the rapid elimination of emergency situations and liquidation of their consequences, as well as inspection activities in order to minimize the economic loss and environmental damage during pipeline operation.

## Theoretical analysis

Oil spill without fire outbreak is the most common result of pipeline failure [2]. However, in specific cases, petroleum fire is likely to happen. Because of oil loss and infrastructure damages, this leads to considerable financial expenses and dramatic degradation of environmental conditions [2]. In order to prevent oil loss, as well as the above-mentioned consequences, it is necessary to find out a damaged section of a pipeline, that represents a technical challenge due to the large burial depth and considerable length of pipeline. Many pipeline leak detection methods have been developed and implemented so far. All these methods can be divided in two groups: leak detection based on pipeline operation specifications (i.e. data provided by automatic process control system); leak detection based on nondestructive inspection tools [2].

The first group of methods is the most widely used due to the simplicity of their implementation. These methods usually do not require the use of a large amount of additional hardware, and they are easily implemented by means of automatics due to the well-studied and relatively simple mathematical models of pipelines and oil transportation processes [6]. Among these methods, the following methods can be named: method of pressure reduction; method of linear balance; method of comparing costs; flow velocity monitoring method. A common drawback of these methods is the low sensitivity that does not allow detecting small-scale leaks [2, 6]. Taking into consideration that measurement gauges of process-dependent parameters (pressure, flow) are permanently installed at special locations (often in pump stations) at both sides of the line pipe section, so that it is impossible to locate the leak in this line section [6]. Moreover, the comparison of costs and linear balance methods does not allow resolving the leak detection issue in principle [6]. Nevertheless, the important advantages of this group of methods are continuous monitoring, reliable and effective leak detection in the case of large scale leak [6]. Since principles of these methods are different, their combined usage allows increasing the reliability of leak detection and improving the accuracy of the decisions. In this connection, Russian leak detection system (SDAs), available on the market, is designed for simultaneous processing of the data coming from different sources and composed of appropriate mathematical models.

Among the methods of non-destructive testing, the following method also can be named: the fiber-optic method, method of acoustical emission, radioactive and ultrasonic methods. The method of acoustical emission is the most widespread [6]. This method is based on the registration of acoustic emission signals by piezoelectric sensors, when sensors are placed on the controlled section of the pipeline. The registered signals indicate the presence of cracks in the pipe wall where the leak may happen [6, 7]. A leak spot is detected by estimating and analyzing indicated acoustic emission pulse damping and the time difference of pulse arrival at sensors. All required operations for leak spot detection are implemented by specialized software [6]. The following advantages of the acoustic emission method are: high sensitivity, flexibility (invariance to pipeline types and petroleum products) and the ability to locate the spots of small-scale leaks [6]. The main disadvantage of this method is the inability to inspect long pipeline sections, which complicates and slows down the inspection of line pipe sections of main pipelines [7]. Due to the low energy of acoustic emission signals, their detection on the background noise is a technical challenge [6], which effective solution significantly influences the quality of pipeline leak detection.

According to the above, when applying the acoustic emission method, the basic informative parameter is the difference in pulse arrival time to sensors. Particularly, in order to evaluate the difference, analysis of the cross-correlation function of the signals should be performed and its maximum should be located [6]. However, practically, due to low-energy of useful signals [7], the maximum of signal correlation function is often not well-marked. Therefore, detection of the leak location may represent a challenge [8].

To resolve this issue, advanced digital filtering of analyzed data can be applied, which allow improving signal-to-noise ratio against the background of the correlation device, which in its turn enables partially to eliminate random emissions in its output. However, this way of analysis cannot be implemented, because the frequency range of the leak acoustic signal is not known a priori and depends on the pressure and flow rate as well as on the shape and size of a leak spot [9, 10]. Determination of the leak signal frequency range is not trivial task that is poorly formalized and that's why it requires experienced and trained experts. Traditionally, the solution of this task is performed by such additional tools as cross-spectrum of the signals and coherence function calculation and analysis. These tools have some disadvantages: cross-spectrum does not give enough information in case of intense or narrow-band noise; due to the presence of several modes of vibration, coherence function could have false peaks that make almost impossible to evaluate frequency range of useful signal identically [9]. In this connection, developing alternative ways of signal spectral information processing and their further adaptation for solving leak detection problems by the correlation method are of great importance.

**Time-frequency correlation approach**

As an alternative method, the analysis based on time-frequency correlation function is considered. The time-frequency correlation function is termed as a correlation function that depends on two variables – response time and frequency. This means that this function represents the correlation between the spectral components of the analyzed signals [8]. Therefore, the method of timeprocedure-frequency correlation function calculation is also based on the decomposition of the original signals into the components localized in a specific frequency range and calculation of correlation functions of signals for each range. The above mentioned idea has become the basis of the algorithm [9] that is described below. Suppose that the problem is to obtain a time-frequency correlation function of signals $x(t)$ and $y(t)$ that are represented by discrete sequences $x_i, y_i \left(i = 0,1,2,\ldots,2^n - 1; n \geq 2, n - algebraicintegervalue\right)$. It should be noted that the signal sample numbers are equal to a degree of two for further application of Fast Fourier Transform (FFT) algorithm. At the initial stage, by the multiplying the spectrum of the signals their cross-spectrum is determined $\left(P_j, j = 0,1,\ldots 2^{n-1}+1\right)$:

$$P_j = F(x_i)\, F^*(y_i) \quad (1)$$

where $F$ – discrete Fourier transform (DFT); $F^*$ – complex conjugate representation of the results of the DFT.

At the second stage, the cross-spectrum separation is performed into $m$ components (where $m$ – number of selected frequency intervals). These $m$ components contain spectral samples that are within the current interval. Mathematically, this procedure reduces to obtaining $m$ vectors $M^k (k = 0,1,2,\ldots,m-1)$ according to the following rule:

$$M_j^k = P_j \cdot w_j^k, \qquad (2)$$

where $w_j^k$ - the window function that is defined in the following way:

$$w_j^k = \begin{cases} 1, \frac{k}{m} < \frac{j}{2^{n-1}+1} \le \frac{k+1}{m} \\ 0, \frac{k}{m} \ge \frac{j}{2^{n-1}+1} > \frac{k+1}{m} \end{cases}, \qquad (3)$$

where $j = 0,1,2,\ldots,2^{n-1}+1, k = 0,1,2,\ldots,m-1$.

At the final stage, obtained vectors $M^k$ are subjected to fast Furier transform:

$$Z^k = F^{-1}[M^k].$$

The obtained complex vector $Z^k$ consists of $m$ vectors and each of them represents correlation function of original signals in the $i$-th frequency interval. The simple change of the notation of the vector $Z^k$ values enables to restore the time-frequency correlation function:

$$r_{xy}(\tau_i, f_k) = Z_i^k. \qquad (4)$$

Since the time-frequency correlation function has two independent arguments, the most obvious way of its presentation is the surface constructed in the three-dimensional space. In this case the leak presence can be detected by peaks of the surface.

The advantages of the time-frequency correlation function in respect with pipeline leak detection issues are determined by its positive influence on the noise tolerance of acoustic emission method, which is particularly important when examining oil pipelines [6]. The capability assessment of the considered pipeline leak detection method was experimentally verified through the inspection of the water pipe section which is 20 mm in diameter and 18 m in length. A leak was located at a distance of 4m from the nearest sensor. The liquid was transported under the pressure of 2 atmospheres. Acoustic signals were recorded with a piezoelectric accelerometer with frequency of 44.1 kHz.

**Experimental part**

Calculation of time-frequency correlation function was made according to the above-mentioned algorithm. For the discrete signal, time interval of discrete samples was 11 seconds. The length of DTF window was 32768 sample units (743 ms). When calculating the time-frequency correlation function, the number of frequency intervals was 23. Final correlation functions were obtained by averaging the correlation functions calculated for different blocks, with their lengths being corresponded to the length of FFT window. The time-frequency correlation function of an experimental signal is represented as a surface in figure 1.

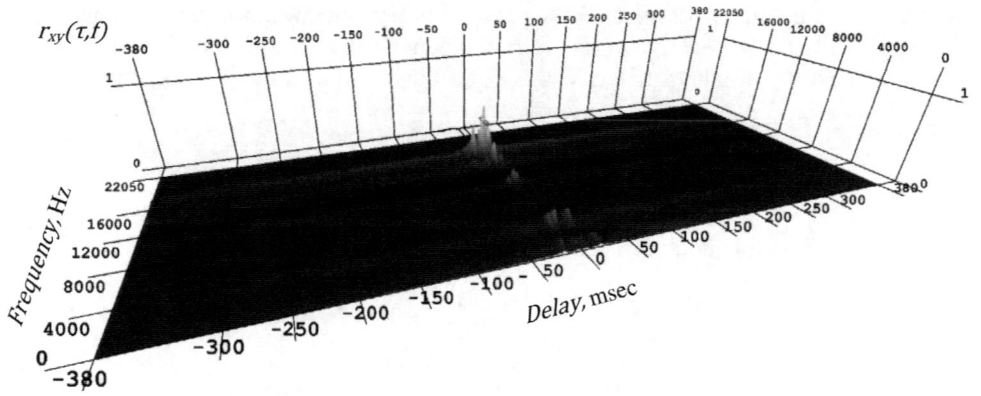


**Figure 1.** Time-Frequency Correlation Function

On the surface, there is clearly distinguishable peak which corresponds to the time delay -9.09 ms. Signal propagation velocity through the pipeline has been experimentally determined in advance, i.e. the value 1100 mps which corresponds to the distance of 3.99 m to the nearest sensor. Thus, the time-frequency correlation function demonstrates high efficiency in the determination of the time delay. Besides, by the location of the surface peak, a specific frequency band signal delay can also be detected – (16.2…17.2 kHz). As a comparison to the time-frequency correlation function surface, the one-dimensional plot of the correlation function is represented in figure 2. As illustrated, it does not contain a single evident peak. Thus, this function does not allow detecting identically the location of a leak.

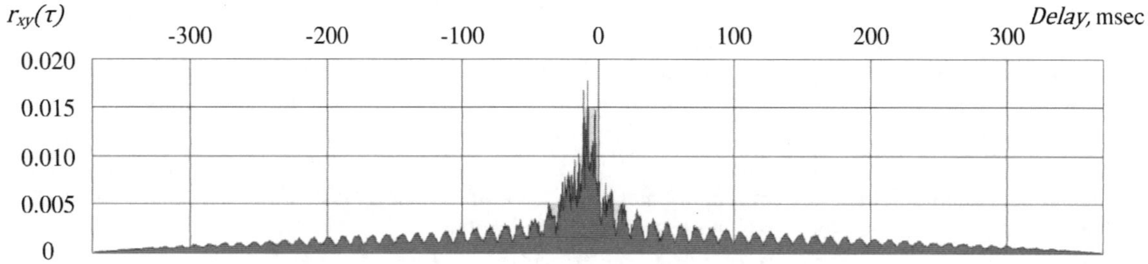


**Figure 2.** One-Dimensional Correlation Function Plot

Pre-filtering of signals in the frequency range of leakage signal can enhance the information content of one-dimensional correlation function. The graph of the one-dimensional correlation function after filter operation is represented in figure 3.

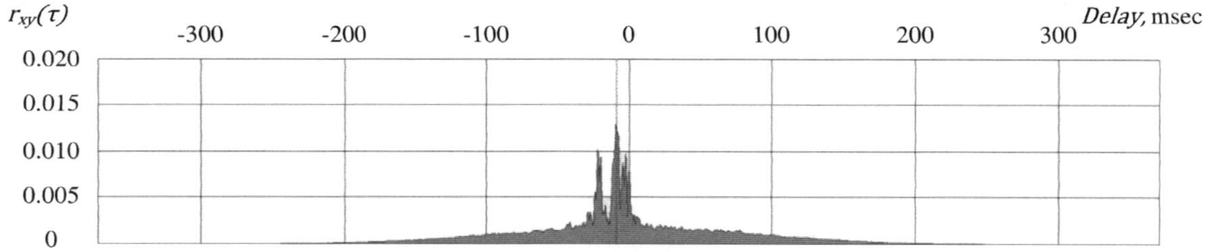


**Figure 3.** One-Dimensional Correlation Function Plot after Filtering

The time-frequency correlation function enables to suppress the noise components in the frequency range of useful signal more effectively, which makes a maximum of the function more pronounced [8]. In addition, this method allows representing processing signal data in more effective and descriptive way. Particularly, through the visual analysis of the plot, difference in time arrival of signals at sensors (to pronounced peak), as well as approximate frequency content of leakage signal, can be determined. The information obtained as a result of the time-frequency correlation function analysis can be used for further research to detect the leak location and set up filters.

The main drawback of the time-frequency correlation function analysis lies in the necessity of a great number of calculations. This may increase pipeline inspection time [8]. A great number of calculations is required because of the calculation of ($m$-1) additional inverse Fourier transforms for each block of analyzed discrete signals. However, this drawback can be partially reduced by the developing and implementing efficient algorithms (including parallel) of computing the fast Fourier transform based on the use of a computer CPU and GPU. Specifically, the use of CUDA technology combined with efficient parallel implementation of Fast Fourier Transform would improve processing speed by more than 80 times in comparison to the non-optimized algorithm.

## Conclusion

Therefore, application of time-frequency correlation analysis could improve the efficiency of pipeline leak detection applying the method of acoustic emission due to the partial elimination of its critical drawbacks. Particularly, implementing the proposed procedure for digital signal processing enables to reduce noise effects and increase the potential differentiation of the useful signal. Above all, simultaneous presentation of temporal and spectral information in one plot (time-frequency correlation function) is more descriptive and simplifies an operator's work.

## References

[1] Information on http://www.oilnews.ru/18-18/formirovanie-sistemy-monitoringa-sostoyaniya-neftepromyslovyx-truboprovodov-v-xanty-mansijskom-avtonomnom-okruge-yugre/.
[2] *Operating engineer oil and gas pipelines and product manual* 2006 (Moscow: Infra-E)
[3] Information on http://docs.cntd.ru/document/499011004
[4] Information on http://plantintegrity.ru/issues/report2/
[5] Information on http://www.gosnadzor.ru/public/annual_reports/
[6] T E Mamonova 2012 Diagnostic methods of linear part of pipeline leaks detection, Problemy informatiki **5** 103-112
[7] A V Jukov and A N Kuzmin 2011 Propagation of acoustic waves in pipelines, V mire NK **53** (3)
[8] V S Avramchuk and V I Goncharov Time-frequency Correlation Method for Improving the Accuracy in Detecting Leaks in Pipelines, Advanced Materials Research, Trans Tech Publications, Switzerland **650** 443–446
[9] V S Avramchuk, V I Goncharov and Chan Viet Tiau 2009 The method of time-frequency correlation analysis of digital signals, Russian Federation Patent, 2405163, 2009118627/28
[10] A L Ovchinnikov, B M Lapshin, A S Chekalin and A S Evsikov 2008 Experience of using a leak detector TAK-2005 for service of the municipal pipelines *Bulletin of the Tomsk Polytechnic University* **312** (2) 196-202

# Toxic elements (As, Se, Cd, Hg, Pb) and their mineral and technogenic formations in the snow cover in the vicinity of the industrial enterprises of Tomsk

**A V Talovskaya[1,a], E A Filimonenko[1], N A Osipova[1], E E Lyapina[1,2] and E G Yazikov[1]**
[1]Tomsk Polytechnic University, Tomsk, Russia
[2] Institute of Monitoring of Climatic and Ecological Systems, Siberian Branch of the Russian Academy of Sciences, Tomsk, Russia

E-mail: [a]talovskaj@yandex.ru

**Abstract.** Snow samples were collected in four industrial areas of Tomsk where brickworks, factories for the production of reinforced concrete structures, machine repair industries and local boilers, petrochemical plant and thermal power station are located. Study of insoluble fraction of aerosols in snow and melted snow water was performed to determine the contents of the emissions from these facilities. The insoluble fraction of aerosols in snow is aerosol particles deposited on snow cover. As, Se, Cd, Hg, Pb concentration was analyzed by inductively coupled plasma mass spectrometry (ICP-MS). Mineral modes of the elements were determined by scanning electron microscope. It was found the snow cover is mainly polluted by As, Se, Cd, Hg, Pb in the thermal power station impact area, by As – in the brickworks impact area, by Se – in the impact area of densely located factories for the production of reinforced concrete structures, machine repair industries and local boilers. The research results show that the mineral modes of As are associated with arsenopyrite, of Pb – with galena in the insoluble fraction of aerosols in snow.

## 1. Introduction

Most toxic pollutants enter the atmosphere from anthropogenic sources such as motor vehicles, factories, power plants and others. While some emissions of metals are the result of natural processes (e.g. volcanic eruptions, dust storms, weathering of rocks, forest fires) there are emissions of anthropogenic origin (e.g. vehicles, factories, power stations). [1]. Industrial emissions of trace elements include nickel, chromium, lead, zinc, copper, mercury, arsenic and some others. Concentrations of these elements are above the natural level [2-5]. Also, presence of heavy metals is the cause of atmospheric solid particles toxicity [6, 7]. For example, the exposure to Ni, Cr, As, Cd may cause cancer diseases, lung damage and/or breathing problems. Heavy metals coming into soil and water from atmosphere destructively affect plant growth and yields [8]. The large cities with diversified industry are not characterized by presence of separate pollutants, but heavy metals associations in the environment, capable to produce a combined effect on the body. However, direct determination of their concentrations at the air is usually costly and labor consuming procedure. The atmospheric deposition is an important process removing aerosol particles and gases from the atmosphere. Snow is one of the main types of precipitations characteristic of middle and high latitudes areas in winter. At subzero temperatures the fallen snow gets accumulated on the ground forming a snow blanket. Snow cover during formation and subsequent accumulation presents the chemical composition of the atmosphere [9, 10]. Snow plays an important role in the deposition of atmospheric contaminants [11]. Even at moderate snowfall a significant portion of elements precipitated from the atmosphere, because the snow is an effective absorber [12]. Snow cover is widely used as a reliable indicator of air pollution [13, 14]. Smaller particles should be given more attention to than larger ones because they have a high relative surface area, which facilitates the adsorption of pollutants [12]. Therefore, the snowpack serves as a reservoir for various pollutants such as heavy metals.

## 2. Materials and methods

### *2.1. Study area*

Tomsk, with a population of 548 000 people, is located in the south-east of the West Siberian Plain (56°29'19", 84°57'08"). It is the administrative center of Tomsk region. The climate is continental with significant seasonal variation in the solar radiation. The wind regime is characterized by an average annual wind speed of 3.1 m/s. The highest wind speeds occur in winter months. The southern (33%) and south-west (15%) winds are predominant in Tomsk. The average annual temperature is negative, -0.5 °C. Annual rainfall is 450-590 mm. Average snow depth is 60-80 cm, snow lasts 178-180 days.

In Tomsk enterprises of petrochemical, energy, engineering, instrument-making, food processing, construction, pharmaceutical and other industries operate. In this paper we present data on trace elements concentrations (As, Se, Cd, Hg, Pb) in snow water and insoluble fraction of aerosols in snow in the vicinity of Tomsk's enterprises. The soluble fraction of atmospheric aerosols accumulated during the winter in the snow attracts attention due to the fact that the biological effects of toxic dust are closely related to its solubility. Having entered the body highly soluble dust is dissolved in the mucus and other biological fluids (blood, lymph), in higher amounts and rapidly distribute throughout the body in dissolved form, exerting a toxic effect [15].

«Area I» that we have chosen in the central part of Tomsk is located in the area of wind mass transfer due to the prevailing wind directions dust emissions from the thermal power station by 100-meter height chimneys.

«Area II» is located in the northwestern part of Tomsk and includes the industrial area, where factories for the production of reinforced concrete structures, machine repair industries and local boilers are operated.

«Area III» is located in the eastern part of Tomsk by the north-east vector from the northern border of the brickworks industrial site.

«Area IV» is located in the north-east direction from the industrial area of one of the largest petrochemical plant in Russia, as well as near the gas power plant.

The background area was chosen at the landfill site observatory «Background» of Institute of Atmospheric Optics (Russian Academy of Sciences, Siberian Branch) (70 km south-west of Tomsk).

### *2.2. Sampling and preparation*

Five snow samples were collected at each of the four studied areas of Tomsk in March 2013 and the distance between adjacent sampling points was 200 - 300 m. There were collected and researched 30 snow samples. Snow samples were collected at the points with undisturbed snow cover for all its depth, the mass of each sample was 15-18 kg. During collection the pit area from which the samples were measured and the time from the formation of a stable snow cover till snow sampling time were fixed.

Snow samples were melted at 20-22°C, and the snow water was filtered through a pre-weighed filter paper. The insoluble fraction of aerosols in snow on the filter were passed through a sieve with a mesh size of 1×1 mm and then weighed with the error of 0.01%. The samples of snow melted water and the insoluble fraction of aerosols were analyzed.

IOP Conf. Series: Earth and Environmental Science **21** (2014) 012042 doi:10.1088/1755-1315/21/1/012042

*2.3. Analysis*
The analysis was carried out in the laboratories of International Innovative Academic Center «Uranium Geology», Geoecology and Geochemistry Department, Tomsk Polytechnic University, Russia. Hg concentration was performed by atomic-adsorption spectrometric method. The modes of element occurrence were determined using scanning electron microscope Hitachi S-3400N with an attachment for microanalysis Bruker XFlash 4010. As, Se, Cd, Pb concentration were studied by mass spectrometry method with inductively coupled plasma in Analytical Centre «Plasma» (Tomsk).

## 3. Results and discussion
Table 1 summarizes the results of analytic research of snow melted water and the insoluble fraction of aerosols in snow.

**Table 1.** As, Se, Cd, Pb concentrations in the insoluble fraction of aerosols in snow and snow melted water in Tomsk industrial areas and background area.

| Area | As | Se | Cd | Hg | Pb |
|---|---|---|---|---|---|
| insoluble fraction of aerosols in snow, mg $kg^{-1}$ | | | | | |
| Background area | 10.0 | 12.2 | 0.30 | 0.08 | 76.7 |
| I[a] | 31.1 | 19.4 | 1.22 | 0.48 | 101.8 |
| II[b] | 10.7 | 7.5 | 1.11 | 0.33 | 70.1 |
| III[c] | 12.4 | 8.8 | 0.67 | 0.11 | 34.4 |
| IV[d] | 16.8 | 11.5 | 1.17 | 0.05 | 71.0 |
| Snow melted water, mcg $l^{-3}$ | | | | | |
| Background area | 0.38 | 0.10 | 0.03 | 0.003 | 1.04 |
| I[a] | 0.94 | 0.12 | 0.05 | 0.012 | 0.81 |
| II[b] | 0.49 | 2.72 | 0.04 | 0.005 | 0.20 |
| III[c] | 2.83 | 0.24 | 0.04 | 0.015 | 0.46 |
| IV[d] | 0.82 | 1.89 | 0.03 | 0.007 | 1.16 |

[a] the central part of city, the thermal power station impact area
[b] the northwestern part of city, the impact area of densely located factories for the production of reinforced concrete structures, machine repair industries and local boilers
[c] the eastern part of city, brickworks impact area
[d] petrochemical plant impact area

Elements concentrations were compared with the background concentration in order to detect abnormal concentrations.

It was observed that the Se concentration exceeds the background l concentration in 27 and 19 times respectively in the snow melted water samples from II and IV area. That is more than other elements. The samples of I and III areas showed that Hg and As are characterized by the highest values above the background concentration.

Hg concentration exceeded the background concentration in 4-6 times in the samples from I and IV area. The concentration exceeded the background one in 3 times in the samples taken in the thermal power station impact area.

Cd concentration exceeded the background one in 2-4 times in the samples taken nearby industrial enterprises, but Se and Pb concentrations were not significantly higher than background values.

Results of electron microscopy scanning show that the mineral modes of As and Pb are associated with sulfides (arsenopyrite and galena) in the insoluble fraction of aerosols in snow (figure 1). The mineral particles size varies from 2 to 12 microns. We have found that Hg is present in the insoluble fraction of aerosols in snow in a physically adsorbed form.

The element composition of fly ashes from Tomsk thermal power station showed higher Hg and As concentrations. The thermal power station uses coals of Kuznetsk region which are of As-U-Ag-Be-Sn-Ge-Zr-Mo geochemical specialization [16].

According to literature data many chemical elements may be emitted during coal burning with the gas phase. Burning of fossil fuels is one of the man-made sources of As, Hg and Se in the urban atmosphere [17].

In addition, Se is one of the specific elements of petrochemical production emissions [13]. It also may form high concentrations of cement kiln dust [18]. It is known that in addition to plant emissions where Cd and its compounds are used in the production process this metal comes with dust emission of plants producing glass, ceramics (bricks) and other as well as along fuel combustion.

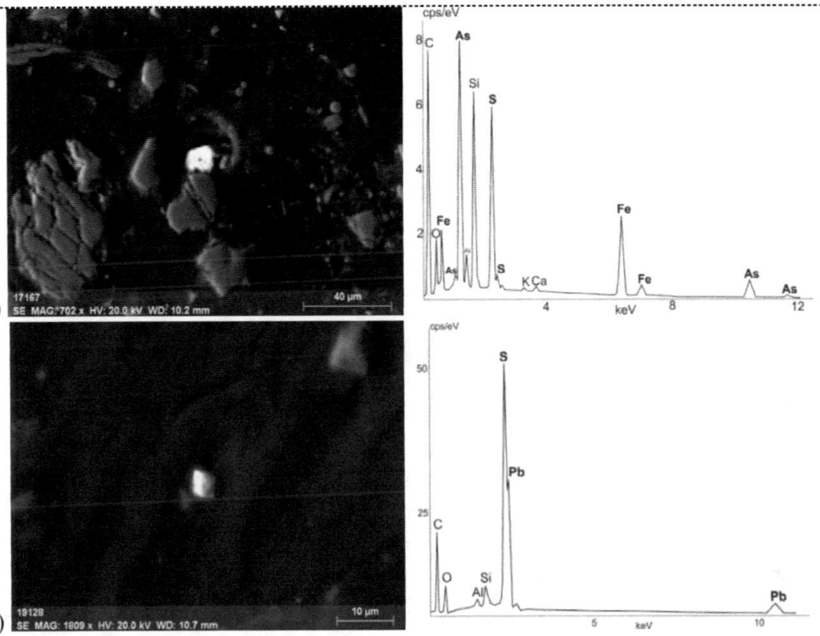


**Figure 1.** Arsenopyrite (a) and galena (b) in photos by the electronic microscope and energy dispersive spectra of the insoluble fraction of aerosols in snow

## 4. Conclusion

Therefore, the studies showed that the highest level of contamination with toxic elements was observed in vicinity of the thermal power station in Tomsk. At the same time the main polluting elements were detected in the petrochemical plant, brick and factories for the production of reinforced concrete structures impact areas. Mineral modes of As and Pb were identified in the insoluble fractions of aerosols in snow.

## 5. References

[1] Moreno T, Querol X, Alastuey A, Viana M, Salvador P, Campa A, Artinano B, Rosa J and Gibbons W 2006 Variations in atmospheric PM trace metal content in Spanish towns: Illustrating the chemical complexity of the inorganic urban aerosol cocktail *Atmos. Environ.* **40** 6791-803

[2] Boyd R S 2004 Ecology of metal hyperaccumulation *New Phytol.* **162** 563-67

[3] Krachler M, Mohl C, Emons H and Shotyk W 2003 Atmospheric deposition of V, Cr, and Ni since the Late Glacial: effects of climatic cycles, human impacts, and comparison with crustal abundances *Environ. Sci. Technol.* **37** 2658-67

[4] Nriagu J O and Pacyna J M 1988 Quantitative assessment of worldwide contamination of air, water and soils by trace metals *Nature* **333** 134-39

[5] Chen B, Stein A F, Maldonado P G, Campa A, Gonzalez-Castanedo Y, Castell N and Rosa J D 2013 Size distribution and concentrations of heavy metals in atmospheric aerosols originating from industrial emissions as predicted by the HYSPLIT model *Atmos. Environ.* **71** 234-44

[6] Lighty J S, Veranth J M and Sarofim A F Combustion aerosols: factors governing their size and composition and implications to human health 2000 *J. Air Waste Manage.* **50** 1565-618

[7] Seaton A, Tran L, Aitken R and Donaldson K 2010 Nanoparticles, human health hazard and regulation *J. of the Royal Soc. Interf.* **7** 119-29

[8] Clemens S 2006 Toxic metal accumulation, responses to exposure and mechanisms of tolerance in plants *Biochimie* **88** 1707-19

[9] Junge C E 1967 Determination of the acid content of aerosol particles *Atmos. Environ.* **5** 165-75

[10] Osada K, Shido Y, Iid H and Kido M 2010 Deposition processes of ionic constituents to snow cover model *Atmos. Environ.* **44** 347-53

[11] Kang J H, Choi S D, Park H, Baek S Y, Hong S and Chang Y S Atmospheric deposition of persistent organic pollutants to the East Rongbuk Glacier in the Himalayas *Sci. Total Environ.* **408** 57-63

[12] Vasic M V, Mihailovic A, Kozmidis-Luburic U, Nemes T, Ninkov J, Zeremski-Skoric T and Antic B 2012 Metal contamination of short-term snow cover near urban crossroads: Correlation analysis of metal content and fine particles distribution *Chemosphere* **86** 585-92

[13] Bosco M L, Varrica D and Dongarra G 2005 Case study: Inorganic pollutants associated with particulate matter from an area near a petrochemical plant *Environ. Res.* **99** 18-30

[14] Hiromitsu S, Tanenori S and Kazuo S 1988 Heavy metal concentrations in urban snow as an indicator of air pollution Himalayas *Sci. Total Environ.* **77** 163-74

[15] Espinosa A F, Rodriguez M T, Barragan de la Rosa F J and Sanchez J C 2002 A chemical speciation of trace metals for fine urban particles *Atmos. Environ.* **36** 773-80

[16] Arbuzov S I, Volostnov A V, Rikhvanov L P, Mezhibor A M and Ilenok S S 2011 Geochemistry of radioactive elements (U, Th) in coal and peat of northern Asia (Siberia, Russian Far East, Kazakhstan, and Mongolia) *Int. J. Coal Geol.* **86** 318-28

[17] Gaffney J S and Marley N A 2009 The impacts of combustion emissions on air quality and climate – From coal to biofuels and beyond *Atmos. Environ.* **43** 23-36

[18] Kunal, Siddique R and Rajor A 2012 Use of cement kiln dust in cement concrete and its leachate characteristics *Resour. Conserv. Recy.* **61** 59-68

IOP Conf. Series: Earth and Environmental Science **21** (2014) 012043 doi:10.1088/1755-1315/21/1/012043

# Epiphytic lichens as indicators of air pollution in Tomsk Oblast (Russia)

**T S Bolshunova[1,2,a], L P Rikhvanov[2,b] and A M Mezhibor[2,c]**
[1]JSC "TomskNIPIneft", Tomsk, Russia
[2]Tomsk Polytechnic University, Tomsk, Russia

E-mail: [a]BolshunovaTS@gmail.ru, [b]Rikhvanov@tpu.ru, [c]Amezhibor@gmail.com

**Abstract.** This research presents the data on concentrations of 28 chemical elements in the ash of lichens growing in Tomsk Oblast where different industrial enterprises are located. In the sites of oil and gas exploration complex the lichens contain higher concentrations of Sc, Fe, Cr, Zn, As, Br, Rb, Ba, and Au in comparison with the background values. In the zone of Tomsk-Seversk industrial agglomeration the lichens accumulate Ca, Sr, Sb, Th, U, and lanthanides.

## 1. Introduction

Tomsk Oblast (Siberia, Russia) has such specific industries as nuclear and coal-fired power energy facilities, petroleum-chemical plant (Tomsk-Seversk industrial agglomeration) and oil and gas exploration complex (west and north-west of the region) which are the sources of various gaseous and particulate pollutants entering the atmosphere. These pollutants contain a wide range of chemical elements including rare and radioactive ones. These peculiarities of the area determine the necessity for control of air condition and analysis of atmospheric emissions.

The use of lichens for observation and investigation of chemical composition of atmospheric fallouts is a relevant trend for the assessment of environment quality because they allow obtaining the long-term averaged characteristics of ecosystems state during quite long life period of lichens. Lichens are found in almost all geographical regions because they are resistant to extreme nature conditions. Today in Russia the use of lichens as bioindicators for the study of the dynamics of atmosphere pollution has been proved in some researches [1,2,3]. As for Tomsk Oblast, the method of lichen-monitoring has not found wide application yet.

The use of lichens for the bio-geochemical research is also explained by wide abundance of forests in Tomsk Oblast (91% of the territory) [4] which are consequently characterized by the abundance of lichen flora. Because of the absence of root system lichens have atmospheric nutrition and their element composition reflect the integrated composition of chemical elements in the air present in gaseous, dissolved, or particulate forms.

Lichens are a symbiosis of mycobiont and photobiont. They have been proved to be excellent bio-monitors because of the particularity to accumulate chemical elements in concentrations exceeding their physiological needs and keep them in their thallus (body) for a long time. Since lichens have atmospheric nutrition, they get substances from wet and dry fallouts absorbing them by the whole thallus surface.

## 2. Data Sets

The samples of epiphytic lichens growing on trees were collected in August and September 2010-2012 in the territories of some oil fields of Tomsk Oblast (32 samples). Moreover, we used the lichen samples collected in 2006 [5] in the impact zone of Tomsk-Seversk industrial agglomeration (8 samples) and in the south and south-west of Tomsk region of Tomsk Oblast (10 samples) used as a regional background. In addition, for the assessment of global geochemical background the lichens from the Alpine region (Semmering, Austria) were sampled in 2012. The samples were collected from adult trees, mainly from warty birch or from coniferous trees at the height ranging from 1.5 to 1.8 m from the ground.

The samples were hermetically packed in plastic bags. In the laboratory we removed bark, needles and other impurities from the samples and dried them. For the analytical research we prepared lichens of the abundant background species: *Evernia Mesomorpha, Usnea Subfloridana, Hypogymnia physodes, Parmelia Sulcata*. In all cases the samples were mixed.

For the quantitative analysis of chemical element concentrations in lichens we used modern nondestructive high-sensitive nuclear-physical method of the neutron activation analysis (INAA) with the irradiation by thermal neutrons performed in the Nuclear-Geochemical Laboratory of the Geoecology and Geochemistry Department on the basis of the research nuclear reactor (Tomsk Polytechnic University). For the INAA the samples were reduced to powder and ashed in a muffle furnace at the temperature of 450-500$^0$C. The sample ash allows decreasing errors in organic analysis and improves the reliability of results due to the element concentration in samples. The samples were weighted before and after ashing. The samples of 100 mg weight were prepared for the analysis. The lichen ash was analyzed by INAA for 28 chemical elements.

Concentrations of chemical elements in the samples were recalculated per a unit of dry mass using ashing coefficient for lichens samples.

The software used in the work included Microsoft Word, Microsoft Excel, STATISTIKA 6.0. The following statistical parameters of element distribution were calculated in terms of the obtained results: mean, standard error, standard deviation. The verification of normal element distribution was performed by the Kolmogorov-Smirnov test.

For the estimation of the ratio of anthropogenic and terrigenous sources of chemical elements we calculated the enrichment factor (EF) according to the formula [6]:

**EF = (xi/Fe) in lichen tissues / (xi/Fe) in lithosphere**

The essence of EF coefficient consists in the estimation of ratio of elements-pollutants and lithogenic elements (iron, aluminum, silicon). The EF calculation is a standard method for the estimation of potential sources of infused matter in plants [7].

The results of epiphytic lichens are presented in table 1.

**Table 1**. Comparative characteristic of the element composition of the epiphytic lichens (mg kg$^{-1}$, recalculated for the dry matter).

| No. | Elements | Area of oil and gas exploration complex, Tomsk Oblast (32 samples) $X_{mean} \pm \delta$ | Area of Tomsk-Seversk industrial agglomeration (8 samples) | Background of Tomsk Oblast (10 samples) | Mean in the lichens of Tomsk region (50 samples) | The Alps, (Semmering, Austria) (1 sample) |
|---|---|---|---|---|---|---|
| 1 | 2 | 3 | 4 | 5 | 6 | 7 |
| 1 | Na | 289±116 | 305±84 | 235±92 | 276 | 88 |
| 2 | Ca | 1700±617 | 3041±1299 | **3650**±1464 | 2797 | 1719 |

| No. | Elements | Area of oil and gas exploration complex, Tomsk Oblast (32 samples) $X_{mean} \pm \delta$ | Area of Tomsk-Seversk industrial agglomeration (8 samples) | Background of Tomsk Oblast (10 samples) | Mean in the lichens of Tomsk region (50 samples) | The Alps, (Semmering, Austria) (1 sample) |
|---|---|---|---|---|---|---|
| 3 | Sc | 0.29±0.1 | **0.36**±0.13 | 0.22±0.04 | 0.29 | 0.06 |
| 4 | Cr | **4.14**±1.6 | 2.14±0.5 | 1.72±0.3 | 2.7 | **3.7** |
| 5 | Fe | 785±263 | **913**±245 | 606±120 | 768 | 401 |
| 6 | Co | 0.43±0.1 | **0.50**±0.13 | 0.29±0.05 | 0.4 | 0.16 |
| 7 | Zn | **45**±14 | 34±0 | no data | 40 | 31 |
| 8 | As | **0.19**±0.04 | 0.08±0 | 0.07±0 | 0.11 | 0.003 |
| 9 | Br | **5.6**±1.8 | 0.89±0.3 | 0.60±0.2 | 2.4 | **5.4** |
| 10 | Rb | **7.2**±2 | 3.12±1.1 | 2.93±0.7 | 4.4 | 1.5 |
| 11 | Sr | 11.4±4 | **32.3**±12 | 1.8±0 | 15.2 | 4.8 |
| 12 | Ag | 0.011±0.007 | **0.019**±0.003 | 0.018±0 | 0.016 | 0.011 |
| 13 | Sb | 0.046±0.02 | **0.133**±0.04 | 0.063±0.02 | 0.081 | **0.113** |
| 14 | Cs | **0.14**±0.04 | 0.13±0.02 | 0.08±0.02 | 0.12 | 0.05 |
| 15 | Ba | **30**±8 | 27±6.6 | 18±5 | 25.0 | 7 |
| 19 | La | 0.9±0.3 | **1.32±0.5** | 0.79±0.15 | 1.0 | 0.009 |
| 20 | Ce | 1.56±0.6 | **2.15±0.7** | 1.30±0.1 | 1.67 | 0.004 |
| 21 | Nd | 0.51±0.2 | **1.19±0** | no data | 0.85 | 0.011 |
| 22 | Sm | 0.14±0.06 | **0.18**±0.06 | 0.10±0.03 | 0.14 | 0.002 |
| 23 | Eu | 0.036±0.01 | **0.040**±0.01 | 0.026±0.007 | 0.03 | 0.02 |
| 24 | Tb | 0.022±0.009 | **0.028**±0.01 | 0.017±0.002 | 0.02 | 0.002 |
| 25 | Yb | 0.064±0.03 | **0.106**±0.03 | 0.065±0.01 | 0.08 | 0.001 |
| 26 | Lu | 0.012±0.008 | **0.014**±0.005 | 0.009±0.003 | 0.01 | 0.002 |
| 16 | Hf | 0.16±0.07 | **0.19**±0.06 | 0.10±0.02 | 0.15 | 0.019 |
| 17 | Ta | 0.028±0.01 | **0.042**±0.01 | 0.024±0.007 | 0.03 | 0.002 |
| 18 | Au | **0.0012**±0.0007 | 0.0002±0 | 0.0002±0 | 0.0005 | 0.0008 |
| 27 | Th | 0.18±0.07 | **0.26**±0.09 | 0.14±0.03 | 0.19 | 0.05 |
| 28 | U | 0.07±0.03 | **0.20**±0.05 | 0.06±0.02 | 0.11 | <0.01 |
| 29 | Th/U | 2.5 | 1.3 | 2.4 | 2.1 | >5 |
| 30 | La/Yb | 14.1 | 12.4 | 12.1 | 12.9 | 19.5 |
| 31 | Ash yield, % | 3.6 | 4.1 | 4.1 | 3.9 | 1.3 |

## 3. Discussion

The results of lichen element concentration analysis (table 1) shows that this biological species from the area of the oil and gas exploration complex is more enriched with Na, Sc, Fe, Cr, Zn, As, Br, Rb, Ba, Au (in 1.3 to 9 times) than the lichens from the background territory of Tomsk region. From the analysis of Table 1 we can also see the peculiarity of the geochemical lichen spectrum taken in the impact zone of Tomsk-Seversk industrial agglomeration generally formed by the enterprises of nuclear-fuel cycle, petroleum-chemical combine, and coal-fired power stations. This peculiarity is determined by the concentrations of lanthanides, actinides (U, Th), Sr, and Sb. Elevated Ca concentration in the background samples can be referred to as a local feature.

Besides, the EF coefficient as an indirect indicator confirms geochemical differences in the lichen composition of the different areas. Comparison of the EF values shows that main elements-pollutants for the area of oil and gas exploration complex are Zn, As, Br, Rb, Ba, and Au; for Tomsk-Seversk industrial agglomeration – La, Ce, Sr, Sb, and U; and background territory is characterized by elevated Ca and Ag concentrations.

When comparing the data from Tomsk region with the background area from the Alps (Austria) we can observe a more considerable difference in the accumulation levels of chemical elements (figure 1). The lichen sample from the Alps contains minimal concentrations of most studied chemical elements. The exceptions are Cr, Br, Sb, and Au. Presumably, elevated Cr, Br and Sb concentrations, which are close to those from the areas of oil and gas exploration complex of Tomsk Oblast, reflect the impact of oil-processing plants or other industrial facilities in Vienna suburbs.

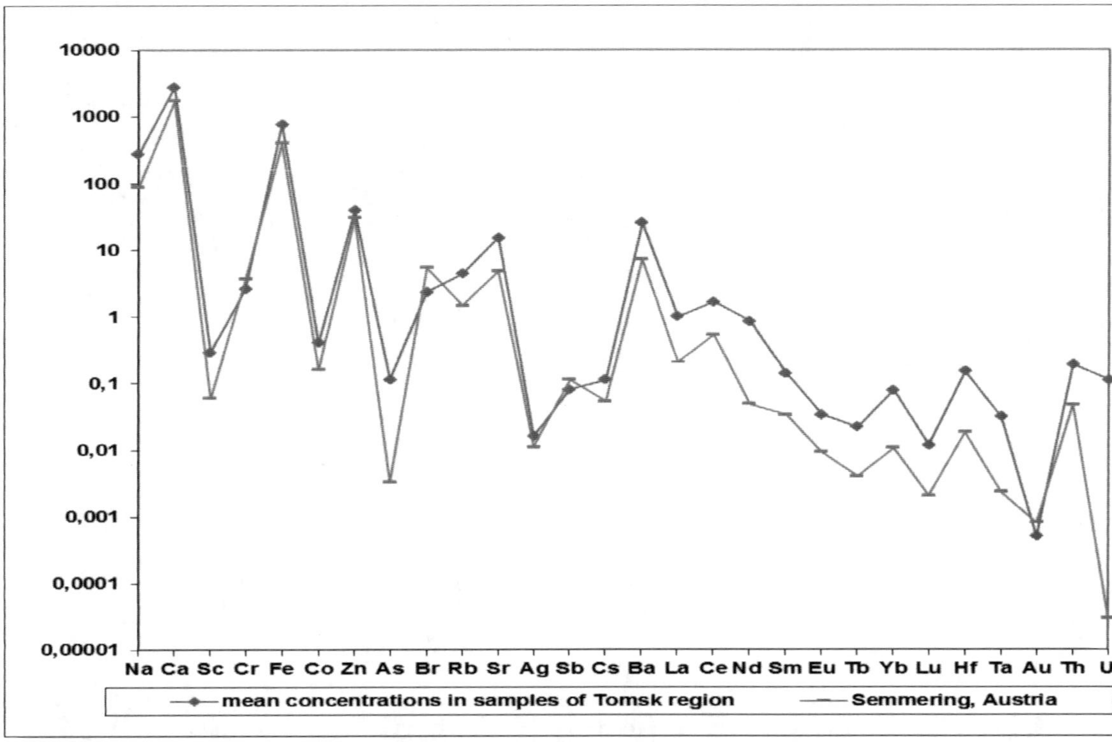


**Figure 1.** Comparison of chemical elements concentrations in the lichen ash from Tomsk Oblast and Alps (Simmering, Austria), mg $kg^{-1}$, recalculated for the dry matter.

## 4. Conclusion

The comparative analysis of lichens growing in Tomsk Oblast allows for conclusion that lichens form the areas of oil and gas exploration complex and Tomsk-Seversk industrial agglomeration have geochemical peculiarities. The geochemical peculiarity of the first area is determined by intense accumulation of such elements as Cr, Zn, As, Br, Rb, Ba, and Au. Elevated concentrations of As, Zn, and Cr were likely to be caused by combustion processes of oil-dissolved gas in flares, oil heaters, diesel power stations and transport, or drilling operations. These elements are known to be accumulated by lichens due to their high biophylity. Oil-field and high mineralized water of oil fields contains elevated concentrations of Br and Au [8] that can explain high concentrations of these elements in lichen ash. Previous investigations of dust aerosols [9] and upland peat [10] in this area showed that Rb, Ba, and Br are indicative elements for the areas of oil and gas exploration complex. Thus, the current research has revealed the fact of atmosphere pollution in the areas of oil and gas exploration complex of Tomsk region.

The peculiarity of the geochemical spectrum of the lichens from the impact zone of Tomsk-Seversk industrial agglomeration formed by the facilities of nuclear-fuel cycle, petroleum-chemical combine, and power stations, is determined by the concentrations of Sr, Sb, lanthanides, and actinides (Th, U).

IOP Conf. Series: Earth and Environmental Science **21** (2014) 012043 doi:10.1088/1755-1315/21/1/012043

The study of the space distribution of these elements in the Tomsk-Seversk industrial zone testifies that their sources are mainly Siberian Chemical Combine facilities.

The results of this study support the concept of good bio-monitoring properties of the epiphytic lichens and allow us to recommend them for using in assessing anthropogenic transformation of natural environment as a main component.

## References

[1] De Bruin M 1990 Les indicateurs biologigues, l'analyse par activation neutronique, et leurs applications a l'etude de la pollution atmospherique par les metaux lourds. AIEA Bulletin **4** 22 –27

[2] Mikhailova I N and Vorobeichik E L 1995 Epiphytic lichenosinusia under conditions of chemical pollution: dose-effect dependencies *Russian Journal of Ecology* **26** 455-460

[3] Valeeva E I and Moskovchenko D V 2002 Trace-element composition of lichens as an indicator of atmospheric pollution in northern West Siberia *Polar Geography* **26** (4) 249-269

[4] Adam A M (Ed.) 2012 Ecological monitoring: Report about the environment state of Tomsk region in 2011. p 166 [in Russian]

[5] Shatilova S S 2007 Geochemical particularities of trace element distribution in lichen ash (Evernia mesomorpha) of Toms district. Problems of geology and subsurface development: Proceedings of the XIth international symposium Tomsk,April, 9-13 2007. 619 – 621 [in Russian]

[6] Voitkevich G V, Kokin A V, Miroshnikov A E and Prokhorov V G (Eds.) 1990 *Hand-book on geochemistry* (Moscow: Nedra) p 480 [in Russian]

[7] Moskovchenko D V and Valeyeva E I 2011 Content of heavy metals in lichens of West Siberian North *Bulletin of ecology, of forestry and landscape* **11** 162-172

[8] Perelman A I and Kasimov N S 1999 Geochemistry of landscapes p 610 [in Russian]

[9] Shatilov A Yu 2001 Matter composition and geochemical characteristic of atmospheric fallouts in the Ob River basin. Dissertation p 205 [in Russian]

[10] Mezhibor A M 2009 *Ecogeochemistry of trace elements in upland peat of Tomsk region* p 160 [in Russian]

IOP Conf. Series: Earth and Environmental Science **21** (2014) 012044 doi:10.1088/1755-1315/21/1/012044

# Toxicity Assessment of Contaminated Soils of Solid Domestic Waste Landfill

**O A Pasko[1] and T N Mochalova[2]**
Tomsk Polytechnic University, Tomsk, Russia

E-mail: [1]oap@sibmail.com, [2]mochalova@green.tsu.ru

**Abstract**
The paper delivers the analysis of an 18-year dynamic pattern of land pollutants concentration in the soils of a solid domestic waste landfill. It also presents the composition of the contaminated soils from different areas of the waste landfill during its operating period. The authors calculate the concentrations of the following pollutants: chrome, nickel, tin, vanadium, lead, cuprum, zinc, cobalt, beryllium, barium, yttrium, cadmium, arsenic, germanium, nitrate ions and petrochemicals and determine a consistent pattern of their spatial distribution within the waste landfill area as well as the dynamic pattern of their concentration. Test-objects are used in experiments to make an integral assessment of the polluted soil's impact on living organisms. It was discovered that the soil samples of an animal burial site are characterized by acute toxicity while the area of open waste dumping is the most dangerous in terms of a number of pollutants. This contradiction can be attributed to the synergetic effect of the polluted soil, which accounts for the regularities described by other researchers.

## Introduction

Human-induced land degradation embraces about 20% and threatens more than 60% of the globe ecosystems [1]. Alongside with the climate change it causes irreversible changes of agro-landscapes and becomes one of the prior issues of land management, economical and environmental activities of the society [2]. The land operational loads result in soil degradation (physical condition, agricultural characteristics) and loss in natural and economic value of lands.

Solid domestic waste landfills (SDW), authorized and non-authorized garbage dumps, animal burial sites, etc. are a source of special concern. Their intensive and extensive grow is caused by urban and rural settlements development [3]. These sites are typically very dangerous for human health. The main risk factor is the pollution of soil and regeneration waters [4] with harmful domestic wastes: paints, pesticides, oil, pharmaceuticals, household chemicals, electronic and electrical equipment, photo chemicals, personal hygiene means, luminous tube lamps, heavy metals, timber treated with harmful substances, etc. [5]. Besides, the SDW landfills are the source of air pollution accompanied by decomposition products, unpleasant smells, methane outbursts and fires, dust, garbage moved by wind, opportunistic pathogenic and pathogenic microflora, helminth eggs, flies, ravens, larines and rodents.

The interaction of SDW with water (snow, rain) produces toxic filtrate water that flows through soil pores and can reach aquifers. Long-term chemical research of soils and filtrate waters conducted in the USA (500 waste landfills, 230 chemical compounds) discovered the dependence of filtrate water's chemical composition on the age of a waste landfill, its geographical location, climate and

IOP Conf. Series: Earth and Environmental Science 21 (2014) 012044 doi:10.1088/1755-1315/21/1/012044

geological characteristics [6]. Such variety of factors can form different chemical transformation models and hinders consolidation of information.

Soil quality is typically evaluated on the basis of chemical analysis that doesn't take into account synergetic and antagonistic effects of the whole polluting complex and doesn't characterize its toxicity level for organisms [7]. Besides, the number of pollutants is much greater than the number of analysis methods used nowadays [8]. It makes it necessary to use biotesting as an additional assessment tool for determining toxic level of soil [9].

The aim of the research is to compare chemical analysis data with biotesting data of SDW landfill conducted in 1995 and 2013.

To achieve the aim the following tasks are to be solved:

1. to identify the soil pollution level on the SDW landfill based on 14 elements chemical analysis
2. to conduct biotesting of the same samples, compare the results and to draw conclusions.

**Research object, tools and methods**

The SDW landfill near village Novomikhailovka was used from 1964 till January 2011. Its total area is 54.3 ha; the animal burial site area is 729 m2. A daily waste volume reached 5400 m3. The first set of tests was conducted to study the dynamic pattern of soil pollution level year-wise (1995, 2011, and 2013). In 1995 The SDW landfill was actively used. In 2011 it was officially closed but it was still being dumped illegally, which implies additional pollution of the area. The second set of tests was conducted to analyze soil chemical pollution level and soil phytotoxicity studying the samples taken in 1995 from different areas of the landfill (recultivated area, animal burial site, open waste dumping area and reference (background) area – northwards from the landfill).

The soil sampling and acid digest preparation were conducted according to the standard techniques. The extractable cupper, cadmium, zinc, chrome and plumbum was determined by means of an atomic absorption spectrophotometer AAS -1 (Germany) (measurement uncertainty (mu) 5-7%). pH was determined by means of pH-meter Mark -903 (mu <0.02) [28]. Nitrate ions were determined by means of a photoelectric colorimeter (mu < 20%). An atomic absorption spectrophotometer AA-7000 "Shimadzu" (Japan) (mu < 20%) was used for arsenic determination. The IR spectroscopy method was applied to determine the rate of petrochemical pollution of the soils. The air samples were tested by means of a gas analyzer (mu < 20%). The confidence coefficient of the measurements is P=0.95.

The experimental part of the research was conducted in 1995 and 2014 in the accredited laboratories of Tomsk Polytechnic University and Tomsk State University, Center of Laboratory Analysis and Technical Measurement of the Siberian Federal Region, Center for Hygiene and Epidemiology in the Tomsk Region and the department of Tomsk specialized inspection of State Ecological Control and Analysis of "Regional Committee for Environmental Protection and Nature". The soil samples were cleaned of mechanic impurities, put into plastic boxes (1 liter) and moistened up to 60% of maximum water capacity.

To determine the soil toxicity level seeds of Cucumis sativus L. (hybrid "Estafeta") were used as a traditional test-object [10]. Their laboratory percentage germination is 80 %. The test was repeated four times with 50 seeds for each set. The seeds were sown 3 cm distance from each other at 1 cm depth. They were watered as long as the substrate dried out. The morpho-physiological method was used to measure seed vigor, germination capacity and their degree of development. The ruler was used to measure 10-day plant biometrics (accuracy up to 1 mm) and the analytical balance "VLR-200" was used for weight measuring. Statistical data processing was carried out by means of software STATISTICA.

**Results and discussions**

The research results prove that petrochemicals and zinc were the most significant pollutants in 1995-2013 (76-134 and 37-62 mg/kg respectively). The second group of pollutants included nitrate ions and heavy metals (mobile forms): copper, plumbum, nickel and chrome with the concentrations over the range of 2-30 mg/kg. The third group consisted of cadmium and arsenic (mobile forms) with their concentrations ranging from 0.005 to 6.225 mg/kg. The maximum general pollution level was registered in 1995, that is the active operational period of the SDW landfill (figure 1). By 2011 the contamination level reduced (5 times) and in 2012 slightly increased (zinc, copper and nickel). This fact can be attributed to the official closure of the SDW landfill and further renewal of illegal waste dumping. The pH value of the salt-water soil digestion was steady and didn't change from year to year (about 6.8).

The comparative analysis of the ultimate composition of the operating landfill's soils shows highly inhomogeneous spatial distribution of the elements over the SDW landfill areas (table 1). The closest to the reference (background) values are those measured in the recultivated area. The most contaminated are the soils under the open waste dumping area. According to the chemical analysis there is an intensive accumulation of the majority of the 24 studied elements. The soils of animal burial site take the in-between position. They are of semi-fluid consistence and characterized by strong smell.

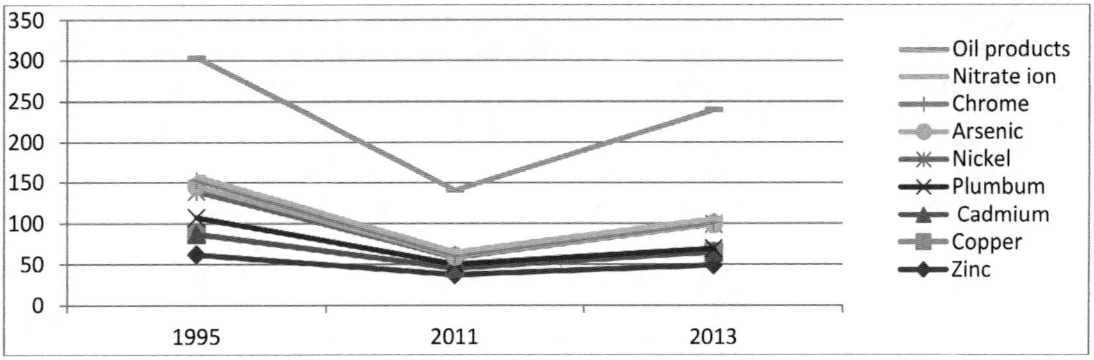


**Figure 1.** Dynamics of soil pollutant (mobile forms) concentrations (mg/kg) on the SDW landfill.

It is necessary to note that even after the operational period of the SDW landfill was finished the air sampling taken in 2013 revealed the following pollutants (mg/m$^3$):

a) at the boarder of the sanitary protection zone: carbonic oxide CO (0.01-0.32); benzol (up to 0.047); toluene (up to 0.08);

b) at the working area of the SDW landfill: carbonic oxide CO (0.37), benzol up to 0.0363); ( toluene (up to 0.005);

c) at the closed area of the SDW landfill: carbonic oxide CO (0.28);

d) garden plots of the citizens (residential area): carbonic oxide CO (2.24); benzol (up to 0.0924);

e) in the Bekkari hollow: carbonic oxide CO (0.35); benzol (0.1199); toluene (0.059); nitrogen dioxide (0.0071).

**Table 1.** Element concentrations (mobile forms) in the soils of the SDW landfill near the settlement Novomikhaylovka during its operating period (1995)

| Area | | Concentration of Elements (mobile forms), mg/kg | | | | | | | | | | | | | |
|---|---|---|---|---|---|---|---|---|---|---|---|---|---|---|---|
| | | Cr | Ni | Sn | V | Pb | Cu | Zn | Co | Be | Ba | Y | Cd | As | Ge |
| 1 | Reference (background) | 62 | 22 | 1,6 | 46 | 15 | 20 | 32 | 6,5 | 0,5 | 220 | 12 | 2,5 | 5,0 | 0,2 |
| 2 | Recultivated | 100 | 40 | 3,2 | 96 | 30 | 33 | 31 | 9,5 | 0,7 | 330 | 20 | 3,7 | 7,3 | 0,2 |
| 3 | Animal burial site | 100 | 41 | 3,0 | 97 | 25 | 16 | 53 | 20,3 | 1,0 | 405 | 20 | 4,7 | 6,5 | 0,3 |
| 4 | The area under open waste dumping | 146 | 41 | 4,8 | 85 | 35 | 25 | 80 | 18,1 | 0,9 | 442 | 21 | 1,7 | 9,3 | 0,3 |

The soil biotesting confirms actual toxicity occurrence. The seeds germinated more slowly and the sprouts were weaker. They were characterized by low morpho-physical characteristics (plant's weight, leaf area) (figure 2). The seeds in the soil from the animal burial site didn't germinate at all, which allows us to make a conclusion about its hyper-phytotoxicity.

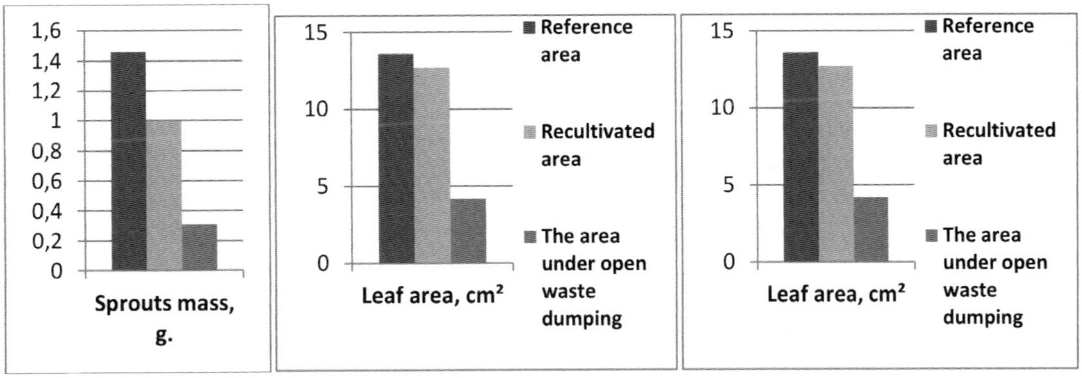


**Figure 2.** Phytotoxicity of the soil samples of the SDW landfill taken from different areas in 1995. SV – seed vigor (the number of sprouts appearing above the soil surface on the 5$^{th}$ day, %; Germination capacity – the final number of sprouts, %).

It is necessary to note that according to the chemical analysis it is the area under the open waste dumping that is the most contaminated one. But the animal burial site appears to be the most phytotoxic one. These differences can be adjusted by taking into account the following:

1. Chemical substances interact with each other in the soil thus strengthening and weakening biological influence. That is why it is difficult to give an integral evaluation of such a complicated multi-elemental composition in terms of the degree and direction of its synergetic effect.
2. Biological objects have selective sensibility when exposed to particular factors. Even slight exposure can cause significant changes in their metabolism.

## Conclusion

The biotesting proves that the soil pollution on the SDW landfill is comparable with the direct soil exposure with diesel and biodiesel [8], as well as with petrochemicals [12]. It allows applying the recommendations developed by the authors mentioned above to the polluted soils of SDW landfills: 1. For effective depollution of the contaminated soils of SDW landfills it is necessary to ensure effective assessment and control of the landfills (areas and distribution of waste disposal) and SDW (content and amount); 2. It is necessary to classify SDW according to the degree of integral toxicity impact on living organisms; decay rate in the natural environment; and particular induced changes in the biosphere; 3. Special attention should be paid to minimization of SDW amount by introducing separate waste collection and waste recycling, thus reducing the landfills' areas with the final disposal of them in the long view.

## References

[1] UNEP Year Book 2011Emerging issues in our global environment, United Nations Environment Programme, Nairobi. Published Februar

[2] Wang G L Qin and G Li et al 2009 Landfill site selection using spatial information technologies and AHP: A case study in Beijing, China *Journal of Environmental Management* **90** 2414–2421

[3] Sumathi V, Natesan U and Sarkar C, 2008 GIS-based approach for optimized siting of municipal solid waste landfill *Waste Management* **28** 2146–2160

[4] Erdogan R, Zaimoglu Z, Yavuz S et al 2008 Applicability of leachates originating from solid-waste landfills for irrigation in landfill restoration projects *Journal of Environmental Biology* **29** 5 779 -785

[5] Kanga J-H, Y Katayama and F Kondo 2006 Biodegradation or metabolism of bisphenol A: From microorganisms to mammals *Toxicology* **217** 81–90

[6] Pavelka C, Heights M, RC Loehr and B Haikola 1993 Hazardous waste landfill leachate characteristics *Waste Management* **13** (8) 43-54

[7] Plaza G, G Naleez-Jaweeki, K Ulfig and R L Brigmon 2005 The application of bioassays as indicators of petroleum-contaminated soil remediation *Chemosphere* **59** 289-296

[8] Repetto G, A Jos, M J Hazen, M L Molero et al 2001 A test battery for the ecotoxicological evaluation of pentachlorophenol *Toxicology in Vitro* **15** 503-509

[9] Cruz J M, Lopes P R, Montagnolli R N et al 2013 Toxicity assesment of contaminatad soil using seeds as bioindicators *Journal of Applied Biotechnology* **10** (1) 1-10

[10] Bagdasaryan A S 2005 Biotestirovaniye pochv tekhnogennykh zon gorodskikh territoriy s ispolzovaniyem rastitelnykh organizmov Author's abstract (Stavropol: SSU) [in Russian]

[11] Pasko O A 2000 Ekologicheskiye aspekty povysheniya produktivnosti tsvetochnykh, ovoshchnykh kultur i kartofelya v tayezhnoy zone Zapadnoy Sibiri *Doctoral dissertation* (Novosibirsk) p 452 [in Russian]

[12] Vasilyev A V and Zabolotskikh V V 2012 Ekologicheskiy monitoring toksicheskogo zagryazneniya pochvy nefteproduktami s ispolzovaniyem metodov biotestirovaniya *Oil and Gas business* 4 242-249 [in Russian]

# The Use of Principle Component Analysis in Type Classification of Air-dry Peat

**N Chukhareva, A Zarubin and O Bulgakova**
Tomsk Polytechnic University, Tomsk, Russia

E-mail: zagtpuru@gmail.com

**Abstract**. The use of principle component analysis (PCA) in the variant of projection on latent values with discriminant analysis (PLS-DA) in type classification of Siberian region air-dry peat by a set of properties is presented. A statistical analysis in principle component space by PCA of different physico-chemical properties of peat such as component composition, concentration of paramagnetic centers and IR-spectra is presented and shows a developed PLS-DA classification model allows estimating peat type by a set of physico-chemical properties with minimum prediction errors.

## 1. Introduction

The study of peat's composition and properties has always been paid great attention to because of its value in the organic part [1–3]. Peat-based materials are widely used for various tasks in the areas of agriculture, industry, as well as resource-efficient solutions for environmental problems [4–5]. But, because peat is an initially difficult natural object, including both organic and inorganic components, there is no uniform approach to choosing the direction of its use at the present time. This is caused by a different degree of influence in botanical composition, type of peat and degradation amount of the organic component on the final product's properties. Therefore, it is important to develop a classification model that contains comprehensive information based on raw material properties. On one hand, the collection of many different characteristics of peat, based on physico-chemical and technical analysis methods, does not always allow excluding part of the incorrectly obtained data considering all of the investigated properties. Not one of their series is possible to identify the specificity of the object. On the other hand, it is impossible to identify objects by baseline characteristics, such as type, if data are not initially.

## 2. Experiment

PCA is widely used in the analysis of the properties and composition of natural objects. So in [6], the authors used the space of the principal components to explain the distribution of peat on the ground, depending on the composition.

The aim of this work is to apply the PLS-DA method to classify the type of air-dry peat by a set of physico-chemical properties.

One of the solutions for the task is to apply PCA to classify objects by a set of different properties [7–9]. This method allows representing the observed data matrix ($X$) as the product of two matrices, plus the residual error, – $E$ [10]:

$$X = SL + E, \tag{1}$$

where $L$ is the intrinsic properties of the data matrix (loadings), and $S$ is the matrix of individual parameters (scores).

The instrumentation of PCA allows the reduction in dimensionality of multivariate data by highlighting the principal components (PC) and interpreting the properties of investigated objects in the new space of PC's [11]. The implementation of PLS in an Excel spreadsheet editor is described by A.L. Pomerantsev [12].

In accordance with the intended aim we put forward following tasks:

- calculation of the scores matrix by physico-chemical properties of different types of air-dry peat;
- building and validation of the PLS-model.

The objects of research: twenty samples of two types (lying and lowland) of air-dry peat of eleven fields in the Tomsk region (Russia) were examined for the set of physico-chemical parameters such as botanical composition [13], elemental composition [14], group composition, functional composition by IR spectroscopy and electron paramagnetic resonance. The set of physico-chemical parameters were presented in the initial data matrix by fourteen variables (table 1).

**Table 1**. The set of physico-chemical parameters.

| Variable | Physico-chemical parameter |
|---|---|
| V1 | lg(S) – logarithm of the concentration of paramagnetic centers (electron paramagnetic resonance) |
| V2 | bitumen content, % on daf |
| V3 | content of humic acids, % on daf |
| V4 | nonhydrolyzable residue content, % on daf |
| V5 | content of water soluble and readily substances, % on daf |
| V6 | fulvic acid content, % on daf |
| V7 | cellulose content, % on daf (the method of determination of group composition) |
| V8 | carbon content, % on daf |
| V9 | hydrogen content, % on daf |
| V10 | nitrogen content, % on daf |
| V11 | the total content of oxygen and sulfur, % on daf (method of determining the elemental composition) |
| V12 | $D_{3400/1620\text{-}1600}$ – ratio of optical densities corresponding to the characteristic lines of the IR spectrum |
| V13 | $D_{2920/1620\text{-}1600}$ – ratio of optical densities corresponding to the characteristic lines of the IR spectrum |
| V14 | $D_{125\text{-}1200/1620\text{-}1600}$ – ratio of optical densities corresponding to the characteristic lines of the IR spectrum |

Data structure of two types of peat (lying – Class1 and lowland – Class2) properties investigation are presented in table 1 and it was divided for calibration and validation sets.

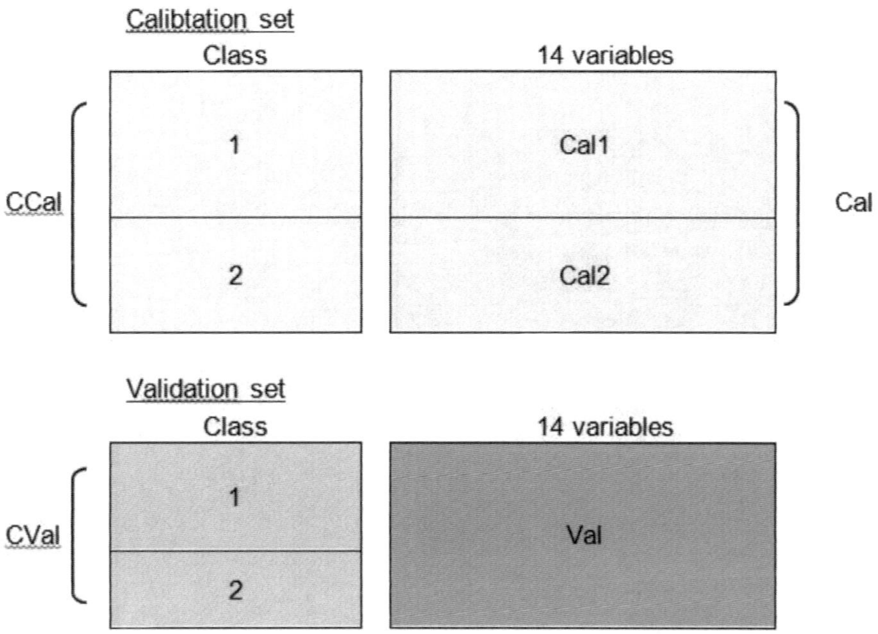


**Figure 1**. Data structure.

### 3. Results and discussion

Based on the data sets, the mathematical model of classification was constructed and validated. The results were obtained for the three latent variables and two of them are shown in figure 2.

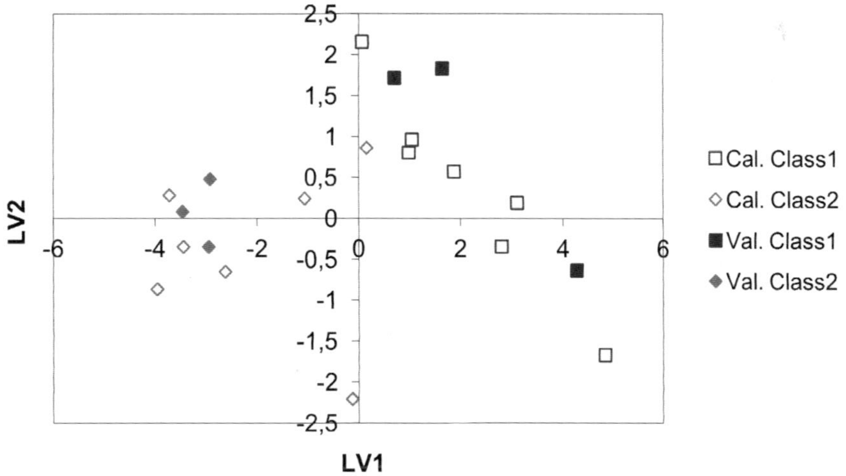


**Figure 2**. Scores of latent variable 1 (LV1) and latent variable 2 (LV2).

The constructed mathematical model with level of discrimination 0.5 allows classifying air-dry peat by type (prediction errors are presented in figure 3).

| | Number of latent variables in model: 1 LV | 2 LVs | 3 LVs |
|---|---|---|---|
| **Training set** | | | |
| Type I errors: | 0 | 0 | 0 |
| Type II errors: | 1 | 1 | 0 |
| **Validation set** | | | |
| Type I errors: | 0 | 0 | 0 |
| Type II errors: | 0 | 0 | 0 |

**Figure 3**. Prediction errors.

The plot of PLS-predictions of three latent variables is presented in figure 4.

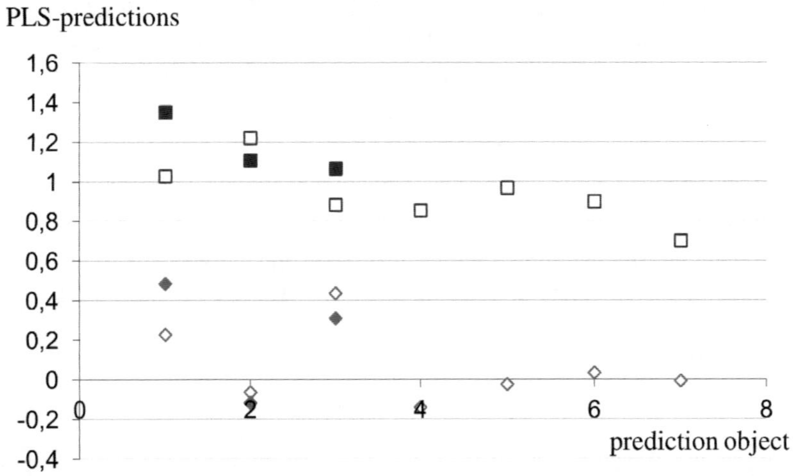


**Figure 4**. PLS-predictions of tree latent variables.

## 4. Conclusion

The principal component analysis in variant of projection on latent values with discriminant analysis was used for estimating changes in the composition and properties of air-dried peat, depending on its type. The PLS mathematical model was constructed and validated with minimum prediction errors. The analysis clearly shows that lowland Siberian peat very easy to discriminate from the lying Siberian peat. With PLS discriminating models using the whole set of 14 physico-chemical parameters, it is possible to achieve excellent discrimination almost all the time. It must be underlined that the statistical model shown in this article is built for Siberian peat only, but it can be extended for the others peat types with model correction.

## References

[1] Arkhipov V S, Preis U P, Bernatonis V K, Maslov S G and Antropova N A 2008 Bituminous peat of the Tomsk region Tomsk: STT [in Russian]

[2] Lishtvan I I and Korol N T 1975 Main properties of peat and methods of their definition Minsk: Science and technology 320 12 [in Russian]

[3] Kujala K, Seppälä M, and Holappa T 2008 Physical properties of peat and palsa formation. *Cold Regions Science and Technology* **52** (3) 408-414

[4] Sokolov B N and Yampolski A L 1986 Socio-economic consequences of industrial development of peat deposits in the USSR In *Socio-economic impacts of the utilization of peatlands in industry and forestry: proceedings of the IPS Symposium, Oulu, Finland, June 9-13, 1986* p 69. International Peat Society, Finnish National Committee.

[5] Shishmina L V, Chukhareva N V and Kravtsov A V 2002 Influence of Peat Humates on Flotation of Coal *Coke and Chemistry* **(2)** 7-9

[6] Muller J, Kylander M, Martinez-Cortizas A, Wüst R A, Weiss D, Blake K, and Garcia-Sanchez R 2008 The use of principle component analyses in characterising trace and major elemental distribution in a 55 kyr peat deposit in tropical Australia: implications to paleoclimate *Geochimica et Cosmochimica Acta* **72(2)** 449-463

[7] Brereton, R G 2007 *Applied chemometrics for scientists* John Wiley & Sons

[8] Wold S, Sjöström M and Eriksson L 2001 PLS-regression: a basic tool of chemometrics. *Chemometrics and intelligent laboratory systems* **58**(2) 109-130

[9] Forina M, Oliveri P, Lanteri S and Casale M 2008 Class-modeling techniques, classic and new, for old and new problems *Chemometrics and Intelligent Laboratory Systems* **93**(2) 132-148

[10] Kowalski B R (Ed) 1984 *Chemometrics: mathematics and statistics in chemistry* No 138 Springer

[11] Wehrens R 2011 *Chemometrics with R: multivariate data analysis in the natural sciences and life sciences* Springer

[12] Pomerantsev A L 2014 *Chemometrics in Excel* John Wiley & Sons

[13] GOST 28245-89 *Turf. Method of determination of botanical composition and degree of composition* Russia: Federal Agency on Technical Regulating and Metrology

[14] Fadeeva V P, Tikhova V D and Nikulicheva O N 2008 Elemental analysis of organic compounds with the use of automated chns analyzers Journal of Analytical Chemistry **63**(11) 1094-1106

IOP Conf. Series: Earth and Environmental Science **21** (2014) 012046 doi:10.1088/1755-1315/21/1/012046

# Political risks of hydrocarbon deposit development in the Arctic seas of the Russian Federation

**Y A Bolsunovskaya, G Yu Boyarko and L M Bolsunovskaya**
Tomsk Polytechnic University, Tomsk, Russia

E-mail: ju_al@inbox.ru

**Abstract**
Nowadays the process of Arctic development has a long-term international cooperation character. Economic and geopolitical interests of both arctic and non-arctic countries meet in the region. Apart from resource development issues, there are problems concerning security, sustainable development and some others issues conditioned by climate and geographical characteristics of the region. Strategic analysis of political risks for the Russian Federation is carried out. The analysis reveals that political risks of hydrocarbon deposits development in the RF arctic seas appear as lack of coordination with arctic countries in solving key regional problems, failure to follow international agreements. Such inconsistency may lead to political risks, which results in strained situation in the region.

## • Introduction

The Arctic region integration into global political and socio-economical processes is an important stage of its development. Today the role of the Arctic region becomes more and more important in global politics and economics. This area is in focus of resource and strategic interests of both arctic and non-arctic countries. The Arctic countries are trying to codify property for a part of the sea bottom, but they encounter opposition of countries-rivals and countries that have no territories in the Arctic region.

Historically, the arctic area was of no economic and political interest for a long time both for Russia and for other countries. Hard climatic conditions made the industrial and transport development of the region impossible. It restricted economical and political interests of the countries and the issue of the arctic region dominion didn't arise.

Global warming resulted in arctic ice melting and changed the situation essentially. Climate change made the arctic regions, rich in natural resources, more accessible and forced many countries, especially Russia, the USA, Canada, Denmark (acting for Greenland and Faeroes) and Norway (the arctic "five") to redefine they interests in the Arctic region. Arctic sea ice loss gave massive prospective of industrial reclaiming of the region, most of all, offshore oil and gas field development and sea lanes economic potential.

The aim of the article is to analyze geopolitical situation and interaction between countries in the Arctic region, to reveal political risks and contradictions during the deposit development in the Russian Federation arctic seas.

- **The Arctic region legal status**

The UN Convention on the Law of the Sea (UNCLOS) was approved in 1982. It established the order to define the boundaries of the continental shelf for the littoral states. According to the Convention state's full sovereignty of territorial waters extends 12 nautical miles from the baseline, aerial domain above it, bottom and mineral resources. There is also an exclusive economic zone (EEZ) that stretches 200 nautical miles from the state's coast where the state has special right over the exploration and use of marine resources but has no right to restrict other navigation. Moreover the Convention empowers coastline states to extend the shelf zone up to 350 nautical miles if it is proved that ocean floor is a part of the continental shelf [1].

The Convention is ratified by all "arctic five" countries, except the USA. Despite the renunciation of the ratification the USA accept many norms of the Convention as norms of international law and seek peaceful solution of possible continental shelf claims. Under these circumstances national interests of the littoral arctic states moved to the forefront, with their interrelation becoming more pragmatic. These states are interested in unanimous defense of their rights and interests in the Arctic region, which makes "arctic five" communication logical and stable.

Most of the arctic resources are within the national boundaries. Nevertheless, territorial debates are still pressing challenge and include not only debates about boundaries of particular economical zones of "arctic five" countries, but also the issues of security and rights concerning passing through territorial and international waters. Iceland, Sweden and Finland are considered to be arctic countries, apart from Russia, Denmark, Norway, the USA and Canada. Moreover, the number of countries claiming to participate in making decisions on key problems of the Arctic is increasing. For example, such non-arctic states as China, Japan, South Korea and others consider the policy of sharing the Arctic among the arctic countries to be shortfall. They regard the arctic resources as the province of all mankind, so their development should be performed within the framework of international collaboration. Such collaboration should imply free access to the arctic resource for any country that has technical and economical potential.

The modern system of international collaboration in the arctic region is of contradictory character. On the one hand, there is a growing business struggle between the arctic states for strengthening their positions in the region. One the other hand, no country has sufficient scientific and technical potential to implement independent arctic projects. Besides, there are some ecological, social and other problems in the region to be solved in international collaboration. All the countries interested in arctic resource development understand the profit potential and solve geopolitical and economical problems from in terms of international collaboration. Therefore, being the biggest arctic state, Russia should have a clear attitude to the perspective of the Arctic development, national project elaboration and international collaboration in the region [2].

- **Research methodology**

Traditionally political risk identification and analysis imply defining regional political uncertainty influencing business activity results. Risk analysis is of strategic nature, the Arctic region having some territorial, climatic, social and other peculiarities.

The research is based on the hypothesis that political risks of the Russian arctic sea deposit development are manifested in inter-country contradictions concerning the Arctic resource development.

The research is based on the analysis of requirement documents, arctic strategies of the states and international non-governmental organizations' reports.

- **Geopolitical position of the Russian Federation**

Geopolitical stance and perspectives of socio-economical development of the Russian arctic region are based on principles of globality (all the processes in the region are influenced by global trends) and sovereignty (the region is a part of the national state system and plays a part in state national security protection).

Russian part of the Arctic is the region of particular state geostrategic interests and long-term economical interests of the society in terms of natural resource development and their rational use, global ecological balance and national security. This fact makes the region to be regarded as a separate object of state politics [2].

No arctic topical issue can be solved without Russia's participation and support. Almost half of the arctic area and waters, limited by Arctic Circle belong to the Russian Federation. The most immense world ocean shelf zone rich in unique resources borders on the Russian arctic coastline. 80% of shelf area girding the Arctic Basin belongs to Russia [3].

Arctic deposits development is not only of resource value for Russia [4, 5], but it also has transport and infrastructural potential (Northern Sea Route development [6]) and social and economic development of the north territories as well. Massive water resources, unique biological marine resources are concentrated in the Arctic region; moreover their sustainability is higher than in the south tropical seas. Besides, it is par5ticularly important to develop a unique life-support and survival system in extreme conditions and to conserve the historical-cultural heritage of north nations.

However, there are risks and uncertainties impeding the Arctic development [7]. In particular, the shelf sea hasn't been thoroughly investigated in terms of geology and geophysics [8], another problem is restrained attitude of petroleum companies towards the region [9].

- **Effective interaction of states in the Arctic region**

Nowadays the most important issues are Russia's presence in the Arctic and its participation in strategic decision-making and problem solving on the arctic resource development.

Experts tend to agree that Russia is a strong geopolitical rival and the leading player in the Arctic region. Particularly, the experts from Council on Foreign Relations, CFR) in their report «The Emerging Arctic» [10] consider that it is not the climate change that is so important, but its consequences that lead to growing activity of the states interested in the region development. At the same time it is noticed that Arctic's development depends on Russia's influence first of all.

In general situation in the Arctic is stable and foreseeable. There is no strong evidence to assume that some arctic issues cannot be solved within the interaction of the states. All interested parties agree that region's countries are able to overcome disagreements appearing between them from time to time and provide the Arctic with strong and structural interaction. Organizing full-fledged and diversified collaboration between the arctic states is a guarantee of finding accurate response for new challenges and menaces [3].

The key points of the international collaboration in the region are identified in the new version of the Russian Federation foreign policy Concept, passed on February 12$^{th}$ 2013. "Russia provides an initiative and constitutive line directed at strengthening heteromorphic international cooperation in the Arctic. Implementing national interests successively, Russia proceeds from sufficiency of present international contract and the legal base for successful problem-solving in the region, including setting the out boundaries of the continental shelf in the Arctic Ocean by means of a dialogue". It is underlined Russia's readiness to mutually beneficial co-operation with non-arctic states "when they respect independence of sovereign right and jurisdiction of the arctic states in the Arctic region" [11].

International cooperation in the Arctic has positive trends; there is development of the following activity fields: research, ecology, technology, fishing industry, shipping and others (figure 1). The arctic states have clear agreements on respecting international law and solving problems by using the cooperation mechanisms. Arctic states community united by historical, cultural, economical and other factors implies some presence of specific interests that can be implemented on the basis of mutually beneficial co-operation. Nowadays collaboration is based on reciprocal conventions or mutual projects of the states (including cooperation with international organization).

Expanding activity in the region makes the arctic states solve complicated tasks. Primarily it is necessary for the countries to coordinate legislation on different aspects of the region's development.

International organizations play a special role in solving disputable issues (table 1). Arctic Council, Council of the Barents Euro-Arctic Region, EC and NATO are the most powerful organizations.

IOP Conf. Series: Earth and Environmental Science **21** (2014) 012046 doi:10.1088/1755-1315/21/1/012046

Interaction experience with these organizations has rather positive trends, despite the fact that their functions, enforcement powers, structure and activity character are heterogeneous, as well as states' attitude to these organizations is different. For example, longstanding cooperation of the arctic states in the framework of the Arctic Council let them define challenges and risks existing in the Arctic, and make effective and practical decisions to minimize the risks.

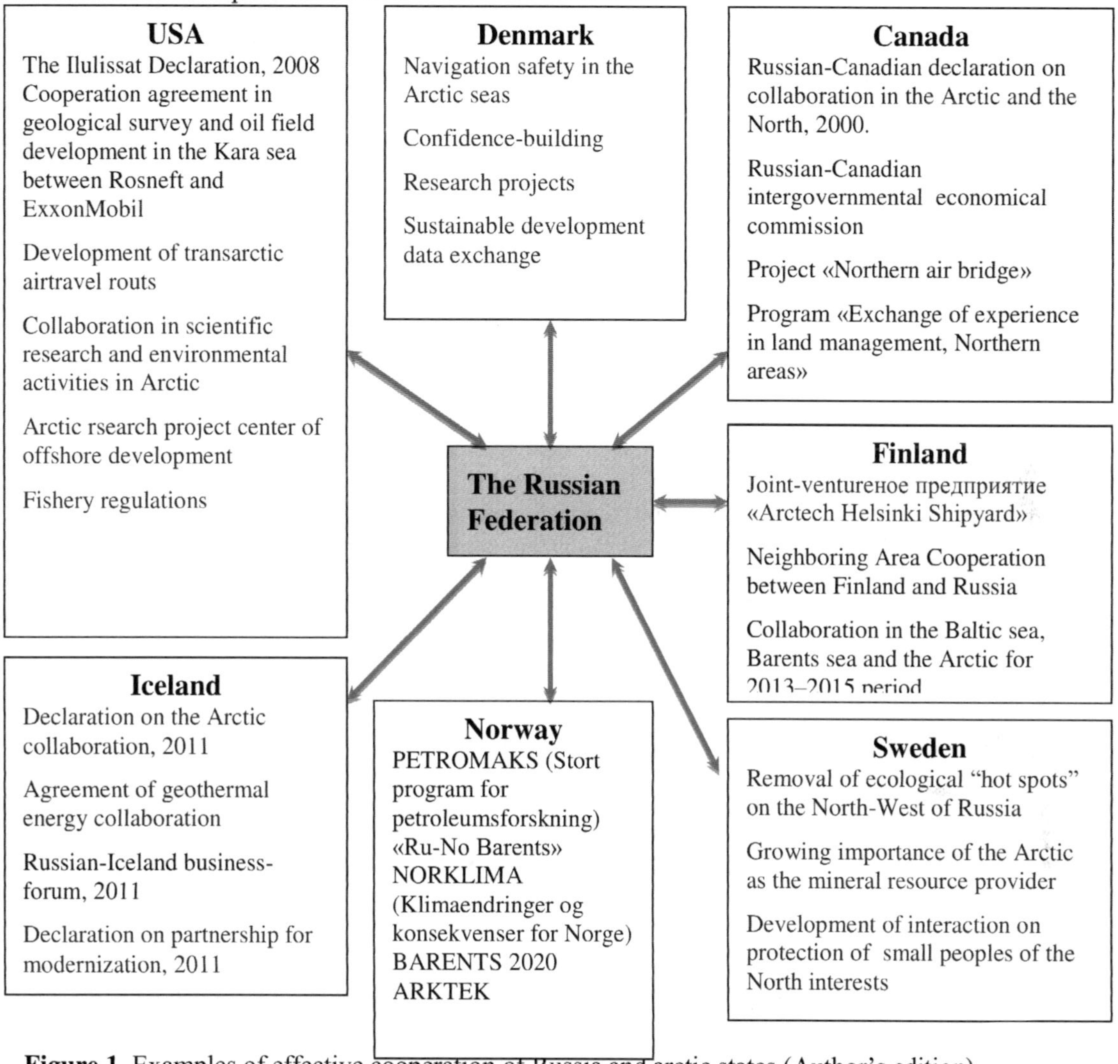


**Figure 1.** Examples of effective cooperation of Russia and arctic states (Author's edition).

**Table 1.** International organization participation of the states interested in the Arctic region.

| | «Arctic five» | UNCLOS | Arctic Council | EU | NATO | UNSC |
|---|---|---|---|---|---|---|
| Canada | + | 2003 | Member States | | + | |
| Norway | + | 1996 | Member States | + | + | |
| USA | + | Not ratified | Member States | | + | Permanent member |
| Russian Federation | + | 1997 | Member States | | | Permanent member |
| Denmark (Greenland and Faroe Islands) | + | 2004 | Member States | + Greenland and Faroe Islands are | + | |

| | | | | | | |
|---|---|---|---|---|---|---|
| | | | | not part of EU | | |
| Sweden | | 1996 | Member States | + | | |
| Finland | | 1996 | Member States | + | | |
| Iceland | | 1985 | Member States | | + | |
| France | | 1996 | Observer | + | + | Permanent member |
| Germany | | 1994 | Observer | + | + | |
| The Netherlands | | 1996 | Observer | + | + | |
| Poland | | 1998 | Observer | + | + | |
| Spain | | 1997 | Observer | + | + | |
| United Kingdom | | 1997 | Observer | + | + | Permanent member |
| People's Republic of China | | 1996 | Observer | | | Permanent member |
| Italy | | 1995 | Observer | + | + | |
| State of Japan | | 1996 | Observer | | | |
| Republic of Korea | | 1996 | Observer | | | Non-permanent member |
| Republic of Singapore | | 1994 | Observer | | | |
| Republic of India | | 1995 | Observer | | | |

The latest strengthening of international organizations' role let us hope that mutual understanding is going to be a key factor of cooperation between countries. International organizations' resources create potential possibilities for Russia to develop cooperation in the Arctic in the longer term. Nevertheless, it is necessary for Russia to stand for its' interests in the Arctic to the full extend.

- **Discussion**

Strategic analysis of the political risks of hydrocarbon deposit development in the Russian Federation Arctic seas shows that the region's sustainable development is not only limited to the protection of resource interests, but requires to solve global geopolitical issues within international process of globalization.

Political risk of disagreement on key arctic issues, uncertainty of the agreements, some disputable issues can have negative consequences for Russian business activities in the Arctic (figure 2). Such risk potential may impede the implementation of the projects concerning field development key problem-solving. International cooperation in the Arctic development is certainly unconditioned for Russia, though its implementation should not contradict Russia's national interests.

The other political risk for Russia is the necessity to protect its' EEZ and the continental shelf as well as to keep national control over the Northern Sea Route. Non-arctic states' initiatives to impose international form of arctic issues management and transform the Arctic Council into intergovernmental organization cause some worries. The "Arctic five" countries consider the present Arctic Council form optimal and international legal base to be sufficient to solve disputable arctic issues. Nevertheless the matters under high discussion, such as freedom of the Northern Sea Route navigation gives Russia ground for straightening its presence in the region, including on disputed territories.

- **Conclusions**

The analysis defines the most significant political risks of hydrocarbon deposit development in the Russian Federation arctic seas.

Considering growing economical and geopolitical interests of many countries in the Arctic, Russia needs to take expedient measures to strengthen its presence in the Arctic. First of all, it is necessary to start developing its legal shelf areas, which will make Russia the owner of the corresponding deposits. This will bring us to productive economical cooperation with other countries.

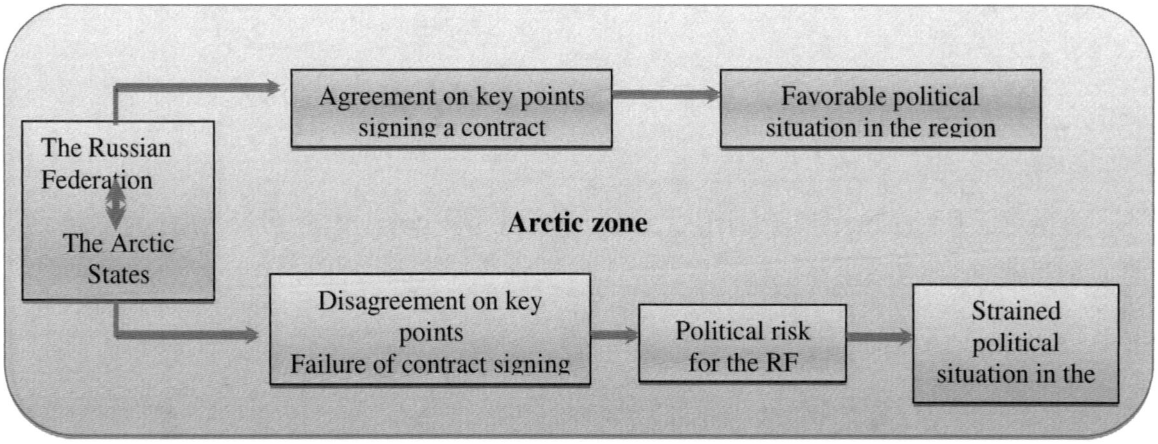


**Figure 2.** Possible options of states' cooperation in Arctic (Author's edition).

Secondly, it is necessary to start the dialogue between the interested parties to strengthen process of decision making in key arctic issues and to make as many agreements as possible. Development of the regional governmental system is of great importance as well. It is necessary to exclude gaps in legislation and to develop common approaches to solving the key arctic issues. Moreover, it is important for Russia to develop its ways of informational influence to position Russia on the world stage as an equitable arctic country.

- **References**

[1] Motyashov V P 2014 Gas and geopolitics: Chance for Russia (Moscow: LENAND) p 296
[2] Bolsunovskaya L M and Bolsunovskaya Y A 2012 Influence of risks on invest potential of Russian Federation arctic continental shelf *Bulletin of Tomsk Polytechnic University* **321**(6) 44
[3] Russia in the Arctic: Opportunities and Peculiarities of International Cooperation URL: http://russiancouncil.ru/en/inner/?id_4=2782#top
[4] Ivanov V L and Suprunenko V I 2013 Oil and gas bearing capacity of arctic shelf of Russia: From scientific forecast to industrial deposits *Gorny Zhurnal* **11** 35-40
[5] Lindholt L and Glomsrød S 2012 The Arctic: No big bonanza for the global petroleum industry *Energy Economics* **34** 5 1465-1474
[6] Bolsunovskaya Y and Boyarko G 2014 Otsenka perspektiv razvitiya Severnogo morskogo puti kak mezhdunarodnoy tranzitnoy magistrali *European Social Science Journal* **1** (4) 529-533 [in Russian]
[7] Bolsunovskaya and Bolsunovskaya L 2012 Vidy riskov pri otsenke investitsionnogo potentsiala arkticheskogo shelfa RF *Gorny Zhurnal* 4 special issue 87-91 [in Russian]
[8] Kaminskii V D, Suprunenko O I and Suslova V V 2011 The continental shelf of the Russian Arctic region: the state of the art in the study and exploration of oil and gas resources *Russian Geology and Geophysics* **52** (8) 760-767
[9] Ermida G 2014 Strategic decisions of international oil companies: Arctic versus other regions *Energy Strategy Reviews* **2** (3-4) 265-272
[10] The Emerging Arctic. URL: http://www.cfr.org/arctic/emerging-arctic/p32620#!/
[11] *Russian Federation foreign policy concept for February 12th* 2013 URL: http://www.mid.ru/brp_4.nsf/0/6D84DDEDEDBF7DA644257B160051BF7F

IOP Conf. Series: Earth and Environmental Science **21** (2014) 012047 doi:10.1088/1755-1315/21/1/012047

# Features of the marketing strategy of oil and gas companies in exploration drilling

**I Sharf [1,a], V Malanina[2,b] and L Kamynina[1,c]**
[1] Tomsk Polytechnic University, Tomsk, Russia
[2] Tomsk State Pedagogic University, Tomsk, Russia

E-mail: [a]irina_sharf@mail.ru, [b]milanskaya@mail.ru, [c] lidia.andreevna19@gmail.com

**Abstract.** The implementation of national and regional programs for the development of new oil and gas provinces of Eastern Siberia poses the challenge of increasing geological exploration. The current drilling service companies' market structure, as well as the strategic task of search and exploration effectiveness requires qualitatively new approaches for choosing a contractor. The proposed strategy to select a contractor based on comprehensive analysis of certain groups of industrial, financial, infrastructural criteria allows not only to optimize the costs of exploration activities, but also to minimize preventively the risks of a poor geological exploration. The authors' SWOT- analysis of the marketing strategy of "Gazprom neft" for choosing a contractor outlined the problem of imperfection of the Russian legislation in the sphere of activities of service companies in the oil and gas sector.

## 1. Introduction

According to the draft Energy Strategy of Russia until 2035, the Program of socio-economic development of the Far East and Trans-Baikal region and a number of regional programs, the intensification of exploration drilling is planned in Eastern Siberia in the Yurubcheno-Takhomskoye zone and Irkutsk region during 2013-2017. In Western Siberia, the largest amount of work may be conducted in new production fields in Yamal, particularly, in the Bolshekhetskaya depression. In the Sakha (Yakutia) Republic further detailed exploration of the Chayandinskoye field is planned to increase the available stocks and long-term filling of "Power of Siberia" pipeline. As a result, by 2020-2022 forecast all-Russian exploratory drilling stabilize at 1.12-1.18 million $m^3$ per year.

According to the development strategy "Gazprom neft" must produce by 2020 100 million tons of oil equivalent (TOE) per year, and proven oil reserves are to be increased up to 2.2 billion tons mainly due to the growth of investment in a geological exploration in new oil and gas provinces of Eastern Siberia. In 2014 the construction of 78 exploration wells with a total progress of up to 225 thousand linear meters is planned, and in 2015 32 billion rubles are to be invested in exploration.

Thus, the company faces the challenge of determining the strategy of choice of a service company according to the type of work - "drilling", which is based on a deep analysis of the exploration drilling market.

## 2. The problem of choosing a service company to conduct exploration drilling

The analysis on such criterion as a linear meterage of penetration shows that leading companies in the exploration drilling market are the intracorporate structures of Russian oil and gas companies, among which the share of OJSC "Surgutneftegas" accounts for 28%, OJSC "LUKOIL" for 23%, OJSC "Rosneft" for 25%. The high share (17%) belongs to the drilling company "Eurasia".

Choosing a contractor for service is difficult by a number of problems, such as:

- long-term payment for the completed orders by the customer;
- severe dumping by low-skilled service providers and small contractors, reasoned by trying to "stay afloat" in the current economic and political conditions;
- gap between Russian and foreign companies in terms of technical equipment, technology, implementation of scientific developments, financial opportunities and skills training;
- lack of experience by many contractors in mining and geological conditions in the new oil and gas provinces of Eastern Siberia;
- substandard study of the issues related to the drilling process, such as the selection of drilling fluid, delivery logistics of personnel and equipment to the place of work, procurement activities.

In addition, the task of "improving the reproductive efficiency of reserves", determined by the state [1, 2] actualizes the improvement of exploration efficiency in a rapidly changing macroeconomic environment, which is an additional parameter in the contractor selection.

The efficiency of prospecting is determined by the following characteristics [3, p.193].

1. Geological:

a) reserves growth in absolute and relative terms;

b) excess reserves to the production of hydrocarbons.

2. Economical:

a) expenses involved in hydrocarbon reserves growth per 1 m of deep drilling;

b) expenses involved in a ton of oil treating.

3. Informational: obtaining the additional information about the structure of the oil and gas-bearing area, deposits, amount of reserves and resources, etc.

## 3. The marketing strategy of "Gazprom neft" in choosing a service drilling company

The problem of selecting a contractor is provided for, in particular, by the implementation of the tender system, which encourages oil and gas companies, and especially small and medium-scale ones to choose the company that will offer the lowest service price that has repeatedly led to the delay in the service work realization and failures in drilling.

As a result "Gazprom neft" has launched a peculiar matrix of exploration. This matrix includes the following successive stages: prospecting, assessment, processing and production, each stage of which is characterized by its specific objectives and goals filled with special content and accompanied by different levels of integration of units of each block, and thus the split-level integration of subsidiaries.

One element of the implementation of this matrix is a marketing strategy for choosing a contractor to conduct drilling operations as the exploration drilling success depends not only on the well-site choosing, but also on a competent technical execution of the work.

Marketing strategy includes a list of systemized software criteria (table 1) to service companies, which are based on the requirements from the customer to the conducted work and on the provisions of normative legal acts of various legal force governing this activity.

According to this marketing strategy for choosing a contractor the work at the facilities of the customer and third-party sites is estimated, technical audit is carried out, human potential is verified. Special attention is paid to working conditions, climatic zoning and distance from the transport and industrial infrastructure. Significant role is played by the cost of well construction set by contractors. Broader criteria is presented in Table 1. As a result, a list of qualified contractors is made up.

**Table 1.** List of criteria for contractors.

| | |
|---|---|
| General information about the contractor organization | Location of head office<br>Average annual volume of work carried out for one half of the current year<br>Licenses and/ or other work permits<br>Company years<br>Work experience with entities within the group of companies OJSC "Gazprom neft"<br>Average annual volume of work carried out for one half of the current year at current prices<br>Affiliations with other oil companies |
| Experience | Experience in the oil and gas industry in Russia<br>Positive reviews from customers for the last 3 years<br>Number of wells, where oil, gas and water show or open well flowing occured for the last 3 years |
| Financial performance (data, ratio) | Nominal capital<br>Receivables and payables<br>Availability of current capital to ensure the production of works<br>Tax due |
| Logistics, commissioning, mobilization | Day-and-night dispatcher service<br>Incurrence of liabilities on transportation of all equipment, materials, chemicals<br>Organization to ensure continued production of well construction |
| Drilling equipment | Availability of drilling rigs (including mobile units) by types and hoisting capacity<br>Average period of rig operation<br>Facilities of the rig<br>Program of mud system, cementing, etc. |
| Providing industrial, environmental, fire and occupational safety | Availability of necessary permits to carry out the activities, as well as other permits, required by legislation in the field of industrial, fire, environmental safety and occupational health. |
| Staff | Total number of office and field drilling engineers eligible to manage mining operations<br>Share of engineers and geologists in personnel with experience in exploration drilling from 3 to 10 years<br>The availability of personnel with experience in formation tester.<br>Availability of personnel training programs<br>The number of available crews |

According to this strategy the work at the customer's sites and third-party sites is estimated, technical audit is carried out, human resources is verified. Special attention is paid to operating conditions, climatic zoning and remoteness from the transport and industrial infrastructure. The cost of well construction by contractors plays a significant role. As a result, a list of qualified contractors is specified.

However, the implementation of this program for the selection of companies is accompanied by a number of difficulties, including:

a) lack of sufficient competitiveness in the market of contractor organizations in operating regions where, in particular, out of 23 contractor organizations, claimed to participate in 2013, were recognized as qualified only 16 (70%);

b) underestimation of the planned cost of some wells, depending on the geological stress and geographical conditions, which led to significant changes in the final cost of the work.

As a result, for 2014 a new system of contracting was suggested by authors that allowed defining the following tasks within the developed strategy.

1. Expand the range of qualified contractors, and as a result in 2014 out of 25 contractors 19 (76%) were recognized as qualified. Positive dynamics is provided by involving contractor organizations conducting their business in the oil and gas provinces of European Russia.

2. Test the planned cost at the preparation stage of investment projects and formation of a business plan aimed at reducing the unit cost of construction of some wells by reducing the project depth and geological stress. In addition, the important point was to compare the average cost of one linear meter of penetration rate, which is affected by many factors specific to the East Siberian petroleum provinces (table 2):

a) "a variety of reservoirs;

b) the complex structure of the development of flow properties of the reservoir" [4];

c) industrial and infrastructural underdevelopment of the East Siberian region;

d) macroeconomic conditions (inflation, economic growth, exchange rates, etc.).

**Table 2.** Average cost of 1 linear meter of penetration rate in OJSC "Gazprom neft", thousand rub.

| Subsidiary organization | 2012 | 2013 | 2014 (plan) |
|---|---|---|---|
| Gasprom neft – Angara, Ltd. | N.A. | 148,6 | 173,05 |
| Gasprom neft – Orenburg | N.A. | 40,91 | 43,23 |
| Messoyahaneftegas, Jsc | 90,85 | 90,85 | 90,85 |
| Tomskneft VNK, Jsc | 41,32 | 41,32 | 41,32 |
| Gasprom neft- Vostok, Ltd. | 45,83 | 45,83 | 45,83 |
| Gasprom neft- Khantos, Ltd. | 38,11 | 38,11 | 38,11 |
| Gasprom neft- Noyabrskneftegas, Ltd. | 41,58 | 41,58 | 41,58 |
| Gasprom neft- Muravlenko, Ltd. (branch) | 43,76 | 43,76 | 43,76 |

3. One qualified contractor is eligible to receive no more than 4 wells in accordance with the criterion of the minimum price. This measure is aimed to maintain competitiveness in the market in the field of drilling.

4. Extend the contractual relations on the basis of the non-competitive selection, taking into account only the positive experience of the previous season.

Thus, this marketing strategy is the most appropriate in the present conditions, since its implementation will optimize the cost of exploration drilling and reduce the probability of risk.

## 4. SWOT-analysis of the marketing strategy of "Gazprom neft"

Taking into account the macroeconomic, macroenergetic, innovation, integration processes as in both the global and national economies, as well as in the oil and gas industry, the SWOT- analysis presented in table 3 was conducted by the authors [5, 6, 7]. This analysis allows not only to determine the further development of cooperation with contractors, which is especially important in the terms of exploring new license areas, but also to identify spheres that require a reconstruction.

**Table 3.** SWOT-analysis of "Gazprom neft" marketing strategy.

| Strengths | Weaknesses |
|---|---|
| 1)Faithfulness of the Customer - by holding full competitive selections in 2009- 2013<br>2) Provides the attractiveness of "Gazprom oil" as a customer.<br>3) The cost of well constructions led to the market level.<br>4) Competitive environment and monitoring of contractors' activities are performed, understanding of the possibilities provided by contractors.<br>5) A framework agreement has been developed and implemented, pricing model has been unified.<br>6) Ratings of contractors have been implemented.<br>7) Implementation of automated selection of contractors. | 1) Significant growth in the exploration drilling in all companies of the Russian Federation with the absence of new drilling contractors causes an increasing demand for drilling rigs and a significant increase in the cost of services.<br>2) Delay of project documentation and approvals from the Customer.<br>3) Inappropriateness of planned costs with a technical project in lots, underestimation of wells planned cost at a formation stage of the exploration program and business plan.<br>4) Poor equipment and undermodernization of Contractors.<br>5) Seasonal work performance. |
| **Opportunities** | **Threats** |
| 1) Additional economic benefits due to the proper formation of lots. To get a discount at the auction by creating of lots, it is necessary to apply the approach to the formation of lots on the basis: one lot – one well.<br>2) Formation of a strategy of commercial negotiations.<br>3) Expanding the range of qualified contractors at the stage of training.<br>4) Ability to attract extra contractors on a competitive selection.<br>5) Ability to preserve (maintain, keep) pricing of existing contracts by some contractors (counteragents). | 1) Unbalanced distribution of volumes and, as a consequence, the disruption of schedule.<br>2) Price increase due to the tightening (toughening) of a standard contract – implementation of the unified scale of penalties and rating of the contractors.<br>3) Changing the production program and the technical component.<br>4) Effect of climate conditions.<br>5) Logistic problems. |

## 5. Conclusion

Thus, a significant role in shaping the weaknesses of the marketing strategy plays the imperfection of Russian legislation, governing service organizations in many areas of oil and gas sector. In addition, organizational and economic measures should be taken in the field of technological and technical renovation of domestic service organizations. It is necessary to improve the legal regulation of relations between oil and gas companies and service companies. Thus first of all, the establishment of effective legislation can significantly improve the skills of contractors as well as the effectiveness of exploration.

## References

[1] Main provisions of the Russian draft energy strategy up to 2035 *Ministry of Energy*

[2] Grinyov V F and Varlamov A G 2014 Povysheniye effektivnosti burovykh rastvorov pri pomoshchi menedzhmenta kachestva *Bureniye i neft* **3** 30-34 [in Russian]

[3] Sharf I V 2012 Performance analysis of the oil companies on the example of Tomsk Oblast *The region's economy* 3 189-196

[4] Grigoriev G A and Borovinskikh A P 2013 Geologo-ekonomicheskaya kharakteristika kak osnova otsenki perspektiv razvitiya neftegazodobychi v Vostochnoy Sibiri *Neftegazovaya geologiya. Teoriya i praktika* **4** [in Russian]

[5] Vorobyov F 2014 Bolshoye budushcheye dlya maloy nefti? *Oil of Russia* **1-2** 20-24 [in Russian]

[6] Valentinov A 2013 Yest li otvet na novyye vyzovy? *Oil of Russia Oil Service* 4-9 [in Russian]

[7] Kotliarov I D 2012 Vnutrennyaya i vneshnyaya sreda firmy: utochneniye ponyatiy *Izvestiya vysshikh uchebnykh zavedeniy. Seriya: Ekonomika, finansy i upravleniye proizvodstvom* **1** 56-61 [in Russian]

# Development of well construction and workover supervising in Russian Federation

**A Sizov[1], G Boyarko[2] and I Shenderova[3]**
Tomsk Polytechnic University, Tomsk, Russia

E-mail: [1]Sizov_aleksey@rambler.ru, [2]Gub@tpu, [3]ruinna-shenderova@yandex.ru

**Abstract.** Despite long history of drilling supervising it still has a number of uncertainties. The period of rapid rise in supervising development at the beginning of the 90's changed in the 2000's. The necessity in the development of this sphere is obvious. The author describes the history of supervising, period of its market condition adaptation. The research also gives principles methods of supervising development and first steps for its position improvement.

## 1. Introduction

Annual spending of Russian petroleum enterprises associated with well construction and workover are on average from 30 to 50 %. But despite the enormous investments some of the constructed objects are still characterized as non-acceptable, as they require minor and capital repair already at the initial stage of operation. [1] Consistently, this leads to the increase in existing wells steady operation maintenance expenditures. Thus, improving the quality of site facilities construction is directly related to the future reserve balance and, as a consequence, results in increasing of enterprises profitability. The main proven tool to improve the quality of drilling at service market is supervising. Despite the long history of RF supervising services, it still needs the development, namely, in the increasing of its efficiency [2,3].

## 2. Research Method

Historical and logical method. The article studies the processes in chronological order starting with the period of supervising initiation and its step by step development in Russian Federation petroleum sphere.

Scientific abstraction method. The research is based on the destruction from the volume of petroleum engineering facilities during investigating period. Thus the dynamics of attraction and demand of supervising service in time is not taken into account. The authors consider supervising service of well construction and workover as a required one on Russian Federation petroleum service market.

System approach method. Supervising is studied from the point of its types. The research characterizes principle differences of supervising services presented on Russian Federation petroleum service market. The supervising service is researched in whole as a variety complex.

Induction and deduction method. Authors in their article give a number of facts revealing the true situation of petroleum supervising in Russian Federation. These facts can be considered as a ground for assuming the necessity of supervising sphere development.

Positive and normative analysis. Taking into account the concrete evidences, the authors present their viewpoint and give the objective estimation of well construction and workover supervising position on Russian Federation petroleum service market. Future of supervising service is attempted to be forecasted on the base of determinants influencing the development of the sphere.

## 3. Domestic Drilling Supervising Development

Need to improve the quality of well construction is a constant factor that led to the creation of drilling contractors market through reorganization of production enterprises. Since the early 90's, during the switchover to market economy, individual entities providing services in facilities construction for mining industry have been generating on Russian petroleum market. Growing competition forced drilling contractors to increase the quality of work that resulted in the considerable growth of collaborative engagement. Contract based cooperation developed relationships between entrepreneurs and stakeholders. In this case, the counterparties were focused on the quality of works done by the contractor.

Work organization experience in the conditions of early market relations was insufficient as well as theory and practice was not prepared for radical changes [4]. As the case may be, the need to improve the quality of oil and gas facilities construction accidentally formed the new market service – supervising. Thus, the main economic problem concerning the improvement of facilities construction quality as well as fundamentals of industrial activity organization was solved. From the perspective of market relations built on a contractual basis, supervising was focused on the effectiveness of customer-contractor system interaction.

In course of time supervising improved its position in the domestic petroleum service market, this is supported by a numbers of annual turnover - 18 billion US dollars [5], but in the Russian legislation there is still no such a concept. In accordance with the Civil Code of the Russian Federation, the customer is entitled to realize control and supervision over the progress and quality of work, compliance with execution deadlines and quality of materials provided by the contractor [6]. For realizing the above mentioned, the customer, without contractor's agreement, may enter into a contract with a definite engineer or specialized organization for provision of equal services. [7]

Industry standards also require the customer to ensure the construction of hazardous production facility to realize control in order to manage the compliance of works with project documentation requirements, technical regulations, results of engineering surveying and requirements of urban land development plan. [8] Lack of standard rules and regulations of legal service rendering for supervisor organization creates uncertainty. Thus, legislation predetermined formation of supervising in the form of technical and technological control, which involves such activities as supervision in monitoring and detecting of nonconformance to technical standards, industrial safety rules and other legal acts regulating the construction of facilities. Supervisor is collecting information about the construction and sends it to the customer in a timely manner. In this case the supervisor does not interfere in the construction works, his authority is limited, and all operational decisions are made by the customer. Basic obligations of the supervisor are clearly set in the agreement.

Starting from the 2000's through the conjuncture of fast developing market, supervising began to change and adapt to the growing needs of customers. Under these conditions a new kind of supervision services emerged – construction management. Apart from technical and technological construction supervision, these services include process control which is intended to improve the quality of work and to reduce cost. Thus, management implies control and estimation of volume and quality of contractor's work. In this case supervisor in addition to basic obligations of technical and technological control can provide the customer with the suggestions of how to improve organization and increase the efficiency of contractor's work, as well as to analyze the actual condition of the object, give suggestions and comments. The supervisor can also perform detailed analysis of working time and coordinate working schedule, at this the customer can make a claim on idleness. Representatives of the supervisor are involved in production meetings and working schedule coordinating. It should be noted that this form of supervising significantly reduces customer's burden, as part of the issues is managed without customer's participation.

Principal differences between the above mentioned forms of supervising can lead to confusion and misunderstanding during signing of Service Agreement. Often customers are trying to shift some of the obligations, not specified by the Agreement, to supervisors. [9] Sometimes, during tendering for the provision of services, supervision over the facilities construction in accordance with technical specification is noted in the customer's tender. Technical specifications set special tender for providing of services which correspond to the construction management. All participants of the economic process has their own visions of supervising services.

An Agreement should be made in terms of the tender documentation, so tender documentation provided by the customer is essential, namely, the draft of the Agreement. The terms of the Agreement after tendering procedure could be changed only in case of re-tendering, accompanied by a temporary loss of the resource. In tenders customer places his own requirements which can often be vague. Duties of the Agreement executives may not be clearly specified. It should also be noted that Russia has an extensive supervising market, but the modern trend, in contrast to the trends of the 90's, has led to the situation when customers are mostly guided now by the price, not the quality of rendered services. In this case, there can emerge such organizations which will agree with the tender terms regardless the value of the contract. The contractor can bear a risk of obligations fulfillment. Thus, market competition displaces supervisors that are not price competitive, but can provide services of better quality.

## 4. Extensive Development of Drilling Supervising in the Russian Federation

The experience obtained in technical and technological well construction, before shifting of the enterprises to market relations, positively reflected in the development of supervising in the early 90's. It is priori known that at the initial stage of supervising development, specialists in well drilling and construction, namely, leading engineers of that time fulfilled supervisor's obligations. At present, in the wage-rate book there is still no such position as "Supervising Engineer", but it is obvious that drilling supervising engineer is a specialist who has experience in drilling. So the reserve of knowledge, skills, and specialists in this field was enough for 15 years. However, even nowadays there is a shortage of professionals.

The situation is critical as in 5 years the current generation of supervisors will retire, but there must be someone who should replace them. The problem is – where to find specialists of next generation and if they be able to provide services without quality loss. The whole complex of factors and problems led to necessity of finding extensive methods of supervising development. Lack of trained specialists resulted in rapid development of large number of training centers, specialists' competence development programs and various supervising courses. In May 30$^{th}$, 2008, firstly in Russia at Gubkin Russian State University of Oil and Gas, drilling supervisors got the certificates in additional education training program "Specialist in Technological Supervision and Control of Wells Construction Process (Drilling Supervisor)". A specialist, who has just graduated from university, is sure not be able to meet all the requirements of the employer as well as to fulfill all the obligations. Method of self development during continuous learning process is the most effective despite the fact that this is costly and risky. It should be also noted that the company would rather invite a professional than spend money and time for young specialists' training.

## 5. Intensive Development of Drilling Supervising in the Russian Federation

At the initial stage of supervising development, before the 2000's, attracting of the external service was profitable for the customer from the economic point of view. Fines and penalties charged to the contractor for violations identified by supervisors covered the losses and justified the risks associated with construction. The general tendency has led to the fact that the system supervisor - contractor started to be intended for the cutting of losses and risk prevention both from supervisor and the contractor. Supervisors timely reacted to possible deviations and irregularities in construction technology facilities, and the contractor coordinated all his works with the supervisor as a legal representative of the customer under the terms and conditions of the Agreement. Ultimately, the supervising organization got profit and contractor performed his obligations set by the Agreement.

Since 2003 the situation has stabilized, the amount of fines reduced and customers has no longer seen the supervising as an object of investments, which in future can cover the costs of engaging a third-party service through penalties to the contractor or the service organization, in accordance with the terms and conditions of the Agreement. Thus, in the course of time, the common viewpoint to the attracting of the supervisor for facilities construction has changed. Supervising started to be considered as an obligatory spending aimed at control of the contractor's actions. Even the contractors refused to undertake working responsibilities without a legal representative of the customer in the face of the supervisor. At the same time, companies were always trying to optimize works and reduce their costs. This fact led to the development of supervising. Overall dynamics led to increasing in supervisors' responsibility and gross in the number of objects under supervision. As a rule, to organize supervising services, Contractor provides Customer with engineers based on the scheme including two interchangeable experts per one object [10].

Well construction process is a continuous process, therefore, supervisor must be constantly present at the facility realizing his obligations, but supervisor cannot be present in two or more places at once. Construction facilities have different geographical location, often at a considerable distance from each other. Such a process of working organization could certainly reduce service quality. Supervisor is certain to be guided by the priority, complexity and risk of technological deviations when planning supervision of a definite object. It is preferable if for example conductor drilling is realized at one object it means that constant presence of supervisor is not urgently required, so supervisor can shift his attention to more complex object where round-trip operations are performed. But processes can be of casual character and here arises a need to look for more intensive methods of supervising development, exactly - methods to improve efficiency of supervising. One of such methods is the implementing of engineering construction management system (hereinafter ECMS) that allows remote monitoring of key parameters of production process in real time. ECMS does not exclude the presence of a field supervisor at the facility, on the contrary, the system aims at the logistics of field service optimization and data gathering.

Under the conditions of the large number of work execution on different objects, senior supervisor must constantly make decisions on the appropriate allocation of field supervisors on the objects. To organize the logistics of field supervisors, senior supervisor must have working schedule and operation data of technological parameters for each object. Key process parameters at well workover such as the weight on the hook, weight on the bit, wrench torque moment, manifold pressure, liquid level in tanks allows for measurement of weight electronic indicator "GIV-1-E" (producer - ZAO "Prompribor", Ekaterinburg) by exchanging information with the computer or other device. Such remote monitoring allows senior supervisor to see the current process at the object in real time and forecast the possible working aspects for the nearest future, which, in its turn allows the prediction of critical or hazardous industrial operations requiring the presence of the field supervisor.

At well construction the presence of supervisor is essential, so the logistics aspect in this context is not relevant. In this case, supervisor has a task to optimize the process. It is mostly seen during construction of wells at great depths (over 4,000 m) in complex geological environment, with complex technological schemes, where customers are ready to cooperate with design institutions concerning changes in the design documentation in order to reduce construction costs [11]. For drilling process monitoring it is recommended to implement a system of drilling parameters control such as "IS MSRV" (Producer - ZOA "AMT" St. Petersburg), "TM-KUB» (Producer OOO "Tomskneftegazinzheniring" Tomsk), Rigsense (Producer MD Totco / National Oilwell Varco, USA). This system monitors the following drilling parameters: the weight on the hook, winch speed, the drilling mud pressure in manifold, the number of pump strokes, rotor speed, wrench torque moment, mud level in each tank, drilling mud parameters, and concentration of combustible gases. Through remote monitoring data can be transferred to the supervisor, who can change the drilling process and predict accidents basing on the data obtained,.

In addition to monitoring of the production parameters there exist an opportunity to install devices for recording manufacturing operations at the facility. Supervisor can not only perform the control over the machines and sensors but also manage the workers.

Besides, ECMS indirectly completes the task in gathering information that helps the supervisor in regular reporting. At rational organization of information flows of Customer and Supervisor internal divisions, complete and timely provision of all geological and technical information can be achieved. Information flows should be based on the source of communication control system parameters, the means of communication and means of information processing - application software on the results of which, strategic decisions are taken. So, operational control can be performed by all involved production units. Results of this control are: cost reduction and cost optimization for facility construction, time reduction for construction, longer service life of materials and equipment, intangible benefits implies productive cooperation with project organizations on a number of issues.

## 6. Conclusion

Supervising as a sphere of activity emerged spontaneously due to the market demand in the effective tool to improve the quality of drilling. In this case supervising should be enshrined in law. Issuing of normative legal documents will help to eliminate the uncertainty and create a common vision of providing Supervision services. Supervising period is characterized by its extensive growth in the 1990's, when supervising was considered as an investment project, where customer can benefit from penalties and claims assessed on supervisor or contractor and the stage of stagnation in the 2000's. The last stage of supervising is characterized by losses, personnel turnover, the growing of service companies' responsibility. So the main extensive method of supervising development under these circumstances can be regarded as a process of teaching and developing of expert staff in terms of best practices exchange with professionals. To the intensive methods the development of engineering constructions monitoring systems can be also referred. The experience obtained in the 1990's allows supervising to exist, but not to develop. So, only the combination of methods applied can give a start for the future of supervising.

## References

[1] Shulyev Y V, Martynov V G, Kulchitskly V V, Sazonov A A, Larionov A S qand Grishin D V 2010 Innovative and educational technologies of drilling supervising *Oil Industry* 3 10-13
[2] Korotchenko A, Zimenkov S, Fedulov N and Pizamov 2013 I Drilling supervising serves in Russia *Drilling & oil* 5 24-26
[3] Korol' A A 2006 Supervising – method of well construction quality improvement *Reporter of the Association of Drilling Contractors* 2 14-16
[4] Belousov V O and Povalikhin A S 2007 Drilling supervising *Petroleum engineer* 2 9511
[5] Moiseeva E 2011 Russian petroleum market *Petroleum serves* **4**(16) 11-13
[6] Article 748 Civil Code of the Russian Federation
[7] Article 749 Civil Code of the Russian Federation
[8] P. 387 Federal norms and rules of industrial safety "Safety rules in petroleum industry" (Resolution №101, 12th march 2013 Federal serves of ecological, technological and nuclear control)
[9] Nesterova T N 2009 Domestic drilling supervising – challenges and experience Part 2 *Drilling & oil*. **6** 62-64
[10] Nesterova T N 2009 Domestic drilling supervising – challenges and experience Part 3 *Drilling & oil*. **9** 60-63
[11] Akhmadullin E A, Kalinin S P and Rozenberger Ye B 2013 Electronic supervising in well construction and workover *Onshore and offshore well construction* **2** 14-18

# Industrial management of peat sorbent production in Tomsk region

**D S Rozhkova[a], I A Hadkevich[b], T S Glyzina[c] and E G Matugina [d]**
Tomsk Polytechnic University, Tomsk, Russia

E-mail: [a]rozhkovads@yandex.ru, [b]hadkevi4@mail.ru, [c]inosine@yandex.ru, [d]emk512542@mail.ru

**Abstract.** The intensification of production development causes the growth of energy consumption and increases the risk of oil leak. This involves the implementation of measures for the effective breakdown elimination. In this case the best method meeting the environmental requirements is the use of appropriate sorbent. The economic research of using the peat sorbent by oil and gas extraction companies in Tomsk region was carried out.

## Introduction

The intensification of production and consumption of goods in modern society has a direct impact on the structure of national economy and the system of economic relations. The orientation on the use of traditional energy resources forms the relationship between the growing demand and the use of energy resources. It is accompanied by the increase of technogenic environmental impact. Not only the enterprises using energy resources sources are the reason of negative impacts are, but also mining enterprises, processing and transporting. In particular, the matter is in the accidental leak of petroleum products. According to the Greenpeace, in Russia oil stock losses during its production and transportation are about 1%; NPO "Centre of Ecology FEC" indicates the amount of 3.5 ... 4.5%. Accordingly, at the current production level of 510 million tons per year, losses amount from 18 to 23 million tons or from 14.2 billion to 17.2 billion dollars. Taking into account that in 2013 the volume of oil production exceeded the forecasted value and amounted 523.2 million tons [1], the losses are of greater amount. However, this is only the direct damage, which should be supplemented by the expenditures connected with the delay of other production operations, the environmental disruption, and the elimination of accident consequences in the current (proper oil spill response) and prospective (compensative measures) periods.

Thus, a prompt emergency response is one of the ways to minimize the damage. There are a lot of sorbents, used in case of oil spill, which are diverse in composition (containing organic, inorganic, synthetic and other components). Depending on the type of feedstock, sorbents can be classified into inorganic (natural or artificial minerals) and organic (caustobiolith, synthetical, organic mineral, natural raw materials of plant and animal origin and recycling waste). But in the aspect of the ecological safety, sorbents of natural origin are the most attractive – sawdust, hay, coal and peat. The disadvantages of synthetic sorbents are - environmental threat, complexity of recycling, and high cost. Despite the hydrophobic properties, they have low selectivity and can absorb oil and water with the same intensity.

IOP Conf. Series: Earth and Environmental Science **21** (2014) 012049 doi:10.1088/1755-1315/21/1/012049

The use of sorbents in a dispersed form (rubber crumb, phenol-formaldehyde powder, polystyrene foam granules) can lead to a secondary pollution by these sorbents, due to poor decomposition in natural conditions (due to the possibility of arbitrary scattering by the wind outside the booms) [2, 3]. The extra arguments for using sorbents of natural origin are their availability in sufficient quantity, renewability, quick and cheap production (in comparison with synthetic sorbent production), and etc. Thus, in Russia peat reserves (68 billion tons) exceed of oil and gas reserves (respectively 31 billion tons and 22 billion tons). Moreover, according to the results of researches 1.5 billion $m^3$ of raw material grow in peat swamps every year.

**Theoretical Bases**

The problem under study is essential for Tomsk region, as oil and gas complex is one of the significant sectors of the local economy (about 30% of GRP is the share of mineral resource extraction (for the year 2011), being at the same time a source of technogenic risk (in terms of accidental leaks in the main oil and gas pipelines and production pipelines). Tomsk region is a unique area due to combination of significant reserves of natural energy resources (by 2030 it is planned to increase the annual oil production up to 20 million tons including the tight oil production up to 5 million tons) as well as a natural sorbent – peat. 1444 of peat deposits are discovered on this territory with total area of 7988209 hectares in commercial value with the ultimate reserves and predicted resources of 30881.4 million tons of peat per 40% of humidity, this accounts 18.07% from the total reserves and predicted resources of peat in Russia. Tomsk region occupies the second place among Russian regions, first is for Tyumen region [4, 5].

The location of peat deposits in the territory of Tomsk region is shown on figure 1.

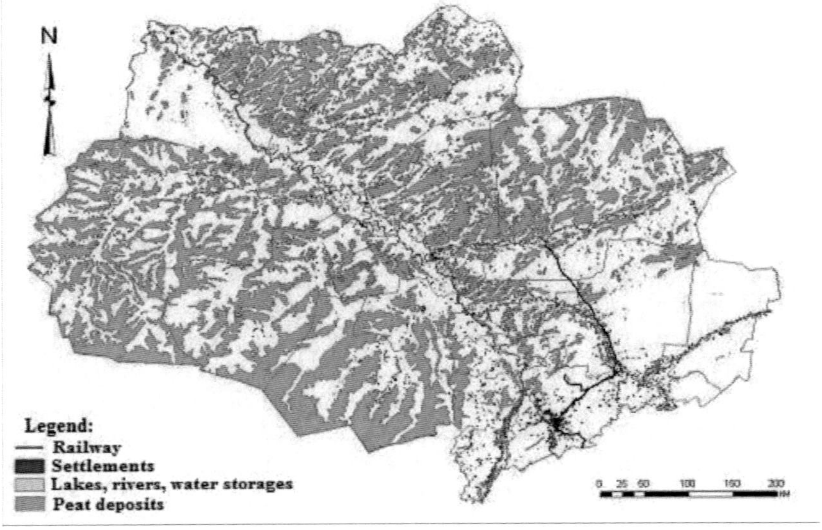


**Figure 1**. Peat deposits in Tomsk region.

**Results and discussions.**

Technical characteristics. According to the intended use of peat as a sorbent, its main characteristic lies in the oil capacity indicator, defined as the amount of oil, absorbed by a unit of sorbent mass. The experiments on sorbent oil capacity of peats with different degrees of a decomposition showed the variability of researched index within 2 ... 8 g/g, which is comparable with the Canadian peat sorbent SpillSorb and Russian synthetic sorbent EcoSorb. Moreover, the peat sorbent allows eliminating completely the accidental oil spill on the ground. If the spills occur on water surfaces, peat is the most effective help on the last stage, as it destroys the oil slick. Comparing oil capacity of Tomsk and Canadian sorbent SpillSorb (comparable as feedstock), widely used by oil companies of Tomsk region, it is worth taking into consideration their identity which conditions the equal consumption rate in oil spill response (table 1).

**Table 1**

**Forecast of the oil spill volume**

| Coefficient (Ratio) | **Breakthrough at the rate of 25% of maximum volume of pumping within 6 hours and the volume of oil between block valves on the damaged section of the pipeline.** | **Perforation at the rate of 2 % of maximum volume of pumping within 14 days** |
|---|---|---|
| Volume of the spilled oil, t | 1287,7 | 4200 |
| Weight of Tomsk Sorbent, t | 161 | 525 |
| Weight of Canadian Sorbent, t | 161 | 525 |
| Consumption rate of a sorbent per 1t of spilled oil, t: | | |
| Tomsk sorbent | 0,125 | |
| Canadian sorbent | 0,125 | |
| Ecosorbo | 0,083 | |

Economic characteristics. It is necessary to compare the sorbent consumption rate for collecting 1 t of oil (table 2).

**Table 2**

**Oil sorbent characteristics comparison**

| **Sorbent** | **Type of Sorbent** | **Oil capicity, g/g** | **Cost, rub/Kg** | **Sorbent** consumption **rate per 1 t of oil, rub** |
|---|---|---|---|---|
| SpillSorb | Peat-sorbent | 4-8 | 284,60 | 35 575 |
| Peat of Tomsk region | Peat-sorbent | 2-8 | 100 | 12 500 |
| Ecosorb | Synthetic sorbent(polyprpylene) | 6-12 | 350 | 29 050 |

* The cost of sorbent is given without transportation costs [6, 7]

When using the peat sorbent of Canadian origin, it is necessary to note significant transport costs, directly affected the expenditures on the oil spill response. Almost a triple excess of the cost of the imported sorbent over the Tomsk analogue (without taking into account the delivery costs) is a strong argument in favor of using local raw materials. It is possible to use a peat sorbent of any deposit, located near the main pipeline. The functioning of the given production will be connected with additional job vacancies (social effect), the development of the territory, and the increased tax payments into the budget.

Despite a wide extension at the oil spill response, sorbents should be appropriately used in order to prevent their inexpedient and excessive expenditure, which creates difficulties in terms of logistics, secondary environmental pollution, gathering, storage and waste disposal. Furthermore, the use of sorbents involves forming an additional infrastructure. The use of peat sorbent implies fire safety measures, careful storage sites planning, and etc.

Thus, the use of sorbent of bog peat origin of Tomsk region is appropriate as having familiar characteristics with the materials, applied now by the enterprises of the region. The competitiveness of a peat sorbent is caused by such characteristics as the geographical proximity of the raw material reserves and consumer companies of a product, minimizing costs of transportation; the low cost of sorbent, (excluding delivery costs), identical rates of oil capacity, and etc.

**References**

[1] RIA, URL: http://top.rbc.ru/economics/10/04/2012/645532.shtml
[2] Zapivalov N P 2011 Peat resources of Siberia – untouched wealth underfoot *ECO* **8** 104-113
[3] Federal State Statistics Service in Tomsk region // URL: http://tmsk.gks.ru
[4] Bernatonis V K and Bernatonis P V 2010 The concept of the development of peat resources in the current economic conditions (on the example of Tomsk Region) *Bulletin of TPU* **1** 161-163 URL: http://izvestiya.tpu.ru/ru/archive/old/issue.html?id=194437&journalId=176237
[5] Chuhareva N V, Rozhkov D S and I A Hadkevich 2012 Study of oil capacity of peat in relation to crude oil and gas condensate *Research Journal of International Studies* **5**(5) 132-134
[6] NPF Ecosorb. URL: http://www.ecosorb.su/catalog/sorbent/sorbent-voloknistyij-ekosorb.html
[7] Delivery of environmental goods in Russia. URL: http://www.ecosorbent.ru/price

# Analysis of Russian Federation budget revenues from the implementation of export duties "60-66-90" scheme

**L Grinkevich[1,a], I Sharf[2,b] and G Gerasimovich[1,c]**
[1] Tomsk State University, Tomsk, Russia
[2] Tomsk Polytechnic University, Tomsk, Russia

E-mail: [a]nasty_saturn@mail.ru, [b]irina_sharf@mail.ru, [c]qweezze@mail.ru

**Abstract**. Innovative development of the national economy provides qualitative transformation of the refining sector. In order to promote the depth of oil refining and the relative increase the refining share in oil extraction the Government of Russian Federation has changed the mechanism for calculating export duty on crude oil, light and dark oil. This research work addresses the nature of financial implications for the budget system in connection with these changes. Different effects are observed for oil extractors and oil refiners. Conclusions are reasoned by calculations.

## 1.Introduction

Government Resolution № 716 of August 26, 2011 [1] established a new scheme of export duties collection, which, in accordance with the rates of export duties on light and dark oil products, calculated from the rate of export duty on oil, was called "60-66-90" (oil, oil products, petrol).

The key provisions of this mechanism are the following:

– changing the coefficient reflecting marginal rate of export duty, from 0.65 to 0.60 (table 1) in the calculation formula for the oil export duty rate;

– unifying light and dark oil products duties at the rate of 0.66 of crude oil duties;

– fixing the export duty on petrol at the rate of 0.9 of oil export duty.

The table below reflects the dynamics of change in formulas for oil export duty calculation in the period from 2002 to 2013.

**Table 1.** Formulas for oil export duties calculating [2, 3, 4].

| World price of "Urals" oil | | 2002 – 31.07.2004 | 1.08.2004 – 30.09.2011 | 1.10.2011 – 31.12.2013 |
|---|---|---|---|---|
| Dollar/t | Dollar/barrel | Oil export duties, dollar/t | | |
| Less than 109.5 | Less than 15 | 0 | 0 | 0 |
| 109.5 to 146 | 15 to 20 | 0.35x(P–109.5) | 0.35x(P–109.5) | 0.35x(P–109.5) |
| 146 to 182.5 | 20 to 25 | | 12.78 + 0.45x(P–146) | 12.78 + 0.45x(P–146) |
| More than 182.5 | More than 25 | 25.53 + 0.4x(P-182.5) | 29.2 + 0.65x(P–182.5) | 29.2 + 0.60x(P–182.5) |

P – world price of "Urals" oil, dollar/t

Note that the period from 1999 to 2013 can be divided into several stages of state customs policy implementation which are described below.

1. 1999 to 2003. - differentiation of oil and oil products export duties, which included the establishment of fuel oil duty at the rate of 50 %, and light oil duty at the rate of 80-120 % of oil duty rate depending on the brand.

2. 2003-2005. - alignment of oil and oil products export duties at the rate of 90%.

3. 2005-2011. - differentiation of oil and oil products export duties, which included the establishment of heavy oil duty at a rate of 40 %, and light oil duty at the rate of 72% of the crude oil duty rate.

4. 2011 - 2013. - export duties calculated according to the scheme "60-66-90" as described above.

The transition to the fourth stage was due to the fact that no significant positive changes in refinery sector were observed which could be explained as follows:

a) quantitative increase in oil extraction and refining was not accompanied by an increase in relative indicators, in particular, the share of oil refining in its extraction has remained almost at the same level: 53.6 % in 2000 and 53.1% in 2013;
b) the depth of oil refining has not changed: 71 % in 2000 and 71.4 % in 2013. It is important to note that the depth of oil refining in advanced economies reaches 90-95 % due primarily to various methods of oil recycling, while in Russia primary refining based on direct distillation of crude oil (catalytic cracking) dominates [5].

Thus, the main goal of innovations in 2011 was a complex influence on the extracting and refining sectors by promoting:

a) extraction of hydrocarbons on the fields with falling production rates, primarily in Western Siberia - the main national oil-extracting region;
b) the growth of oil recycling, and consequently, enhancing the modernization of refineries [6, 7, 8 ].

In addition, changes in the state export policy were reasoned by the following socio-economic factors:

1. Growth in production and exports of gas oil and fuel oil (end-products of primary oil refining). At the same time periodic petrol deficit in some country's regions occurred as a result of simultaneous decrease in production caused by the closure of several plants for modernization and increase in petrol exports. Note that oil, vacuum gas oil and straight-run petrol are used primarily as raw materials for foreign refineries;

2. Decline in the share of mini refineries, which focused only on primary processing;

3. Compliance with the terms of the World Trade Organization (WTO), which limits the rate of oil products export duties at the rate of 90 % of maximum oil duties;

4. Maintenance of high average refining margins (the difference between the cost of oil barrel and a basket of oil products produced from it), which made the increasing of the refining depth economically unfeasible;

5. Inconsistency of motor fuels to European standards, primarily on environmental safety.

## 2. Calculation of the financial implications for the Russian Federation budget system from the implementation of export duties "60-66-90" scheme

Certainly, the introduction of incentives entails certain positive or negative financial consequences for the country's budget system, and the main objective of this research is to determine them. The authors examined financial implications for the Russian Federation budget system from the implementation of export duties "60-66-90" scheme.

Calculations were carried out taking into account the fact that the export and export duty rates are stable (table 2, figure 1).

**Table 2.** Export duty rates for light and dark oil products in the Russian Federation, 2010-2013 [1, 4].

| | 2010 to 30.09.2011 | 1.10.2011 - present |
|---|---|---|
| Light oil products | 0.67 x $D_{c.o.}$ | 0.66 x $D_{c.o.}$ |
| Dark oil products | 0.467 x $D_{c.o.}$ | 0.66 x $D_{c.o.}$ |

$D_{c.o.}$ – export duties for crude oil.

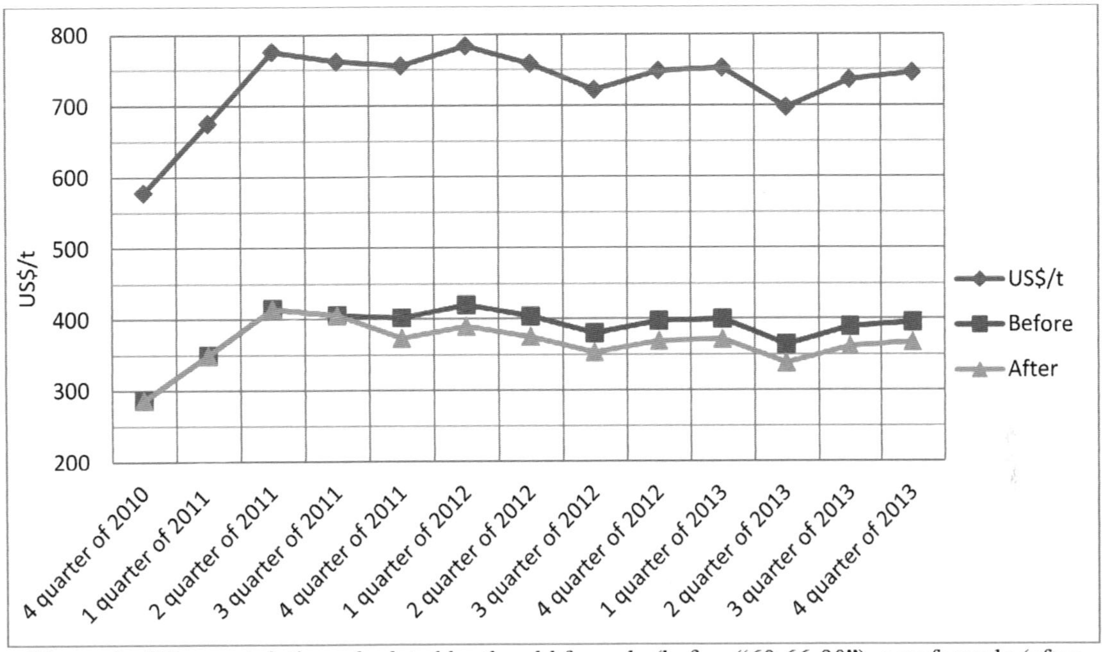


**Figure 1.** Oil export duties calculated by the old formula (before "60-66-90"), new formula (after "60-66-90") and oil prices (Table 2) [9].

Having calculated the difference in revenues from oil exports, obtained after the implementation of levying export duties "60-66-90" scheme, and the revenues that would have gone to the state budget without the implementation of new calculating mechanism, following results were received. Budget revenues declined by a substantial amount – 13359.8 million dollars, that is the estimated amount of losses for the country's budget system from reforming customs policy during 2012-2013 (table 3).

**Table 3.** Calculations of the Russian Federation budget revenues from crude oil export duties.

| | Oil price | | Oil export duties, US$/tn | | | Export | Revenues from oil duties, million US$ | | |
|---|---|---|---|---|---|---|---|---|---|
| | US$/bbl | US$/tn | Before | After | Δ | million tn | Before | After | Δ |
| 4 quarter of 2010 | 79.14 | 577.72 | 286.09 | 286.09 | 0.00 | 63.70 | 18224.21 | 18224.21 | 0.00 |
| 1 quarter of 2011 | 92.48 | 675.10 | 349.39 | 349.39 | 0.00 | 59.10 | 20649.10 | 20649.10 | 0.00 |
| 2 quarter of 2011 | 106.23 | 775.48 | 414.64 | 414.64 | 0.00 | 62.60 | 25956.24 | 25956.24 | 0.00 |
| 3 quarter of 2011 | 104.33 | 761.61 | 405.62 | 405.62 | 0.00 | 57.50 | 23323.20 | 23323.20 | 0.00 |
| 4 quarter of 2011 | 103.54 | 755.84 | 401.87 | 373.21 | -28.67 | 65.30 | 26242.26 | 24370.30 | -1871.96 |

| | | | | | | | | | |
|---|---|---|---|---|---|---|---|---|---|
| 1 quarter of 2012 | 107.3 | 783.29 | 419.71 | 389.67 | -30.04 | 58.80 | 24679.15 | 22912.83 | -1766.32 |
| 2 quarter of 2012 | 103,92 | 758.62 | 403.68 | 374.87 | -28.81 | 59.90 | 24180.16 | 22454.69 | -1725.47 |
| 3 quarter of 2012 | 98.88 | 721.82 | 379.76 | 352,. 9 | -26.97 | 59.10 | 22443.85 | 20850.15 | -159.70 |
| 4 quarter of 2012 | 102.53 | 748.47 | 397.08 | 368.78 | -28.30 | 62.20 | 24698.37 | 22938.20 | -1760.16 |
| 1 quarter of 2013 | 103.08 | 752.48 | 399.69 | 371.19 | -28.50 | 57,40 | 22942.18 | 21306.33 | -1635.85 |
| 2 quarter of 2013 | 95.5 | 697.15 | 363.72 | 337.99 | -25.73 | 58,40 | 21241/39 | 19738.62 | -1502.78 |
| 3 quarter of 2013 | 100.83 | 736.06 | 389.01 | 361.34 | -27.68 | 59,70 | 23224.10 | 21571.72 | -1652.37 |
| 4 quarter of 2013 | 102.14 | 745.62 | 395.23 | 367.07 | -28.16 | 61,20 | 24188.03 | 22464.88 | -1723.15 |
| Total in 2012 | | | | | | | 96001.53 | 89155.87 | -6845.66 |
| Total in 2013 | | | | | | | 91595.71 | 85081.55 | -6514.16 |
| Total 2012-2013 | | | | | | | 187597.24 | 174237.42 | -13359.8 |

Calculations of Russian Federation budget revenues from the change of light oil (gasoline, diesel) and heavy oil (fuel oil) products export duties are presented in table 4.

**Table 4.** Calculations of oil products export duties and total export revenues in Russian Federation budget, 2012-2013.

| Year | | 2012 | 2013 | Total 2012-2013 |
|---|---|---|---|---|
| Oil price, dollar/tn | | 753.05 | 732.8 | |
| | Before | 400.06 | 386.9 | |
| Oil export duties, dollar/tn | After | 371.53 | 359.4 | |
| | Δ | -28.53 | -27.50 | |
| | Petrol | 3.20 | 4.33 | |
| Export, million tn | Diesel | 36.74 | 42.1 | |
| | Fuel oil | 44.70 | 64.4 | |
| | Petrol before | 360.05 | 348.21 | |
| | Petrol after | 334,38 | 323.46 | |
| | Δ Petrol | -25.67 | -24.75 | |
| | Diesel before | 268.04 | 259.22 | |
| Oil products export duties, dollar/tn | Diesel after | 245.21 | 237.20 | |
| | Δ Diesel | -22.83 | -22.02 | |
| | Fuel oil before | 186.83 | 180.68 | |
| | Fuel oil after | 245.21 | 237.20 | |
| | Δ Fuel oil | 58.38 | 56.52 | |
| | Petrol before | 1152.17 | 1507.75 | |
| | Petrol after | 1070.01 | 1400.58 | |
| | Δ Petrol | -82.16 | -107.17 | -189.33 |
| Revenues from oil products export duties, million dollars | Diesel before | 9847.73 | 10913.29 | |
| | Diesel after | 9009.00 | 9986.29 | |
| | Δ Diesel | -838.73 | -927.00 | - 1765.73 |

| | | | |
|---|---|---|---|
| Fuel oil before | 8351.16 | 11635.94 | |
| Fuel oil after | 10960.87 | 15275.94 | |
| Δ Fuel oil | 2609.72 | 3640.00 | 6249.71 |
| Total revenues from oil products export duties, million dollars | 1688.83 | 2605.83 | 4294.66 |
| Total revenues, million dollars | -5156.83 | -3908.83 | -9065.15 |

The results presented in Table 4 demonstrate that the positive effect for the Russian budget system from the new system of oil products export duties calculation was obtained only for dark oil products, namely oil. Budget revenues estimates 6249.71 million dollars.

Analysis of complex effect, which is calculated by summing up the results obtained for the export of crude oil, dark, light oil and petrol, shows a negative value in the amount of 9065.15 million dollars. These losses of federal budget occurred in 2012-2013 from the implementation of export duties "60-66-90" scheme, i.e. increase in revenues of 6249.71 million dollars from selling of dark oil does not cover losses in the amount of 13359.8 million dollars by reducing the crude oil export duty.

## 3. Conclusion

Thus, further study is required for financial and tax incentives, stimulating the effective development of fuel and energy complex of oil refining segment, in order to find a compromise regulation of relationship between business and the state, which is especially important in terms of innovative development of the national economy.

## References

[1] Russian Federation Government Resolution № 716 of August 26, 2011 " O vnesenii izmeneniy v postanovleniye Pravitelstva Rossiyskoy Federatsii ot 27 dekabrya 2010 g. N 1155" [in Russian]

[2] Federal Law № 409-FL " O tamozhennom tarife " (as amended on December 6, 2011) [in Russian]

[3] Russian Federation Government Resolution № 276 of March 29, 2013 " O raschete stavok vyvoznykh tamozhennykh poshlin na neft syruyu i otdelnyye kategorii tovarov, vyrabotannykh iz nefti, i priznanii utrativshimi silu nekotorykh resheniy Pravitelstva Rossiyskoy Federatsii " [in Russian]

[4] Russian Federation Government Resolution № 1155 of December 27, 2010 " O raschete stavok vyvoznykh tamozhennykh poshlin na otdelnyye kategorii tovarov, vyrabotannykh iz nefti " [in Russian]

[5] Mastepanov A M 2013 O nekotorykh problemakh, stoyashchikh pered neftegazovym kompleksom Rossii *Oil Industry* **5** 14-15 [in Russian]

[6] Antonova Z G 2013 Perekhod Rossii na innovatsionnyy put razvitiya *Bulletin of Tomsk Polytechnic University* **6** 26-32 [in Russian]

[7] Kapustin V 2014 Neftekhimii ne khvatilo «rychaga» *Oil of Russia* **1-2** 34-39 [in Russian]

[8] Cui N, Lei Y and Fang W 2011 Design and impact estimation of a reform program of China's tax and fee policies for low-grade oil and gas resources *Petroleum Science* **8** (4) 515-526

[9] Export of Russian crude oil during 2000-2012 *The Central Bank of the Russian Federation*

**Institute of Physics**
Dirac House, Temple Back
Bristol BS1 6BE UK

CURRAN ASSOCIATES INC
proceedings.com